Chemical Pollutants: A Threat for Environment

Chemical Pollutants:
A Threat for Environment

Edited by **Ralph Britton**

SYRAWOOD
PUBLISHING HOUSE

New York

Published by Syrawood Publishing House,
750 Third Avenue, 9th Floor,
New York, NY 10017, USA
www.syrawoodpublishinghouse.com

Chemical Pollutants: A Threat for Environment
Edited by Ralph Britton

International Standard Book Number: 978-1-68286-170-7 (Hardback)

Printed in the United States of America.

Contents

Preface   IX

Chapter 1   **The influence of water and sewage networks on residential water consumption**   1
Falah A. Almottiri and Falah M. Wegian

Chapter 2   **A methodology of finding dispersion coefficient using computational fluid dynamics (CFDs)**   5
Rouzbeh Abbassi, Faisal Khan and Kelly Hawboldt

Chapter 3   **Introducing sand filter capping for turbidity removal for potable water treatment plants of Mosul/Iraq**   12
S. M. Al-Rawi

Chapter 4   **Radon and radium concentrations in 120 samples of drinking springs and rivers water sources of northwest regions of Mashhad**   21
A. Binesh and H. Arabshahi

Chapter 5   **Prioritization of micro watersheds on the basis of soil erosion hazard using remote sensing and geographic information system**   27
Vipul Shinde, K. N. Tiwari and Manjushree Singh

Chapter 6   **Atmospheric pollution from the major mobile telecommunication companies in Tanzania**   34
M. Kasebele and W. J. S. Mwegoha

Chapter 7   **Geoelectric study of major landfills in the Lagos Metropolitan Area, Southwestern Nigeria**   48
Oladapo, M. I., Adeoye-Oladapo, O. O. and Adebobuyi, F. S.

Chapter 8   **Regional model for peak discharge estimation in ungauged drainage basin using GIUH, Snyder, SCS and triangular models**   60
Majid Dabbaghian Amiry and Mohammadi A. A.

Chapter 9   **Evaluation of the water quality status of Lake Hawassa by using water quality index, Southern Ethiopia**   71
Adimasu Woldesenbet Worako

Chapter 10   **Study of geosynthetic clay liner layers effect on decreasing soil pollution in the bed of sanitary land fills**   79
Rouhollah Soltani Goharrizi, Fazlollah Soltani and Bahador Abolpour

Chapter 11   **Study of heavy metals pollution and physico-chemical assessment of water quality of River Owo, Agbara, Nigeria**   85
Kuforiji Titilope Shakirat and Ayandiran Tolulope Akinpelu

Chapter 12   **Performance and kinetic evaluation of phenol biodegradation by mixed microbial culture in a batch reactor**   93
Sudipta Dey and Somnath Mukherjee

Chapter 13   **Comparison of spectrophotometric methods using cuvette tests and national standard methods for analysis of wastewater samples**   103
Sonya Dimitrova, Nadejda Taneva, Kapka Bojilova, Vesela Zaharieva, Svetlana Lazarova, Mariana Koleva, Rumen Arsov and Tony Venelinov

Chapter 14   **Removal of phosphorus from Nigeria's Agbaja iron ore through the degradation ability of *Micrococcus* species**   110
O. W. Obot and C. N. Anyakwo

Chapter 15   **Industrial sludge based adsorbents/ industrial by-products in the removal of reactive dyes – A review**   116
A. Geethakarthi and B. R. Phanikumar

Chapter 16   **Assessment of health risks associated with wastewater irrigation in Yola Adamawa State, Nigeria**   125
Hussaini I. D., Aliyu B., Bassi A. A. Abubakar S. I. and Aminu M.

Chapter 17   **Influence of suspended matters on iron and manganese presence in the Okpara Water Dam (Benin, West Africa)**   130
Tomètin A. S. Lyde, Mama Daouda, Sagbo Etienne, Fatombi K. Jacques, Aminou W. Taofiki and Bawa L. Moctar

Chapter 18   **Water quality assessment of a wastewater treatment plant in a Ghanaian Beverage Industry**   140
Emmanuel Okoh Agyemang, Esi Awuah, Lawrence Darkwah, Richard Arthur and Gabriel Osei

Chapter 19   **Flow system, physical properties and heavy metals concentration of groundwater: A case study of an area within a municipal landfill site**   148
Adebisi, N. O., Oluwafemi, O. S., Songca, S. P. and Haruna, I.

Chapter 20   **Study on the effects of vegetation density in reducing bed shear stress on the downstream slope of earthen embankment**   157
H. M. Rasel, M. R. Hasan and S. C. Das

Chapter 21   **Contamination of groundwater due to underground coal gasification**   164
R. P. Verma, R. Mandal, S. K. Chaulya, P. K. Singh, A. K. Singh and G. M. Prasad

Chapter 22   **Design of a waste management model using integrated solid waste management: A case of Bulawayo City Council**   173
Bupe Mwanza and Anthony Phiri

Chapter 23   **Modelling dynamics of organic carbon in water hyacinth Eichhornia Crassipes (Mart.) Solms artificial wetlands**   182
Aloyce W. Mayo

Chapter 24   **An integrated process investigation of self-sustaining incineration in a novel waste incinerator: Drying, pyrolysis, gasification, and combustion of sludge-plastic blends**   192
Jianzhong Zhu, Lieqiang Chen, Jun Fang and Buchang Shi

Chapter 25   **Water quality index of fresh water streams feeding Wular Lake, in Kashmir Himalaya,**
                India   **203**
                Sayar Yaseen, Ashok K. Pandit and Javaid Ahmad Shah

                Permissions

                List of Contributors

Preface

The widespread uses of chemicals and hazardous toxics in industrial activities have resulted in polluting and degrading ecosystems. A wide range of chemicals like pesticides, fertilizers and oil spills contaminate natural resources and have adverse effects on environment and health. The book presents a detailed account on disposal and treatment of chemical pollutants. It provides various tools and techniques to detect, test and reduce the ill effects of toxic chemical compounds. Students and researchers will find this book beneficial.

This book is the end result of constructive efforts and intensive research done by experts in this field. The aim of this book is to enlighten the readers with recent information in this area of research. The information provided in this profound book would serve as a valuable reference to students and researchers in this field.

At the end, I would like to thank all the authors for devoting their precious time and providing their valuable contributions to this book. I would also like to express my gratitude to my fellow colleagues who encouraged me throughout the process.

Editor

The influence of water and sewage networks on residential water consumption

Falah A. Almottiri and Falah M. Wegian

Civil Engineering Department, College of Technological Studies (Kuwait) P.O. Box: 34 Ardia, 13136 Kuwait.

In this study a real case is used in order to evaluate the effect of the factors controlling the residential water demand in one of the suburb of the state of Kuwait. The study investigates the effect of having or not having water and sewage networks on the water demand. The importance of this study arises from the difference between the two studies residential areas in which one of them contains complete and modern water and sewage networks. The other area under investigation is adjacent to the first area but is lacking similar networks yet. These two residential areas have the same controlling factors for water demands, such as economical conditions, climatic conditions, standard of living, and social life. The method presented is based on a sensitivity analysis of the effect of networks availability on the per capita daily water demand. The analysis of data from both residential areas showed a greater water demands for area without networks, which is contradicting the expectations and experiences.

Key words: water demand, residential water, sewage, network.

INTRODUCTION

The community in Kuwait has unique social life that differs from that in other oil countries in the Gulf region. All Kuwaiti citizens have high income (GDP per capita is USD 56,000, World Fact Book) that minimizes the effect of the water price on the consumption rate. The common luxurious life of citizens requires the use of several faucets in houses. This lavish style of life makes Kuwait one of the top water consuming countries per capita.

Water consumption in Kuwait is about 460 liter/capita/day(lpcd) that is about one and a half times the consumption of an American person and three times the consumption of a European person (Al-Shawaf, 2008). On the other hand, the available water resources in Kuwait is very limited that the demand/resource ratio for water is higher than 20 (ESCWA, 2009). The efficient performance of water resources management depends largely on the accuracy of future water demand estimation, which in turn depends on the evaluation of the factors affecting water consumption rate in the city. There are many important factors playing an important role in the estimate of the per capita water consumption rates.

These factors include, but not limited to, climatic conditions, economical conditions, geographical locations, composition of the community, and the existence of water and wastewater pipe networks system with installed water meters.

Many studies dealing with residential water demand focused on the water consumption modeling for urban and rural areas. Whitford (1972) has developed a forecasting model based on the use of a formal decision tree to include some of the important controlling factors. Saunders and Warford (1976) determined the most important variables affecting municipal water demand in some metropolitan areas. Narayanan et al. (1987) examined the feasibility of seasonal water pricing considering metering costs.

Mukhopadhyay et al. (2001) collected data from 48 household over a period of 56 weeks to estimate the consumption inside a house in Kuwait. The consumption ranges from 182 to 2018 lpcd with an average of 814 lpcd. The collected data and the followed analysis considered few parameters that affect the water consumption including number of people in residence, number of rooms, number of bathrooms, existence, size, and watering of gardens, and income level of the occupant. The neural network analysis conducted through the study presents the watering of the attached garden is a prime

*Corresponding author E-mail: falah13@hotmail.com.

reason for increasing water consumption.

Keshavarzi et al. (2006) investigated the consumption in rural area of Ramjerd, Fars District, Iran over a 5 year period. The consumption is found dependent on the household size, age of household's head, garden size, greenhouse size, and garden watering frequency. Another study was conducted by Milutinovic (2006) in which he examined several numerical models for the water consumption in Kuwait considering factors such as the household size, family income, existence of a garden, water price, and temperature. Nazerali (2007) conducted a comparison between water resources management in Kuwait and that in Singapore. He elaborated that some factors need to be considered with respect to water management in Kuwait such as the economy of the country, lifestyle of the rural population, and environmental change.

However, none of the available literature examined the effect of water and sewage network availability on the water consumption rate in Kuwait. A comparison is conducted in this study between the water consumption of an established area with water and sewage network and another newer area without similar networks. The comparison considered the area of houses, number of people in the house, age of head of household, and house garden.

RESEARCH OBJECTIVE

The target of this study was to evaluate the effect of the availability of water supply network with installed meters and sewers on the water demand. The investigation was conducted by making a comparison between two adjacent residential areas in one of Kuwait suburb called Sabah Al-Naser located about 25 km from the center of Kuwait city (Figure 1). This modern residential area is a perfect model for applying this study. The location of the studied residents gives the opportunity to evaluate the effect of water and sewage networks on the municipal water consumption rates, particularly the household water consumption rates.

The other important controlling factors affecting residential water consumption rates are considered to have the same effect on both residential areas. It is presumed that factors such as climatic conditions, economical conditions, standard of living, habits and traditions, to have the same effect on both residential areas under investigation.

It was hypothesized that the availability of water and sewage networks in the city will increase the daily per capita consumption rate. However, the efficiency of the existence of such network in Kuwait was the prime target of the study.

METHODOLOGY

It was aimed in this program to evaluate the relationship between household consumption rates and the effect of water and sewage networks. This examination is made by comparing two sets of data collected from two different residential areas.

However data collected from residential area with water and sewage networks (Area-A) were made by day to day direct water meter readings for one hundred and twenty individual houses. The other part of this residential area with no water and sewage networks (Area-B) was getting its water by tankers from water filling stations, and also transporting its wastewater by tankers as needed. A sample of about one hundred and twenty houses was taken. The data collected included, area of house, number of people living in each house, and monthly water consumption.

Each house's daily water consumption rates in Area-B were calculated depending on the size of tankers, the frequency of water supply to these houses, and the volume of tanks placed over each house. Moreover, a monthly bill of water supply by water tankers in houses in Area-B houses was taken to estimate average water daily consumption rates.

In Area-A, houses were of two floors and a basement. Houses were categorized into three groups depending on the area of the house. The numbers of persons living in these houses was considered. There are small variations of number of people living in these houses with a total average of 10 persons in the house. The distribution of number of people living in a household is:

1. 40 houses with a 750 m^2 area with an average of 12 habitants.
2. 40 houses with a 600 m^2 area with an average of 10 habitants.
3. 40 houses with a 500 m^2 area with an average of 8 habitants.

Houses categorization was selected to have the same number and same area for both areas being considered. A time period of one year was selected to follow up the consumption rate of the houses based on monthly basis. Data collection of water consumption started on June 2003 to May 2004. A full record of water meters' reading has been recorded for each month in order to estimate the daily per capita consumption rates.

For Area-B in which there were no networks, an estimation of water consumption for each house has been conducted by considering the effective capacity of the tankers, the volume of tanks located on the top of each house, and average frequency of filling the tanks. Then, the per capita consumption rate was calculated by dividing the total daily consumption of water by the number of people living in each house.

It is worthy to note that all the houses selected in Area-B were getting water from tankers according to contracts that guarantee steady water supply for each house every day on a monthly payment. Also, it is important to keep in mind that all the residents of selected houses are Kuwait national families with high income and they own their houses.

RESULTS AND DISCUSSION

It was found that for similar houses in both areas A and B, the average per capita consumption rate for all the categories in Area-A is less than the corresponding value of Area-B. The average daily water consumption per capita for the houses of 500 m^2 area is 491 lpcd for Area-A, while the corresponding rate in houses of the similar area and number of habitants is about 605 lpcd in Area-B. Also, it was found that a lower water consumption for houses of 600 m^2 area in Area-A than those in Area-B of the same area and number of habitants. These are found to be 418 lpcd for Area-A houses and 523 lpcd for Area-B houses. In addition, a greater water consumption was found in houses of 750 m^2 area in Area-B compared to

Figure 1. Location of Sabah Al-Nasser region in Kuwait.

Table 1. Water demand in each area of our study (liters per capita per day).

Area of house (m^2)	Residential area (Part A)	Residential Area (Part B)
500	491	605
600	418	523
750	473	568

that of corresponding houses of the same area and number of people in Area-A. There are 568 lpcd for houses in Area-B and 473 lpcd for houses in Area-A (Table 1).

The higher values obtained for water consumption in area without water and sewage networks may be referred to several reasons. The existence or not of the networks is not a prime factor in the consumption difference since the common habits for residents in both areas are the same. All residents do not care about the price of consumed water whether through a fixed network or delivered by tankers. The possible reasons for that difference include the social characteristics of the residents in each area. The residents of the new area are new families with lower ages and lower number of household size. These families with frequent travels abroad in vacations, work, or study have more modernized style of life.

The young age of household parents and children make the family more dynamic more consuming for water. Also, the small number of persons in the house leads to higher rate of water consumption per person since some fixed activities is not dependent on the number of persons such house cleaning and garden watering. Moreover, the luxurious modernized style of living consumes more water because of using en suite rooms, utilizing new machinery for washing and cleaning, having frequent faucets inside and outside bathrooms and kitchens to facilitate cleaning activities.

Conclusions

The conclusion of this study is contradicting the initial hypothesis, since it is usually presumed that the existence of water and waste water networks for a region will result in an increase of water consumption rates. The study compared two areas in the same district with similar house areas, very close income rates, same climatic conditions, and all houses with gardens. The diffe-rences were limited to the existence of water and sewage networks or not and the age of the family members.

The following points are the possible explanation for the obtained results:

1. The water pricing is very low due to the governmental subsidizes and high income rate. Hence, water cost has no effect on the residents' response to water consump-tion for both areas.

2. Similar to the effect of the pricing factor, the existence of the water and sewage networks has a very limited effect on the consumption since water is required to be available all the time based on contracts with water delivery companies in Area-B without networks.

3. The young age of household parents and children in Area-B makes the family more dynamic more consuming for water. Also, the small number of persons in the house leads to higher rate of water consumption per person since some fixed activities is not dependent on the number of persons such house cleaning and garden watering.

4. Luxurious modernized style of living, which is more applicable in Area-B, consumes more water because of using en suite rooms, utilizing new machinery for washing and cleaning, having frequent faucets inside and outside bathrooms and kitchens to facilitate cleaning activities.

5. Kuwaiti citizens in bad need of awareness of the demand/resource problem in Kuwait to participate in water management conservation process, specially for houses with new families.

REFERENCES

AlShawaf M (2008). Evaluating the Economic and Environmental Impacts of Water Subsidies in Kuwait. Thesis, Louisiana State University, LA, USA.

Economic and Social Commission for Western Asia (ESCWA) (2009). Water Development Report 3 – Role of Desalination in Addressing Water Scarcity. United Nations, New York, USA.

Keshavarzi AR, Sharifzadeh M, Kamgar HAA, Amin S, Keshtkar Sh, Bamdad A (2006). Rural domestic water consumption behavior: A case study in Ramjerd area, Fars province, I.R. Iran. Water Research, 40: 1173 – 1178.

Milutinovic M (2006). Water Demand Management in Kuwait, Thesis, Massachusetts Institute of Technology, USA.

Mukhopadhyay A, Akber A, Al-Awadi E (2001). Analysis of Freshwater Consumption Patterns in the Private Residence of Kuwait. Urban Water, 3: 53-62.

Nazerali NA (2007). Sustainable Water Resources Development in Kuwait: An Integrated Approach with Comparative Analysis of the Case in Singapore. Thesis, Massachusetts Institute of Technology, USA.

Narayanan R, Beladi H, Roger D, Hansen A, Bishop B (1987). Feasibili-ty of Seasonal Water Pricing Considering Metering Costs. J. Am. Water Res. Ass., 23(6): 1091-1099.

Saunders RJ. Warford JJ (1976). Village Water Supply: Economics and Policy in the Developing World. Baltimore: The John Hopkins University Press.

Whitford PW (1972). Residential Water Demand Forecasting. Water Resourc. Research. 8(4): 829-839.

World Fact Book, CIA, www.cia.gov.

A methodology of finding dispersion coefficient using computational fluid dynamics (CFDs)

Rouzbeh Abbassi*, Faisal Khan and Kelly Hawboldt

Faculty of Engineering and Applied Science, Memorial University of Newfoundland, St.John's, NL, Canada, A1B 3X5, Canada.

The treatment efficiency of waste stabilization pond is directly related to its hydraulic regime. The hydraulic efficiency of the pond is dependent on parameters such as the pond geometry, the location of inlet and outlet and the inlet flow velocity. Poorly designed or specified hydraulic parameters may lead to short circuiting and dead regions within the pond. This in turn impacts the dispersion coefficient. Drogue and tracer studies are often used to get actual dispersion coefficients; however, tracer studies can be costly and are therefore not practical to do frequently. The objective of this paper is to obtain the actual dispersion coefficient using computational fluid dynamic (CFD) approach (using Fluent). The CFD results are validated using an actual tracer test.

Key words: Stabilization pond, modeling, computational fluid dynamic, residence time distribution, dispersion coefficient

INTRODUCTION

A waste stabilization pond (WSP) is a simple and cost effective method for treating wastewater (Khan and Ahmad, 1992). According to many studies (Arceivala, 1981; Polprasert and Bhattarai, 1985; Chien and Liou, 1995), the dispersed flow model may predict the transport of contaminants more reliably than the idealized continuous stirred tank reactor (CSTR) or plug flow reactor (PFR) models. This model is a strong function of dispersion coefficient, which is in turn dependent on the hydraulic regime of the pond. Therefore, identifying the hydraulic performance of the pond is required to obtain the actual amount of dispersion coefficient.

Poor hydraulic considerations and design of the WSP reduces the treatment efficiency of this system (Shilton and Harrison, 2003; Shilton and Bailey, 2006). In fact, the treatment efficiency of the WSPs is a function of the numerous physical parameters that may affect fluid movement in a pond (Piondexter and Perrier, 1981; Thackston et al., 1987; Muttamara and Puetpaiboon, 1997; Salter et al., 1999; Torres et al., 1999; Shilton and Harrison, 2003; Aldana et al., 2005; Abbas et al., 2006; Agunwamba., 2006; Fyfe et al., 2007):

1. Pond geometry (including the influences of baffles)
2. Inlet size and position
3. Outlet position and design
4. Flow rate
5. Temperature/density effects
6. Wind shear stress and its variation over time

Pond geometry is one of the important factors that affect the hydraulic performance of the basins (Marecos et al., 1987; Torres et al., 1999). L/W ratio is the most important factor that affects the hydraulic performance of the basins (Piondexter and Perrier, 1981).

The placement of inlet and outlet impacts the hydraulic efficiency of WSP. Waste water can be discharged at the surface, mid depth and bottom of WSPs. The position of outlet may also be diverse in variety of different ponds.

The effect of inlet and outlet locations on short circuiting was evaluated by Agunwamba (2006). Short circuiting is the phenomena where the retention time of a particle in the pond is shortened due to flow conditions; in essence it decreases the treatment efficiency of the pond. Different inlet/outlet positions showed that short circuiting is highly related to the location of inlet/outlet. Minimum hydraulic efficiency occurs when the inlet and outlet are in front of each other and improves significantly if the inlet and outlet are positioned on the opposite

---

*Corresponding author. E-mail: rabbassi@mun.ca

corners of the pond (Persson and Wittgren, 2003). The presence of baffles in the pond reduces short circuiting. When baffles are present, shifting the outlet toward the baffles may reduce short circuiting (Safieddine, 2007). Inflow jet produces short circuiting within the pond depending on the flow velocity (Fyfe et al., 2007). The influence of the inflow jet reduced as the flow heads to the outlet.

Considering the hydraulic behaviors of WSPs, an accurate method of predicting the dispersion coefficient has been sought in a number of research studies. Tracer tests are widely used for tracking the flow motion in WSPs. The determination of the dispersion coefficient of the WSP using tracer studies have been evaluated by many researchers (Marecos et al., 1987; Moreno, 1990; Uluatam and Kurum, 1992; Pedahzur et al., 1993; Salter, 1999; Shilton et al., 2000; Vorkas and Lioyd, 2000). It should be noted that tracer tests are costly in time and finances. The second way for calculating the dispersion coefficient is using empirical equations. The simplest proposed by Arceivala (1981) is based on the pond width. Polprasert and Bhattarai (1985) developed an empirical formula based on the pond geometry and retention time. Other researchers used an empirical formula based on the pond geometry and retention time, but with different correlation factors (e.g. Liu, 1977). Agunwamba et al. (1992) have stated that the shear stress of the wind also affects the hydraulic behavior of the basin and axial dispersion coefficient res-pectively. Some of the empirical equations to obtain the dispersion coefficient can be seen in Table 1.

Empirical equations reduce the cost of actual tracer studies and may be a suitable option for predicting the dispersion coefficient. The empirical equations, unlike actual tracer studies may solve the problem of predicting the dispersion coefficient for the WSPs to be constructed in the future. Although, these equations are themselves defined based on different actual tracer tests, they may not be applicable in all WSPs with diverse hydraulic conditions. Therefore, the fluctuations of hydraulic parameters of the ponds and their effect on dispersion coefficients may not completely evaluated using empirical equations.

Use of CFD is another option to obtain the dispersion coefficient. These programs have the ability to model the various conditions of the pond. For the WSPs that have not been constructed yet, these models give the designer the ability to predict the hydraulic behaviors and dispersion coefficients respectively. Using CFD for finding the dispersion coefficient has the following advantages:

1. Includes the effect of pond's characteristics such as ponds geometry, inlet size and position.
2. Includes parameters such as temperature and viscosity.
3. Includes surrounding environmental parameters such as temperature fluctuations and wind.

4. Considers the effect of hydraulic behavior of the basins.

These programs are case sensitive and the lack of complete description of different parameters in the accurate way would cause uncertainty in the results. Furthermore, the user must be aware of the CFD model limitations, assumptions and working knowledge of actual ponds to prevent misinterpretation results. In this paper, a methodology is discussed using CFD (Fluent) to obtain the dispersion coefficient. The validation of the methodology is done using actual tracer study. The result of this approach is compared with the ones found by using empirical equations.

A METHODOLOGY OF FINDING DISPERSION COEFFICIENT USING CFD

The Fluent CFD model used in this paper is a commercially available computer package which is produced by Fluent Inc. in USA. Fluent solves a finite volume form of the conservation equations for mass and momentum (Fluent, 2003). The methodology is presented to simulate the stimulus response techniques (Levenspiel, 1972) to obtain tracer concentrations in different time steps and to draw residence time distribution (RTD) respectively. As the first step, the model should be meshed using Gambit (A.I.F., 2002). The sensitivity analysis is performed to ensure the mesh-independency of the numerical simulation. For this purpose, the simulation is repeated with different meshes (consecutive smaller meshes) until the differences between solutions become negligible. After meshing the model, two steps are undertaken for the modeling, started by steady state simulation. This work is solved the three momentum components (u,v,w) and the two turbulence components (K and $\mathcal{E}$) (Fluent, 2003).

After completion of the steady state simulation, particles with the same density and size are injected to the influent. Next, it is possible to carry out a transient simulation of particle movement with respect to the time. For this purpose, the solvers for pressure, momentum and turbulence are turned off and the results of steady state simulation are used. Based on the values stored from the steady state run, the simulation then stimulated through a series of time steps solving for the dispersion of the particles.

For a low surface fraction of dispersed second phase (particle), an Eulerian - Lagrangian approach was used. This allows the effects of turbulence modulation (effect of particles on turbulence) to be neglected. The Lagrangian approach divides the particle phase into a representative set of discrete individual particles and tracks these particles separately through the flow domain by solving the equations of particle movement. Assumptions regarding the particle phase included the following: (i) no particle rebounded off the walls/surfaces (ii) no particle coagulation in the particle deposition process and (iii) all particles are spherical solid shapes. Trajectories of individual particles can be tracked by integrating the force balance equations on the particle (Fluent, 2003):

$$\frac{du_p}{dt} = F_D(u - u_p) + g_x(\rho_p - \rho)/\rho_p + F_x \qquad (1)$$

Where F_D (Drag force) is calculated according to the following equation:

$$F_D = \frac{18\mu}{\rho_p D^2_p} \frac{C_D \, \text{Re}}{24} \qquad (2)$$

Table 1. Empirical equations for determining the Peclet number (UL/D).

Name	Condition	Formula
Liu (1977)	Large width to depth ratio	$d = \dfrac{0.168 * (\tau.v)^{.25} * (W + 2Z)^{3.25}}{(LWZ)^{1.25}}$
Polprasert and Bhattarai, (1985)	Waste stabilization pond	$d = \dfrac{0.184 * [\tau.v(W + 2Z)]^{0.489} W^{1.511}}{(LZ)^{1.489}}$
Arceivala (1981)	For pond width greater than 30 m	$D^{**} = 16.7\,W$
Arceivala (1981)	For pond width less than 30 m	$D = 2W^2$
Murphy and Wilson, (1974)	The volume over 300000 m³	$d = K\tau / L^2$
Nameche and Vasel (1998)	Stabilization pond and lagoon	$\dfrac{1}{d} = 0.31(\dfrac{L}{W}) + 0.055(\dfrac{L}{Z})$
Agunwamba et al. (1992)	Stabilization pond	$d = 0.102(\dfrac{u^*}{u})^{-0.8196} (\dfrac{H}{L}) (\dfrac{H}{W})^{-(0.981+1.385\frac{H}{W})}$

*. τ : Retention time, **. The unit of D in Arceivala's equation is m²/h.

And the Reynolds number is defined as:

$$Re = \frac{\rho D_p |u_p - u|}{\mu} \tag{3}$$

When the flow is turbulent, Fluent uses mean fluid phase velocity in the trajectory equation (Equation 1) in order to predict the dispersion of the particles. The amount of the particles in each time step at the outlet position is monitored until the end of the transient simulation. These concentrations versus time help to draw RTD. The methodology of drawing RTD is demonstrated in Figure 1 Integrating the RTD with Levenspiel formula leads to obtain the dispersion coefficient. The dispersion number, D/UL may be calculated from the dimensional variance which is defined as (Levenspiel, 1972):

$$\sigma^2 = \frac{\sum t_i^2 C_i}{\sum C_i} - \left[\frac{\sum t_i C_i}{\sum C_i}\right]^2 \tag{4}$$

The amount of variance is calculated according to concentrations in each time step (Equation 4). The dispersion number (d) (Levenspiel, 1972) is determined using the following equation:

$$\sigma_i^2 = \frac{\sigma^2}{t^2} = 2d - 2d^2(1 - \exp(\frac{-1}{d})) \tag{5}$$

The variance and t are used to estimate "d" by a process of trial and error using Excel Solver in Microsoft Excel.

COMPUTATIONAL FLUID DYNAMIC APPROACH: A CASE STUDY

A methodology for determining the dispersion coefficient is

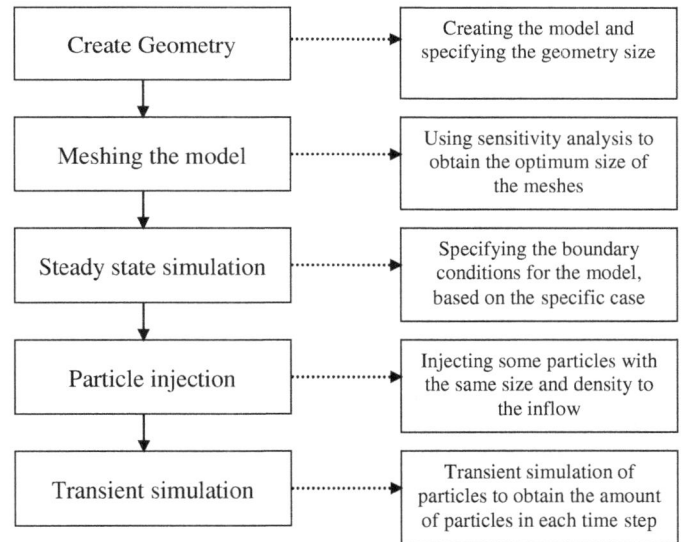

Figure 1. The CFD Simulation methodology to draw RTD.

proposed by the combining of CFD approach methodology and Levenspiel's formula (Levenspiel, 1972). Testing and validation of the method is assessed by the field's data. The name and the place of the basin used as a case study in this paper will not be disclosed herein, due to the confidentiality of the data.

Flow domain and mesh

A two dimensional model was developed for this study. The model created and meshed using Gambit (version 2.4.6). The whole surface was divided to 711819 homogenous quadrilateral cells (0.3

Table 2. Geometry and flow parameters of the basin.

Parameter	Units	Value
Length (L)	m	≈500
Width (W)	m	≈100
Inlet width	m	0.45
Inlet velocity in x-direction	m/s	4.63
Inlet velocity in y-direction	m/s	0
Fluid Density	Kg/m^3	998.2
Fluid viscosity	Kg/(m.s)	0.001

Table 3. Inputs to CFD tool (Fluent).

Models	Two dimensional
	Pressure based, Steady state
	Standard k-epsilon turbulence model
Solution Control	Second order upwind discretization
Materials	Liquid water (H2O), Solid particles
Operating conditions	Operating pressure: 101325 Pa
	Gravity: Off
Boundary conditions	Inlet: Velocity inlet (V=4.63m/s)
	Walls: No slip boundaries
	Outlet: Outflow
	Discrete phase condition at walls: reflect,
	normal constant 0.5, tangential constant 0.8
	Discrete phase condition at inlet and outlet: escape
Convergence limit	Scaled residuals: 1.0E-04

x 0.3m). The parameters related to this model are presented in Table 2.

Initial and boundary conditions

The governing equations were solved in combination with the proper initial and boundary constraints. The inlet boundary was specified by inlet velocity (V = 4.63 m/s). The no slip boundary condition was chosen for the walls. For discrete phase boundaries, the outflow was chosen as an escape boundary and the walls as reflective boundaries. The boundary conditions that were picked for this case are presented in Table 3.

Modeling flow and solid phase particles

Fluent solves the equations of turbulent flow in a two-dimensional geometry to obtain the water velocity. The standard K-$\mathcal{E}$ approach is a widely used, robust, economical model, which has the advantages of rapid, stable and reasonable results for many flows (Marshall and Bakker, 2003). In this case study, the standard K-$\mathcal{E}$ model is used. After running the model for 5,000 iterations and obtainning acceptable convergence, the unsteady particle tracking is used for tracking the solid particles within the basin. For this reason, 5,000 spherical particles with the same size are injected at the inlet at the same time. Particle diameters (100 μm) and density (1020 Kg/m^3) is selected based on previous investigation (Gancarski, 2007). This size and density is an acceptable option for

modeling the particle as a drogue in the basins. After injection, this model is run for 104 time steps, 1800 s each and the amount of the particles escaped from the basin in each time step is calculated.

RESULTS AND DISCUSSION

The data received from actual tracer studies from the field are plotted, as are illustrated in Figure 2. This is one of the typical RTD for this basin between 12 RTDs that draw during different months of the year, but the final value of 'd' is calculated based on the concentrations mean value. Tracer concentration versus time shows the existence of short circuiting in the basin. The maximum concentration of the tracers received approximately four hours after injection which is less than the actual retention time. The existence of short circuiting in the pond was previously reported by other researchers as well (Vorkas et al., 2000; Moreno, 1990). The amount of 'd' using actual tracer test is calculated which is 0.6. The process for calculating this coefficient can be seen in Table 4. Concentrations in different time steps are calculated using un-steady particle injection in Fluent, as demonstrated in Figure 2. Integration of this calculation with Levenspiel's formula as mentioned previously shows the amount of 'd' is 0.5. The summary of this calculation

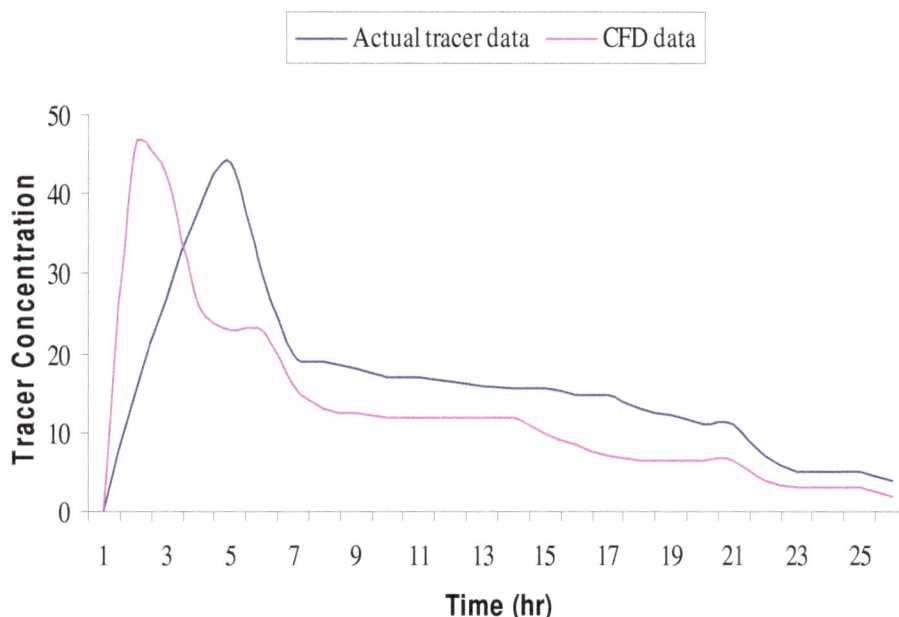

Figure 2. Comparison of CFD versus one typical tracer results.

Table 4. Summary of variables to obtain "d" using actual tracer study.

Function	Value	Function	Value
$\sum C_i$	712.27	$\dfrac{\sum t_i{}^2 C_i}{\sum C_i} - \left[\dfrac{\sum t_i C_i}{\sum C_i}\right]^2$	2.1E+10
$\sum C_i t_i$	4.9E+7	σ^2	2.1E+10
$\sum t_i{}^2 C_i$	1.49E+13	$\sigma_i{}^2$	0.61
$\dfrac{\sum t_i{}^2 C_i}{\sum C_i}$	2.1E+10	d	0.6

* i = Different time steps.

can be seen in Table 5.

Comparing 'd' obtained using actual tracer study and the CFD modeling, it is confirmed CFD is a suitable option for calculating 'd'. The dispersion coefficient was also calculated using empirical equations outlined previously (Table 6). Although, these empirical equations were a good predictor of dispersion coefficient in their own cases, they are not a comprehensive technique to obtain a dispersion coefficient. Some of these equations (e.g. Agunwamba et al., 1992) was claimed to be simple, accurate and economical in comparison to the use of actual tracer studies, however, there are serious limitations on using these equations for different actual modeling scenario. For the given WSP, the dispersion coefficient found by Arceivala (1981) has a better prediction of actual dispersion coefficient.

The result of finding dispersion coefficient using the CFD, predicts the actual dispersion coefficient better in this case. The use of CFD gives opportunity to consider various hydraulic parameters and their effect on dispersion coefficient. Fluctuations of hydraulic behavior and critical conditions of these fluctuations may be determined by using CFD. The effects of these fluctuations can be seen in obtaining dispersion coefficient as well.

Conclusion

The dispersion model is highly dependent on dispersion coefficient which itself is based on hydraulic performance of the WSP. The result of comparing CFD analysis with

Table 5. Summary of variables to obtain "d" using Fluent.

Function	Value	Function	Value
$\sum C_i$	4457	$\dfrac{\sum t_i{}^2 C_i}{\sum C_i} - \left[\dfrac{\sum t_i C_i}{\sum C_i}\right]^2$	2.6E+10
$\sum C_i t_i$	2.31E+08	σ^2	2.6E+10
$\sum t_i{}^2 C_i$	1.27E+14	$\sigma_i{}^2$	0.75
$\dfrac{\sum t_i{}^2 C_i}{\sum C_i}$	2.84E+10	d	0.5

* i = Different time steps.

Table 6. Comparing different methods to obtain dispersion coefficient.

Methods	Calculated D	Difference from tracer
Actual tracer test	0.8	---
CFD based approach	0.67	0.13
Liu (1977)	2.18	1.38
Arceivala (1981)	0.58	0.22
Polprasert and Bhattarai (1985)	1.4	0.6
Nameche and Vasel (1998)	0.1	0.7

actual tracer test shows using CFD is a suitable option to determine the dispersion coefficient for use in dispersion model. The value of 'd' found by using CFD for the WSP used as a case study is 0.5 which is approximately similar to the one found by actual tracer tests which was 0.6. Modeling the tracer using a methodology discussed in this paper (particle injection in CFD) can supplement the actual tracer studies, although, CFD analysis is case sensitive and ignoring the description of hydraulic conditions of the WSP entirely, leads to misleading results. Using CFD in turn reduces the frequency of actual tracer tests. Empirical equations are another option for finding the dispersion coefficient in the design stage. This research shows that the using of these equations for the specific case should be done precisely.

ACKNOWLEDGEMENT

Authors gratefully acknowledge the financial support provided by Natural Science and Engineering Research Council (NSERC) and Inco Innovation Centre (IIC).

REFERENCES

Abbas H, Nasr R, Seif H (2006). Study of waste stabilization pond geometry for the wastewater treatment efficiency. Ecol. Eng., 28: 25-34.

Agunwamba J, Egbuniwe N, Ademiluyi J (1992). Prediction of the dispersion number in waste stabilization ponds. Water Res., 26(1): 85-89.

Agunwamba JC (2006). Effect of the location of the inlet and outlet structures on short circuiting: Experimental investigation. Water Environ. Res., 78: 580- 589.

A.I.F (2002). Using software Gambit 2.0 and Fluent 6.0 for simulation of heat mass transfer problems. Progress report.

Aldana GJ, Guganesharajah K, Bracho N (2005). The development and calibration of a physical model to assist in optimizing the hydraulic performance and design of maturation ponds. Water Sci. Technol., 51(12): 173-181.

Arceivala SJ (1981). Waste water treatment and disposal. Marcel Dekker INC., New York and Basel.

Chien YS, Liou CT (1995). Steady state Multiplicity for Autocatalytic reactors in a Nonideal Mixing of CSTR with two Unpremixed Feeds. Chem. Enginee. Sci., 50(22): 3645-3650.

Fluent (2003). User manual 6.2, Fluent Inc.

Fyfe J, Smalley J, Hagara D, Sivakumar M (2007). Physical and hydrodynamic characteristics of a dairy shed waste stabilization pond system. Water Sci. Technol., 55(11): 11-20.

Gancarski P (2007). CFD modeling of an oxidation ditch. MSc thesis, Cranfield University, UK.

Khan MA, Ahmad SI (1992). Performance evaluation of pilot waste stabilization ponds in subtropical region. Water Sci. Technol., 26(7): 1717-1728.

Levenspiel O (1972). Chemical Reaction Engineering. Department of Chemical Engineering, Oregon State University, USA, ISBN 0-471-53016-6.

Liu H (1977). Predicting dispersion coefficient of stream. J. Environ. Eng., 103(1): 59-69.

Marecos Do Monte MHF, Mara DD (1987). The hydraulic performance of waste stabilization ponds in Portugal. Water Sci. Technol., 19(12): 219-227.

Marshall EM, Bakker A (2003). Computational fluid mixing; Fluent Inc, Lebanon, New Hampshire, USA, www.Fluent.com.

Moreno MD (1990). A tracer study of the hydraulics of facultative stabilization ponds. Water Res., 24(8): 1025-1030.

Murphy KL, Wilson AW (1974). Characterization of mixing in aerated lagoons. J. Environ. Eng., 100(5): 1105-1117.

Muttamara S, Puetpaiboon U (1997). Roles of baffles in waste stabilization ponds, Water Sci. Technol., 35(8): 275-284.

Nameche T, Vasel JL (1998). Hydrodynamic studies and modelization for aerated lagoons and waste stabilization ponds. Water Res., 32(10): 3039-3045.

Pedahzur R, Nasser AM, Dor I, Fattal B, Shuval HI (1993). The effect of baffle installation on the performance of a single cell stabilization pond. Water Sci. Technol., 27(7-8): 45-52.

Persson J, Wittgren HB (2003). How hydrological and hydraulic conditions affect performance of ponds. Ecol. Eng., 21: 259-269.

Piondexter ME, Perrier ER (1981). Hydraulic efficiency of dredged material impoundments: A field evaluation. Symposium on Surface Water Impoundments, 2: 1165-1174.

Polprasert C, Bhattarai KK (1985). Dispersion model for waste stabilization ponds. J. Environ. Eng., 111(1): 45-59.

Safieddine T (2007). Hydrodynamics of waste stabilization ponds and aerated lagoons, PhD thesis, Department of Bioresource Engineering, McGill University, Montreal, Canada.

Salter HE, Boyle L, Ouki SK, Quarmby J, Williams SC (1999). Tracer study and profiling of a tertiary lagoon in the United Kingdom. Water Res., 33(18): 3782-3788.

Shilton A (2000). Potential application of computational fluid dynamics to pond design. Water Sci. Technol., 42(10-11): 327-334.

Shilton A, Harrison J (2003). Development of guidelines for improved hydraulic design of waste stabilization ponds. Water Sci. Technol., 48(2): 173-180.

Shilton A, Bailey D (2006). Drouge tracking by image processing for the study of laboratory scale pond hydraulics. Flow Measurement and Instrumentation, 17: 69-74.

Thackston EL, Shields FD, Schroeder PR (1987). Residence time distributions of shallow basins. J. Environ. Eng., 113(6): 1319–1332.

Torres JJ, Soler A, Saez J, Leal LM, Aguilar MI (1999). Study of the internal hydrodynamics in three facultative ponds of two municipal WSPs in Spain. Water Res., 33(5): 1133-1140.

Uluatam SS, Kurum Z (1992). Evaluation of the wastewater stabilization pond at the METU treatment plant. Inter. J. Environ. Stud., 41(1-2): 71-80.

Vorkas CA, Lioyd BJ (2000). The application of a diagnostic methodology for the identification of hydraulic design deficiencies affecting pathogen removal. Water Sci. Technol., 42(10): 99-109.

3

Introducing sand filter capping for turbidity removal for potable water treatment plants of Mosul/Iraq

S. M. Al-Rawi

Environment Research Center (ERC), Mosul University, Mosul, Iraq. E -mail: alrawism@yahoo.com.

Sand filter capping had been tried as an alternative for the currently practiced rapid sand filters. A 1000 m³/day capacity pilot plant constructed similarly to full scale water treatment plants was used for this purpose. Four levels of sand material with respect to grain size and thickness were used. Capping filter was represented by one level of anthracite coal with one thickness. In order to optimize the most efficient combination(s) of the two materials for maximum filter performance in terms of quality and quantity, four filtration rates were tried. A series of forty test runs and experiments exceeding 120 using different filter configurations were tried. Filters were operated in pairs and subjected to the same conditions. Filters consisting of 20 cm anthracite coal (0.91 mm E.S.) over 40 cm of sand (0.69 mm E.S.) appeared to be the best fit configuration among tried filters. Considerable economical consequences could be achieved using sand filter capping. This was reflected in reducing filter numbers or increasing the plant capacity by two folds in the minimum. Economic revenue was gained through reduction of disinfection doses as well as reduction in filter sand material. Runtimes of filters were increased by 2 - 3 folds indicating capability of more furnished water productivity and less amount for backwash need. Above all the water produced was of very good quality that met the most stringent specifications and promoted health. Capping sand filters were proven to suitably operate under varying conditions of influent turbidity and filtration rates.

Key words: Drinking water filtration, rapid sand filtration, dual-media filtration, multi-media filters, water treatment, filter media, gravity filtration.

INTRODUCTION AND LITERATURE REVIEW

The recent constructed water treatment plant in Mosul city (north of Iraq) dated back to 1980s. The designs of this plant and most water treatment plants in the city were based on concepts developed in the early 1960s, when many aspects of the treatment processes were not fully understood. Since then many changes had taken place including guidelines, consumption patterns and life style.

Besides, water quality underwent a great change due to water impound to the north of the city. Turbidity - as example- was reduced greatly as this impoundment acted as a huge sedimentation tank. Hardness —on the other hand- increased due to geologic formation of the area of the impoundment. Released water from the dam was loaded with nutrients e.t.c. Further, misconceptions of treatment works as well as lack of specialized personnel complicated these problems (Al-Rawi 2009).

Sand had been traditionally used as the filter medium in conventional water treatment plants because of its wide availability, low cost and the satisfactory results that it had given. Sand filters remained the predominant method

of filtration in most developing countries and Iraq is no exception.

However the grading of sand that might occur in backwashing of rapid filters, leaving the finer sand on top, could restrict the capacity of conventional rapid sand filters. The floc particles removed in filtration might concentrate on the topmost layers of the filters leaving most of the thickness of the filter unused.

Besides, these filters were proven to be inefficient in satisfying the recent stringent quality guidelines (Al-Rawi, 1996; Peavy et al., 1987). Further, the design of these filters was intended to treat high turbidity water and could not tackle low turbidity water.

Based on the above facts, most current existing plants turned incapable of tackling such proper water treatment and provision of sufficient water for various demands. These plants acted merely as a passing-through units. Complains of the public increased, supplied water was not healthy and tasted bad. Many cases of contamination were recorded. Diseases - particularly among infants – did

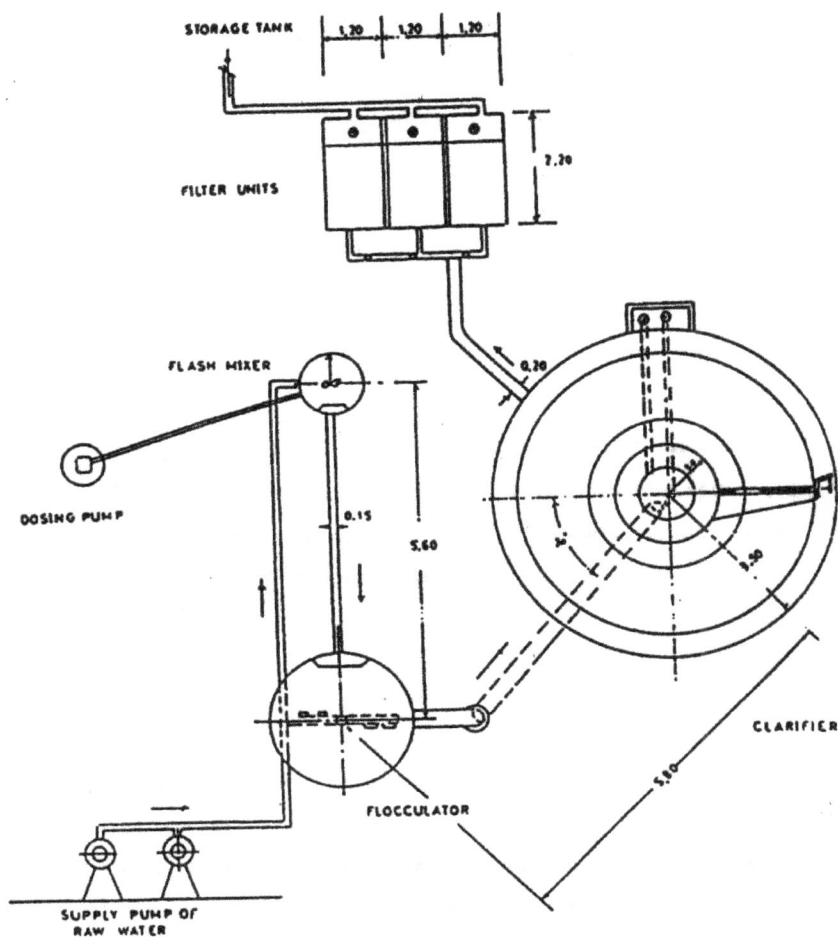

Figure 1. Pilot plant used in the experiments.

not become uncommon (personal communications, 2009).

In order to encounter this situation at present time an additional technology in the water treatment methodology must be -however- introduced to cope with society needs. Such situation suggests upgrading of the existing potable water treatment plants by establishing effective operation and improving unit processes. This will be reflected on plant production and improve the quality of the treated water.

Dual and Mixed media filters have gradually replaced conventional gravity sand filters in many treatment plants in developed countries (Hindricks, 1974; Al-Ani and Al-Baldawi, 1986; Hammer, 1986; Culp, 1974). The idea behind these filters is that the lighter media of larger size occupy the upper layers of the filter, allowing greater penetration of the floc. Floc that escapes the upper layer is caught in the finer sand at the bottom of the filter (Schulz and Okun, 1984).

Usually the thickness of upper lighter layers equals twice that of the bottom heavier layers. However, from view of cost, the developing countries may use sand filter capping. Here the thickness of the upper layers is often half that of the sand layer.

Sand filter capping is the process in which the top portion of a rapid sand filter is replaced with anthracite coal in order to achieve the improved performance typical to sand/anthracite filter (Tillman, 1996). If an increase in capacity is desired, a larger amount of sand is replaced as such practice will increase filter run as well as it will increase the amount of fines.

This paper aims to study this technology and revealing some of the merits that can be gained if introduced in water treatment plants.

MATERIALS AND METHODS

This technique (capping filters) is still unknown in Iraq and its use is restricted on experimental works in laboratories. The distinct feature in this study is that the used pilot plant is very large compared to those usually adopted in labs. Its capacity is 1,000 m³/day and is built similarly to a full scale water treatment plant. Figure 1 shows a

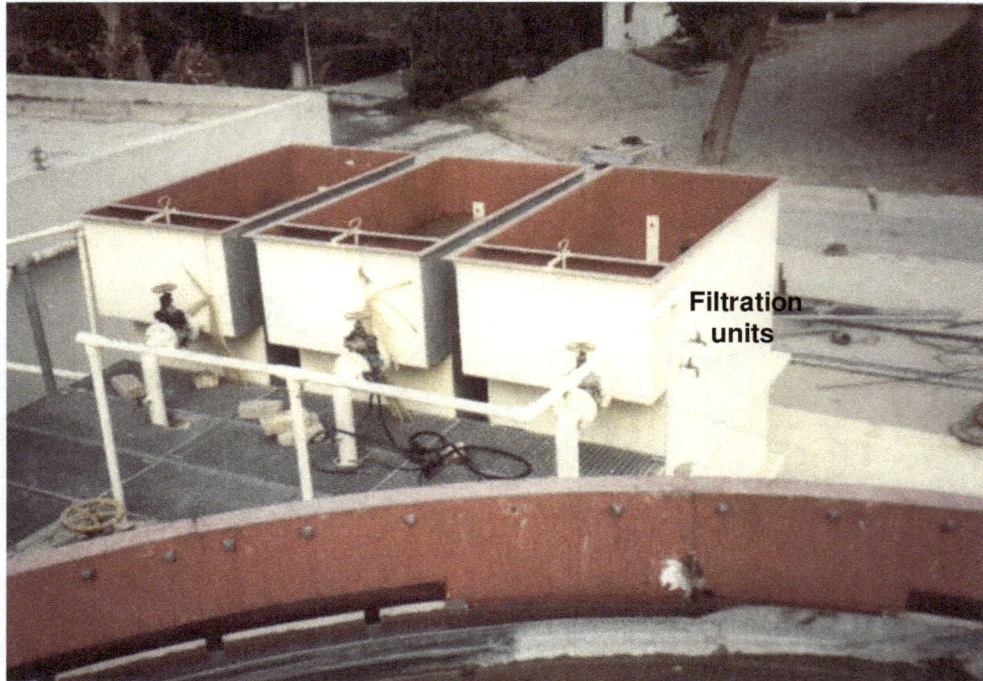

Photograph 1. The used filter units.

Photograph 2. Sedimentation tank and filter units of the used pilot plant.

sketch of the used plant. Photographs 1 and 2 show some close illustration of plant units and Table 1 lists its components. It is worthy to mention here that (2) of the (3) filters shown in the photograph were employed for the purpose of this paper. The third filter unit was employed as a slow sand filter.

Four types of sand materials with different effective size were

Table 1. Units of the pilot plant

Unit	Description
Raw water pipe	150 mm diameter conveys Tigris raw water to storage tank for maintaining constant rate of supply
pumps	Two centrifugal pumps (one stand by) of 45 m³/hr capacity.
Rapid mix basin	One cylindrical tank of 90 cm diameter and 120 cm high. This unit is provided with a stirring propeller of constant rotational speed.
Flocculation tank	One cylindrical tank 250 cm diameter and 230 cm high. It is provided with a paddle made of two blades.
Sedimentation tank	One tank of 600 cm diameter and 400 cm high. It is equipped with de- slugging system (scraper and flush outlet pipe).
Filters	2 identical filter units each of dimensions 1.0 m x1.50 m with 3.75 m high . the inflow rate is controlled by a weir.

Table 2. Media characteristics

Sand type	Effective size mm	Uniformity coefficient	Specific gravity	% loss by weight
No.1	0.92	1.32	2.677	3.3
No.2	0.98	1.58	2.588	3.61
No.3	0.69	1.79	2.617	1.95
No.4	0.46	2.86	2.644	1.72
Anthracite coal	0.91	1.33		

used One type of anthracite coal was used as a capping layer. Table 2 shows the characteristics of these media.

Measurements for turbidity were taken by Hatch turbidimeter type 2100A where measurements were given in NTU. Measurements being determined with each sampling time by appropriate apparatus.

The size distribution of sand and anthracite were made by sieving a sample of known weight through a series of standard sieves. The weight of the material in each sieve was recorded and the accumulative percentages of material retained on each sieve were estimated. Note in the Tables that the loss by weight of sand Nos. 1 and 2 are not meeting authorized guidelines (Degremont, 1991).

Plan of experimentation

The first task in the experimental program was to select the proper type of sand filter capping media as regards to size and thickness to be used. The anthracite coal layer available at time of study was of one type only. The sand media prevailing at water treatment plants of the country was imported from Karbala governorate and was of one size. For this reason some mix trials were made to obtain another additional type with respect to gradation.

The next step was to compare the behavior of a set of different sand capping filters. Every two filters were made to run parallel, having the same thickness and subjected to the same field conditions (Table 3).

Later the best pair of sand filter capping types that would appear as to as to conform to the objectives and goals of the study would be tried under changing conditions of flow rates to check their suitability under such conditions. Measurements were replicated so as to statistically analyze the results.

RESULTS

For the comparison of sand filter capping investigated in this study, considerations of different turbidity standards were made. These turbidities were recommended by the customary level of (5) NTU adopted by World Health Organization (WHO, 1995) and (Iraqi standards, 1991) as well as (1) NTU adopted by U.S. Environmental protecttion Agency (EPA) standards (Hindricks, 2006).

More than forty test runs were conducted throughout the period of study. Some grab runs were made to check the adaptability of sand filter capping to tackle uncoagulated suspensions.

The process variables investigated in the study included some raw water characteristics (pH, temperature and turbidity), size and thickness of media, filtration rate, head loss and filter runtimes. Bacterial populations on standard plate count basis were estimated for the conducted test runs.

After getting the results, it was found that filters using 10 cm of anthracite did not show an efficient performance in removing turbidity. Besides, the runtimes were relatively short that made operation unfeasible for backwash and demands needs. These results were excluded.

Turbidity could be the most variable of the water quality parameters of concern in drinking water sources

Table 3. Run Configuration as cited in the literature

Experiment group set	Run No.	Media thickness cm[1]	Sand size mm[2]	Experiment group set	Run No.	Media thickness cm	Sand size mm
	1	20x20	0.92		23	10x20	0.69
	2	20x20	0.98		24	10x20	0.46
	3	20x30	0.92		25	10x30	0.69
	4	20x30	0.98	C	26	10x30	0.46
A	5	20x40	0.92		27	10x40	0.69
	6	20x40	0.98		28	10x40	0.46
	7	20x50	0.92	D	29	10x20	0.92
	8	20x50	0.98		30	10x20	0.98
	9	30x50	0.92		31	10x30	0.92
	10	30x50	0.98		32	10x30	0.98
	11	20x20	0.69		33	10x40	0.92
	12	20x30	0.46		34	10x40	0.98
	13	20x30	0.69	E	35	10x50	0.92
	14	20x30	0.46		36	10x50	0.98
	15	20x40	0.69		37	10x40	0.69
B	16	20x40	0.46		38	10x40	0.46
	17	20x50	0.69	F***	39	20x40	.
	18	20x50	0.46		40	20x40	
	19	20x60	0.69		41	20x40	
	20	20x60	0.46		42	20x40	
	21	30x60	0.69		43	20x40	
	22	30x60	0.46		44	20x40	

1. The first figure denotes anthracite thickness(cm) while the second figure denotes sand thickness(cm)
2. for all filters anthracite coal effective size = 0.91 mm
*** various filtration rates were applied to best filter configuration in terms of performance

(Montgomery, 1985) and filter performance was based on how much removal of it could be achieved. Besides, turbidity and various harmful organisms usually could occur together in surface water, any removal of turbidity would mean removal of organisms (Conley and Pitman, 1960). Further, turbidity would exert a significant chlorine demand. The World Health Organization (WHO, 1985) strongly recommended that from public health point of view, it was vitally important to produce safe, pathogenic-free drinking water using chlorine as a disinfectant that turbidity be kept low, less than (1) nephelometric turbidity unit NTU. Focus would be made on sand filter capping using 20 cm anthracite (Table 4).

Relative performance of all tried filter configurations based on the percent removal of turbidity of filters using sand No.3 and 4 were shown in Table 5.

One of the factors that affected filter performance was - intermix- (Table 6). This intermix was a function of (D_{90}/D_{10}) of the media. Intermix was increased as this ratio approaches (3).

The effects of increasing sand thickness on filter performance was also dealt with (Table 7).

DISCUSSION

The author et al published a series of papers concerning various aspects of studied parameters. Most of the concerned variables were covered in depth. Readers are requested to refer to such works elsewhere (Al-Layla et al., 1989; Al-Rawi, 1987; Al-Ani et al., 1988). Focus in this paper will highlight the advantages and economic merits drawn by application of sand filter capping in Mosul treatment plants.

Table 4 listed the results obtained throughout the period of study. These results were encouraging. They were so good when compared with the above standards. Turbidities less than the stringent guideline were obtained.

Sand filter capping consisting of 40 cm of 0.69 mm of sand overlain by 20 cm of 0.91 mm anthracite coal showed superiority among tried filter configurations in terms of quality and runtime. Filter of 40 cm of 0.46 mm of sand and 20 cm of 0.91 mm anthracite came next in this comparison. The latter filter showed higher percentage removal of turbidity for some period of operation (98.6%) as shown in Table 5. The above judgement is

Table 4. Percent removal of turbidity in function of runtime*.

Sand type No.	Coal thick-ness cm	Sand thick-ness cm	24 h		36 h		44 h	
			Effluent ntu	Percent removal	Effluent ntu	Percent removal	Effluent ntu	Percent removal
No.1	20	20	1.33	60.80	1.55	56.71	1.78	53.08
	20	30	0.98	75.01	1.17	61.31	1.70	67.84
	20	40	0.81	63.95	0.92	51.57	0.88	44.21
	20	50	0.45	80.43	0.78	61.84	3.26	64.96
No.2	20	20	1.60	52.94	1.79	50.45	1.92	49.77
	20	30	1.21	69.87	1.37	54.64	2.03	60.05
	20	40	0.84	62.36	1.00	47.36	0.93	41.07
	20	50	0.51	78.31	0.78	61.84	3.20	65.65
No.3	20	20	1.37	60.28	2.15	59.76	3.73	61.92
	20	30	1.05	78.90	3.65	90.88	5.25	81.40
	20	40	0.20	98.00	0.5	96.50	0.80	95.50
	20	50	0.49	92.00	0.6	89.65	0.73	83.59
No.4	20	20	1.48	56.95	2.45	53.34	3.37	52.61
	20	30	1.13	77.30	4.05*	90.00	3.6	83.35
	20	40	0.14	98.60	0.8	94.30	1.6	91.06
	20	50	0.50	91.85	0.59	89.65	0.73	83.59

*__Note:__ the indicated runtimes are selected as to cope with max. Operation of one of the tried filters category of compared filters and for the purpose of comparison. It does not imply that they are the max. runtimes of various filter configurations.

Table 5. Relative percent efficiency for turbidity removal in terms of sand type in sand filter capping (%).

Run length h	Filters using sand 1	Filters using sand 2	Filters using sand 3	Filters using sand 4
12	72.3	68.8	100.0	100.00
24	66.6	65.6	100.0	100.50
36	53.6	49.3	100.0	97.03
44	46.7	43.4	100.0	95.50

Table 6. Intermix probability.

Filter type	Filter with sand No.1	Filter with sand No.2	Filter with sand No.3	Filter with sand No.4
D_{90}/D_{10}	1.68	1.602	2.27	3.41

Table 7. Effect of increasing sand depth upon filter performance.

Configurations cm x cm		Percent removal of turbidity	
		12 h	24 h
Coal x sand	20 x 50	92.80	92.01
	20z60	98.51	96.00
	20 x 50	91.20	91.85
	20 x 60	98.60	96.00

based on the fact that the runtime of filter using sand No.3 is about 1.8 times that of filter using sand No. 4. Further the increase in turbidity removal of the latter filter is very slight compared to that of earlier filter. Both types of filter produce turbidity less than (1) NTU.

Based on original data of influent turbidity received by the filter units, it was noted that most of the incoming water to the filters could be described as "tough water".

Such water was hard to tackle as far as turbidity removal was concerned. Obviously a filter could remove the maximum raw turbidity when the incoming turbidity was high. On the contrary, the filters could remove very little when received turbidity was low.

As stated earlier, sand filter capping were proven to tackle quite good with such situation. Again filters using sand Nos. 3 and 4 showed superiority among other tried filters. The effluent turbidity was so low where influent turbidity to filters amounted to 10 ntu.

Some other quality merits of sand filter capping was the probability of intermix that might occur among anthracite coal and sand grains. Intermix provided more uniform decrease in pore size with thickness. This allowed more efficient use of storage space in the media and consequently provided longer runs (Cleasby, 1975). A decrease in the uniformity coefficient would reduce the amount of intermix since larger particles would not be present. This strongly confirmed in Table 6.

By referring to original results obtained by this work showed that tried sand filter capping could also tackle situation of relatively high influent turbid water to filter units. Some influent turbidity to filter units amounted to (16) NTU. The latter value was relatively high for conventional current rapid sand filters. However, it was quietly well tackled by used sand filter capping. This situation highly confirmed the flexibility and reliability, etc. of the sand filter capping and gave a good indication of proper purification.

In the light of above findings, the encouraging results of highly reduced effluent turbidity are reflected on health. Consumers will be very satisfied by the water they are drinking. The quality is palatable, wholesome and satisfies the most stringent standards in terms of turbidity.

The upper layer of sand filter capping contributes a lot in achieving these excellent results. The angularity of the anthracite results in a large void ratio and allows more solids holding capacity than the most rounded sand. The latter yields a consistency better filtrate quality throughout a filter run. The capping media system more fully utilizes the entire bed for storage of filtered solids since water passing through a capping filter contacts the coarser anthracite offering ample void ratio to effect removal and storage of the suspended solids.

The lower uniformity coefficient anthracite has less oversized and undersized particles resulting in a very uniform bed. The low uniformity coefficient anthracite coal requires much lower backwash rate which results in a substantial savings to the treatment plant due to the reduced monthly/annual wash water requirements.

From the other side, the clear quality means that the addition of disinfectants will be minimum. This in turn suggests a good signs from health points of view. Reduced turbidity also means that the added disinfection can operate effectively and efficiently. In other words, an economic revenue may be achieved through reducing costs of added disinfectants as the finished water is almost clear.

On the contrary, an elevated record of turbidity necessitates addition of considerable amounts of disinfection. Excessive disinfectant doses may give rise to Trihalomethane (THM) to occur. The latter is very carcinogenic and may be fatal to organisms.

It is well accepted that the plant production of finished water is a function of its filtration rate. Keeping this in mind, operation of filtration rate at 8.3 m/h may add another positive point. Such rate will enable water supply authority to save in two aspects. First: treatment plants can produce at least two folds the amount currently achieved when using 4 m/h rate. This will avoid the idea of constructing new treatment plant to satisfy required demands. Second, if the produced amount of water is thought to be sufficient, then a reduction by 50% or more in filter unit numbers can be saved. Both of above cases will help to reduce the cost of water treatment plants.

Further, it is seen that increasing filtration rate above 8.83 m/h showed a decline in filter efficiency (Figures 2 and 3). This was attributed to increased collision between removed particles and the media. At the same time it increased the hydraulic shear force which tended to push the particulate matter deeply in the filter where it ultimately emerged with effluent. However, turbidity of effluent for all rates was kept below (1) NTU.

Further, sand filter capping may reduce the amount of sand media as some of this media replaced by anthracite. Current plants use as much as 100 cm of sand as a filter medium. The selected configuration indicated that 40 cm of sand could be used in sand filter capping. This means that 60% of sand can be saved. Generally the accumulated amounts of saved sand from many operating filters will be considerable.

Filter runtime were approved to be longer. This would be reflected on producing more clear water. The breakthrough of impurities would be uniform and lasts longer. The latter encouraged gradual head loss buildup and consequently longer runtimes. This was attributed to the fact that the whole filter thickness was utilized in the filtration process compared to the 10 - 15 cm in the current conventional filters. Running filters for 83 h (more than 3 folds the current runtime in existing Iraqi plants) meant uniform head loss, more water and less backwash. These advantages collectively added to the economy of the process.

Longer runtimes reduced the backwash amount of water. The capping media also required a lower backwash rate to achieve fluidization with filtration rates of 2 - 3 times higher than those of rapid sand filters.

Using sand filter capping does not need further experience or skilled labor. It is simple, easy to operate and needs no complications.

All the above facts suggest the shining or positive aspects of introducing this technology. A question may arise concerning the extra cost incurred by using anthracite. This is quite reasonable. However, this will be taken care by one of the many achieved advantages of using sand filter capping. Irrespective of all mentioned

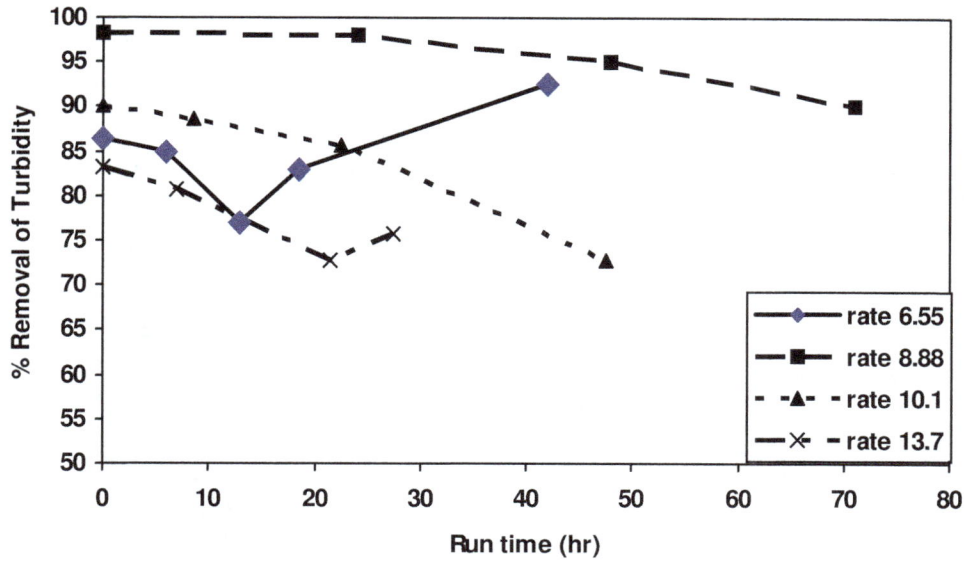

Figure 2. Capping filter performance throughout operation runtime using 20 cm anthracite and 40 cm sand No.3.

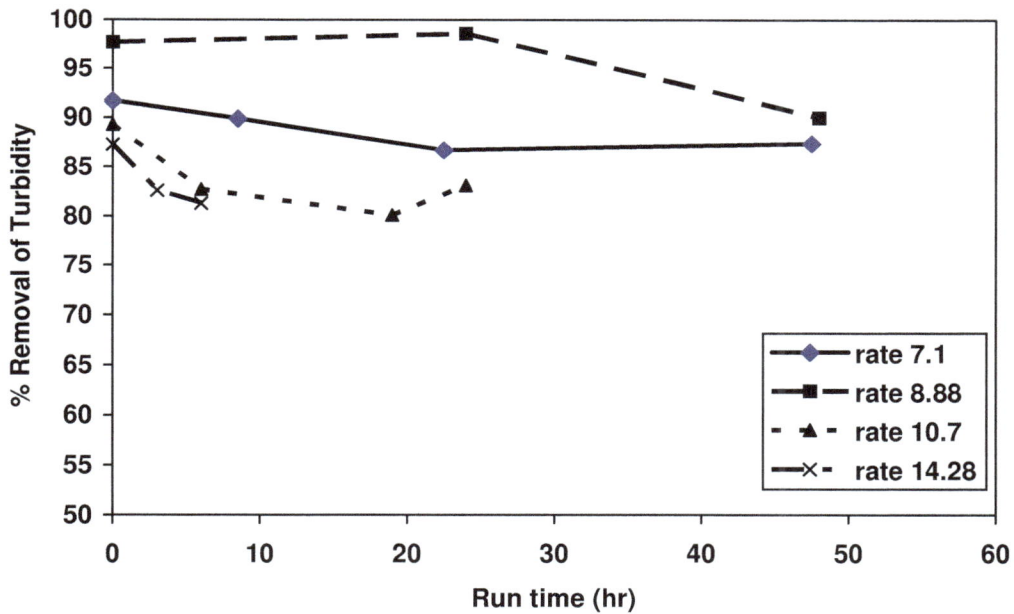

Figure 3. Capping filter performance throughout operation runtime using 20 cm anthracite and 40 cm sand No.4.

merits the provision of clear aesthetic and palatable water is solely sufficient and invaluable.

Conclusion

The following conclusions are drawn:

1. A sand filter capping consisting of 40 cm of 0.69 mm sand and 20 cm of 0.91 mm anthracite coal was found to show superiority in turbidity removal among more than 120 filter configurations tested. This filter achieved 90 - 98% of turbidity removal over 83 h of service time. Influent turbidity as much as 16 NTU had been successfully handled in the filter.

2. The whole thickness of the media was utilized in the filtration process as opposed to the top layers of conventional rapid sand filters.

3. Produced effluent satisfied most stringent quality standards, offers a satisfaction for consumers as well as it was healthy.

4. Intermix that might occur in sand filter capping provided reliability and flexibility towards shock loads of turbidity,

5. A considerable revenue could be gained through reduction of sand thickness and reducing disinfection doses.

6. A great economical burden could be achieved through operating at higher rates. Consequently more water was produced which avoids construction new plants and/or reducing number of filter units by 50% in the plants and thus reducing the plant cost.

7. Another merit concerning plants located downstream an impoundment represented by adopting direct filtration mode of purification. This is because the received turbidity is considerably low as raw water stays long in the impoundment upstream.

8. The sand filter capping could operate for much longer periods of time (three or more times as long at the same filtration rate), before backwashing was necessary because the bed could hold more turbidity.

9 .The extra cost of importing anthracite coal could be substituted excessively by any of the listed merits.

10. The study highly stressed on necessity of introducing this technique in the country water treatment plants for the various merits gained.

REFERENCES

Al-Ani MY, Al-Baldawi MF (1986). The effectiveness of using single and dual media in filtration" proceeding 4th Scientific Conference SRC, Iraq.

Al-Ani MY, Al-Layla MA , Al-Rawi SM (1988). Selection of Size and Thickness of Dual Media Filters in Treating Tigris River Water. Aqua J. UK. 6: 328-337

Al-Layla MA, Al-Ani MY, Al-Rawi SM (1989). Studies of Dual Media Filtration of Tigris River Used for Drinking Water, Mathematical Relationship. J. Environ. Health Eng. Part A Environ. Sci. Eng. USA. 24 (2): 99-110.

Al-Rawi SM (2009). Use of Dual Media Filtration for Drinking Water Treatment EPCRS symposium on new trends in water purification, 23/3/2009.Mosu/Iraq.

Al-Rawi SM (1996). Water Filtration Units, Status and Perspectives. Arab J. Sci. Tunisia (Arabic text). No.26

Al-Rawi SM(1987).Turbidity Removal of Drinking Water by Dual Media Filtration. MSc. thesis Mosul University m 1987, Iraq. p.196

Cleasby J (1975). Intermix of Dual Media and Multi-media Granular Filters. J. AWWA. 76: 4.

Conley WR, Pitman RW(1960). Innovation in Water Clarification. J.AWWA, October. 52: 10

Culp GL (1974). New Concepts in Water Purification. Van Nostrand .

Degremont (1991). Water Treatment Handbook. 6th. Ed , 1991. Paris.

Hammer M (1986). Water and Waste Water Technology 2nd ed. Wiley Pub. Book, USA, 536pp.

Hindricks DW (1974).Process Design for Water and Waste Water Treatment. Class note, CE, Colorado State University, USA.

Hindricks DW (2006). Water Treatment Unit Processes, Physical and Chemical. CRC,Press USA. 1266pp.

Iraqi Environmental Council(1991). Water Specifications. Iraq(Arabic text). p. 108

Montgomery JM (1985). Water Treatment, Principles and Design. John Wiley Inc, New York, USA. p.696.

Peavy HS, Donald RR, George T (1987). Environmental Engineering. McGraw Hill Book 2nd ed. 700pp.

Personal Communications with some pediatrics, (2009).

Schulz RC, Okun DA (1984). Surface Water Treatment for Communities in Developing Countries. Wiley Interscience Publication, USA 299pp.

Tillman MG (1996). Water Treatment Troubleshooting and Problem Solving. Lewis Publishers 1996 USA. p.156

WHO(1995). Guidelines for Drinking Water Quality"(volume 2) Geneva.

WHO(1985).Guidelines for drinking Water Quality. Vol.2 , Geneva.

4

Radon and radium concentrations in 120 samples of drinking springs and rivers water sources of northwest regions of Mashhad

A. Binesh[1] and H. Arabshahi[2]*

[1]Physics Department, Payam Nour University, Fariman, Iran.
[2]Physics Department, Ferdowsi University of Mashhad, Mashhad, Iran.

Radon makes up approximately half of the total dose of radiation were received naturally. The majority of it comes from the inhalation of progeny of [222]Rn and is prominent in a closed atmosphere. The continuous measurement of the levels of [222]Rn concentration in different geographical areas is of great importance, particularly in living places. In this study, the concentration of radium and radon in 120 samples of drinking, springs and rivers water sources of northwest regions of Mashhad city have been measured. Solid state nuclear track detectors were used for measuring the concentration. The average value of radon and radium concentrations in the studied area is found to be 30.2±5.1 and 18.4±2.2 Bq m^{-3}, respectively. The dose rate due to radon, radium and their progenies received by the population in the studied location between 0.1-0.5 mSv y^{-1}. The arithmetic and geometric mean concentrations are 0.2±0.05 and 0.2 mSv y^{-1}, respectively. The results show no significant radiological risk for the inhabitants of the studied regions.

Key words: Radon and radium concentrations, water sources.

INTRODUCTION

There is much concern these days on the part of the public and government organizations about natural radiation and the environment, particularly for dwellings (Folger et al., 1994). Due to relatively higher doses found as a consequence of elevated radon concentrations, some countries are now passing legislation to deal with the problem. This is true particularly in cold climate countries where the energy crisis is a serious problem and where houses were built more hermetically so as to minimize ventilation conditions. Radon contributes most to the effective dose received by a population from natural sources. It has been estimated that radon and its progeny contribute three-quarters of the annual effective dose received by human beings from natural terrestrial sources and are responsible for about half of the dose from all sources. Radon emanates to a certain degree from all types of soil and rocks (Al-Kazwini et al., 2003). The presence of [222]Rn in the biosphere is due to its semi-disintegration period of 3.8 days, which allows it to diffuse from the earth's crust into the atmosphere (Khan, 2000). The radiological importance of radon does not depend on the concentration of radon gas itself, but on its short-lived decay progenies, such as polonium, bismuth and lead. During breathing, radon is exhaled, but the progenies, being material particles, may deposit on to the lungs, tracks of breathing, etc. (Kearfott, 1989). Some factors that influence the diffusion of radon from soil into the air are the existence of uranium and radium in soil and rock, emanation capacity of the ground, porosity of the soil and rock, pressure gradient between the interfaces, soil moisture and water saturation grade of the medium. Radon can enter to the body via respiring, drinking and eating. The alpha emitted by this radon and other radiation emitted from its decay products increase the absorbed dose in respiratory and digestion systems (Kendal et al., 2002). Nearly 50% of annually radiation dose absorption

*Corresponding author. E-mail: arabshahi@um.ac.ir.

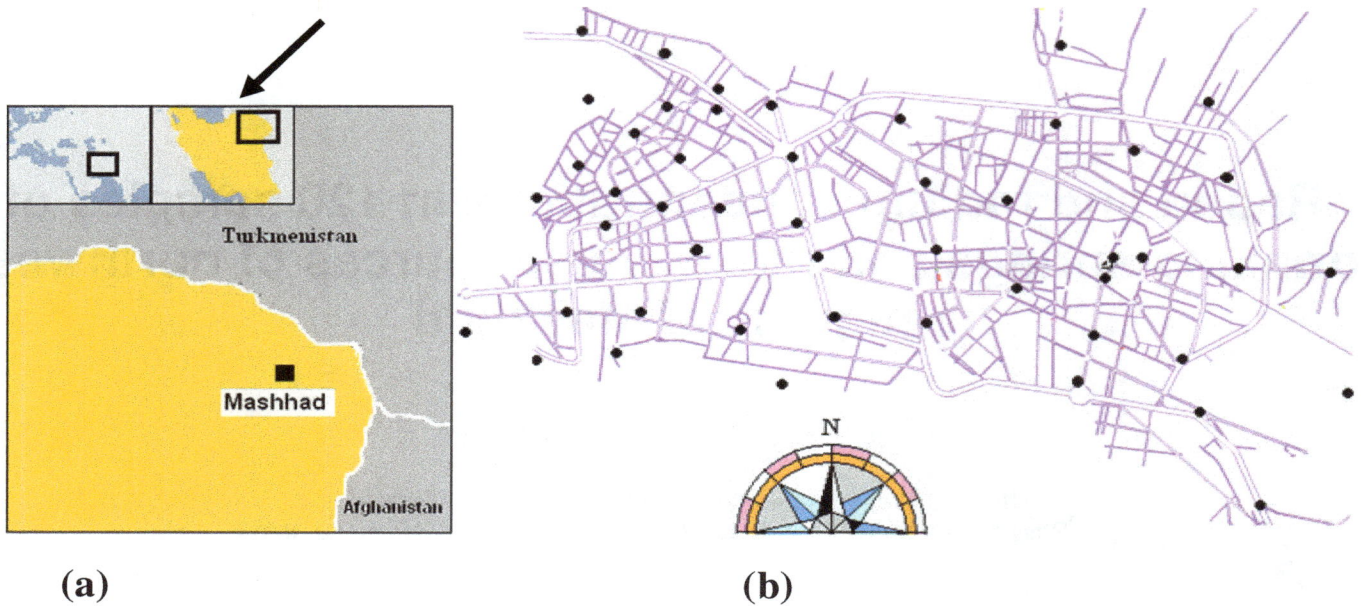

(a)   **(b)**

Figure 1. (a) Mashhad location in Iran. (b) The map of Mashhad city and ● shows the sampling sites of Zoshk, Shandiz and Torghabeh.

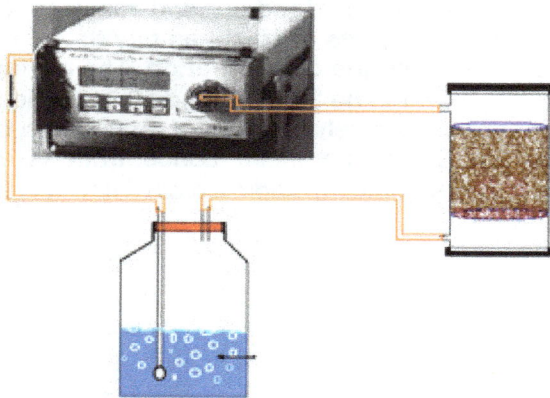

Figure 2. The PRASSI system set up for radon measuring in the water samples

the human respiratory tract system to deliver radiation dose.

MEASUREMENT METHOD

In this study, radon was measured in the water samples using PRASSI system (Savidou et al., 2001). A total of 120 samples including 38 samples of drinking water, 56 river water samples and 26 samples of spring water were tested. Figure 1 shows the sampling sites. Radium in the water samples were measured keeping the water samples in the bottles for 35 days to let radon reach the equilibrium with radium whereby we obtained radium concentration in the samples. Figure 2 shows the system set up of measurement including bubbler and drier column. PRASSI pumping circuit operates with constant follow rate at 3 liters per minute in order to degassing the water sample properly. Its detector is a scintillation cell coated with ZnS(Ag) 1830 cm^3 volume. The sensitivity of this system in continuous mode is 4 Bq/m^3 during the integration time 1 h. Numbers shown by the device is based on Bq/m^2. Using relationship Equation 1, radon gas density is calculated based on $\frac{Bq}{l}$.

$$Q_{Rn}(\frac{Bq}{l}) = Q_{PRASSI} \times \frac{Vtot(m^3)}{V(l)} \times \left[\exp(\frac{Ln2}{3.8 \times 24}t) \right]$$

(1)

Where Q_{PRASSI} value recorded by the device, V_{tot} is the total volume of air connections, V is volume sample and within the brackets is a correction factor in the delay measurement.

Radon in water samples

The third column in Table 1, radon concentration samples

of human is due to radon which is one of the main cancers cause at respiratory and digestion systems (Li et al., 2006). Radon in water can enter the human body in two ways. Firstly, radon in drinking water or mineral drinks can enter the human body directly through the gastro-intestinal tract and irradiate whole body which the largest dose being received by the stomach (Kusyk et al., 2002). Assuming an average consumption of 0.5 L of water per person per day, and stomach dose per Bq of radon is 5 nGy/Bq, with the consider 0.12 for stomach tissue weighting factor and 20 for quality factor of α radiation, the annual equivalent dose per Bq of radon concentration in water is about 2.19×10^{-6} μSv/(year Bq l). Secondly, radon can escape from household water and became an indoor radon source, which then enter

Table 1. Radon and radium concentration data of different water samples.

Sample No.	Water sample	$Q_{Rn}(\frac{Bq}{l})$	$Q_{Ra}(\frac{Bq}{l})$
1	Zoshk River	0	0
2	River 10 km before Zoshk	0	0.23
3	River 2 km before Zoshk	0	0
4	River 8 km before Zoshk	0	0.24
5	River 1 km after Zoshk	0	0.15
6	River 4 km after Zoshk	0.30	0
7	Zoshk spring water	0.33	0
8	Zoshk drinking water (No. 1)	0.32	0.045
9	River 1.5 km after Zoshk	0.38	0.09
10	Torghabeh drinking water (No. 1)	0.54	0
11	River of Shandiz waterfall (No.1)	0.56	0.68
12	River 2.3 km after Torghabeh	0.58	0.08
13	River 2.5 km after Zoshk	0.6	0.05
14	River 1.3 km after Zoshk	0.66	0.099
15	Zoshk drinking water (No. 2)	0.94	0
16	Shandiz waterfall	1.04	0.18
17	River 2.8 km after Abrdh	1.180	0.17
18	River 0.8 km after Zoshk	1.299	0.018
19	River 1.8 km after Zoshk	1.3	0
20	River 2.7 km after Torghabeh	1.35	0
21	Shandiz drinking water (No.1)	1.4	2.19
22	River 2.5 km after Zoshk	1.54	0.056
23	Shandiz drinking water (No. 2)	1.96	0
24	Torghabeh drinking water (No. 2)	1.641	0.163
25	River 2.3 km after Zoshk	1.763	0
26	Zshk drinking water (No. 3)	1.853	0.13
27	Upper Torghabeh drinking water (N0.1)	1.937	0.308
28	River 0.7 km after Zoshk	2.241	0.096
29	Zshk spring water (No. 1)	2.3	0.036
30	River 2.7 km after Zoshk	2.352	0
31	Shandiz drinking water (No.3)	2.412	0.492
32	River 0.8 km after Zoshk spring water	2.435	0.14
33	Lower Torghabeh drinking water (No.1)	2.476	0
34	Shandiz drinking water near the Mosque	2.476	0
35	Shandiz drinking water (No. 4)	2.63	0.854
36	Upper Torghabeh drinking water (No. 2)	2.698	0.07
37	River 5 km after Torghabeh	2.85	0
38	River 1.7 km after Zoshk	2.873	0.208
39	Lower Torghabeh drinking water (No. 2)	2.87	0
40	Lower Abrdh spring water	3.049	0.24
41	Shandiz drinking water (No. 5)	3.153	0.66
42	River of Shandiz waterfall (No.1)	3.215	0137
43	Lower Torghabeh drinking water (No. 3)	3.24	0.491
44	River 1.3 km after Zoshk	3.269	0
45	River beginning Zoshk	3.418	0.07
46	River 5.5 km after Torghabeh	3.492	0
47	Shandiz drinking water (No. 6)	3.619	0.787
48	River at Zoshk	3.76	0
49	River 5.9 km after Torghabeh	4.012	0.013
50	River 2.4 km after Torghabeh	4.17	0.25

Table 1. Cont.

51	River 0.5 km after Zoshk	4.2	0.133
52	Shandiz Drinking Water (No.7)	4.231	0
53	River 1.5 km after Zoshk	4.237	0.051
54	Upper Torghabeh drinking water (N0. 3)	4.254	0
55	Upper Abrdh drinking water (N0. 4)	4.375	0
56	River 2.6 km after Torghabeh	4.729	0
57	River 1.2 km after Zoshk	4.87	0
58	Lower Torghabeh drinking water (N0. 4)	4.895	0.3
59	Shandiz drinking water (No. 8)	4.98	0
60	Lower Torghabeh drinking water (N0. 5)	5.051	0.1108
61	River of Shandiz waterfall (No. 2)	5.05	0.316
62	River 3.5 km after Abrdh	5.081	0.059
63	Lower Abrdh spring water	5.130	0.244
64	River 0.1 km after lower Torghabeh	5.255	0
65	River 1.6 km after Zoshk	5.431	0.057
66	Upper Torghabeh spring water	5.441	0.044
67	River 0.2 km after Zshk	5.453	0.29
68	Torghabeh Drinking Water (N0. 3)	5.482	0
69	River 4 km before Torghabeh	5.579	0.133
70	River 5 km before Torghabeh	5.675	0
71	River 0.5 km after Torghabeh	5.66	0.094
72	Zoshk spring water (No. 2)	5.727	0
73	Upper Torghabeh drinking water (No. 5)	6.141	0.087
74	Lower Torghabeh drinking water (N0. 6)	6.574	0.047
75	Torghabeh drinking water (N0. 4)	6.907	0.288
76	Spring water 1 km after Zoshk	7.02	0
77	Lower Torghabeh drinking water (N0.7)	7.15	0.24
78	River 2.8 km after Zoshk	7.13	0
79	Torghabeh drinking water (No. 5)	7.77	0.24
80	River 0.2 km after lower Torghabeh	7.587	0.093
81	Lower Torghabeh spring water (No. 1)	7.631	0.132
82	River 2.9 km after Zoshk	7.867	0.291
83	Zoshk Spring Water (No. 3)	7.895`	0
84	River 4.5 km after Torghabeh	7.969	0
85	Torghabeh drinking water (No. 6)	8.131	0.178
86	Zoshk drinking water (No. 4)	8.155	0.058
87	Zoshk drinking water (No. 5)	8.310	0
88	Zoshk spring water (No. 4)	8.327	0
89	River 0.4 km after Zoshk	8.356	0
90	Zoshk drinking water (No. 6)	8.603	0.054
91	Lower Torghabeh drinking water (No. 8)	8.630	0.437
92	Zoshk spring water (No. 5)	9.034	0.183
93	Zoshk spring water (No. 6)	9.056	0.280
94	River 2.5 km after Torghabeh	9.931	0.0189
95	River of Shandiz waterfall (No. 3)	10.124	0
96	Qelqeli spring water	10.402	0.083
97	Zoshk drinking water (No. 7)	10.721	0.0014
98	Lower Torghabeh drinking water (No. 9)	10.729	0
99	Zoshk Drinking Water (No. 8)	10.915	0.0052
100	Lower Torghabeh drinking water (No. 10)	10.992	0.022
101	Shandiz drinking water (No. 9)	11.199	0
102	Spring Water 0.5 km after Zoshk	11.360	0.127

Table 1. Cont.

103	River 1 km before Zoshk	11.434	0.207
104	Lower Torghabeh drinking water (NO. 11)	11.595	0.096
105	River 2 km after Zshk	11.778	0.433
106	Zoshk spring water (No. 7)	13.055	0.133
107	River 1 km after Zoshk	13.058	0.091
108	Zshk spring water (No. 8)	13.761	0.0026
109	Zshk spring water (No. 9)	14.43	0.183
110	Spring water 0.1 km after Zshk	14.577	0
111	Spring water 2 km after Zshk	14.863	0.207
112	Zshk drinking water (No. 9)	15.755	0
113	River 0.5km before Zshk	16.324	0
114	Spring water at Zshk	16.344	0
115	River of Shandiz waterfall (No. 4)	17.363	0.354
116	Upper Abrdh drinking water (No. 6)	17.879	0.207
117	Lower Abrdh spring water (No. 2)	18.445	0.047
118	River 1.5 km after Abrdh	18.578	0
119	Spring water 0.7 km after Zshk	21.495	0.01
120	Spring water 1.5 km before Zshk	31.881	0.66

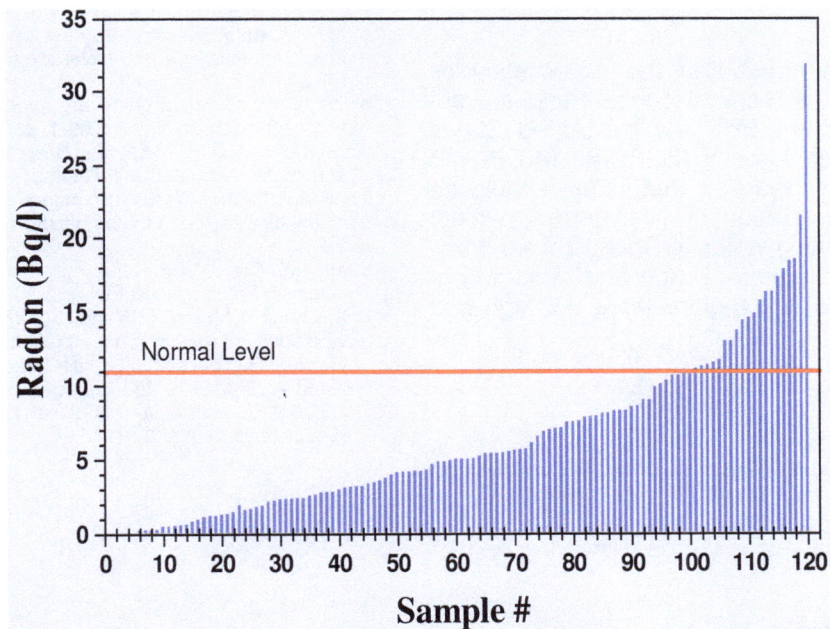

Figure 3. The histogram of radon gas concentration in 120 water samples of Shandiz, Zoshk and Torghabeh regions.

that have been ordered from low to high, is listed. Also, the radon gas density results are shown in histogram of Figure 3. As it can be seen, only 81/12% of the samples, the last 19 samples in Table 1 have concentrations higher than 11 Bq/l, particularly the sample number 120 that related to the spring in the village of Zoshk has concentration about 30 Bq/l.

Radium in water samples

Figure 4 shows the histogram of radium concentration in different water samples as well as the data listed in fourth column of Table 1. The radium concentration of samples were less than 1 Bq/l, except sample number 21, drinking water of Shandiz region is about 1.87 Bq/l.

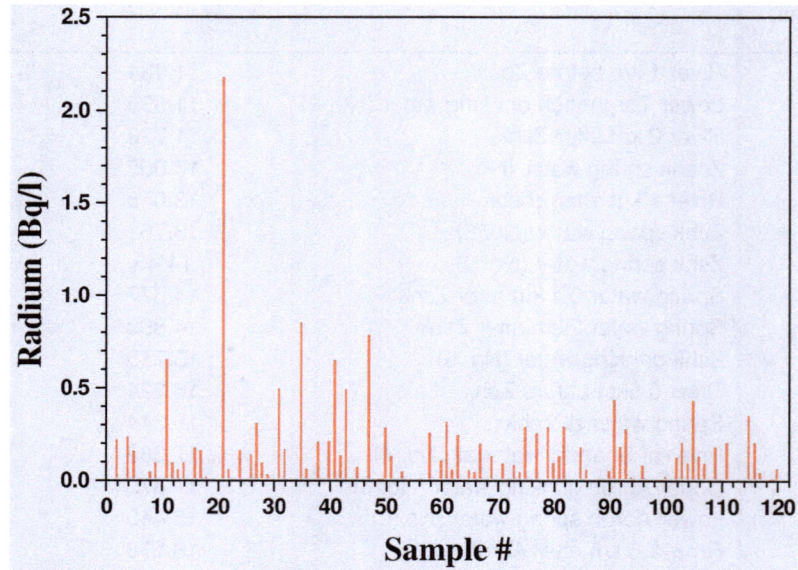

Figure 4. The histogram of radium concentration in different water samples.

Conclusion

Results of radon concentration in the water samples showed that only 14.67% sample concentrations were higher than the normal 11 Bq/l, set by United States Environmental Protection Agency (USEPA). 148 Bq/l is limit amount of action or reaction that radon should be reduced. Radium concentration of all samples, except sample number 21, drinking water of Shandiz were small and less than 1 Bq/l. Therefore, radon and radium concentration in the water of the regions were not high and these were appropriate.

REFERENCES

Al-Kazwini AT, Hasan MA (2003). Radon concentration in Jordanian drinking water and hot springs. J. Radiol. Prot., 23: 439-448.

Folger PF, Nyberg P, Wanty RB, Poeter E (1994). Relationship between [222]Rn dissolved in groundwater supplies and indoor [222]Rn concentrations in some Colorado front range houses. Health Phys., 67: 244-252.

Kearfott K J (1989). Preliminary experiences with 222Rn gas Arizona homes. Health Phys., 56: 169-179.

Kendal GM, Smith T J (2002). Dose to organs and tissues from radon and its decay products. J. Radiol. Prot., 22: 389-406.

Khan AJ (2000). A study of indoor radon levels in Indian dwellings, influencing factors andlung cancer risks. Radiation Measure.,, 32: 87-92.

Kusyk M, Ciesla KM (2002). Radon levels in household waters in southern Poland. NUKLEONIKA, 47: 65- 68.

Li X, Zheng B, Wang Y, Wang X (2006). A study of daily and seasonal variations of radon concentrations in underground buildings. J. Environ. Radioactivity, 87: 101-106.

Savidou A, Sideris G and Zouridakis N (2001). Radon in public water supplies in Migdonia Basin, central Macedonia, northern Greece. Health Phys., 80: 170 - 174.

5

Prioritization of micro watersheds on the basis of soil erosion hazard using remote sensing and geographic information system

Vipul Shinde*, K. N. Tiwari and Manjushree Singh

Agricultural and Food Engineering Department, Indian Institute of Technology, Kharagpur, India-721302, India.

Degradation of agricultural land by soil erosion is world wide phenomenon leading to loss of nutrient-rich surface soil, increased run off from more impermeable subsoil and decreased water availability to plant. Thus estimation of soil loss and identification of critical area for implementation of best management practice is central to success of soil conservation programme. In this study universal soil loss equation (USLE) interactively with raster-based geographic information system (GIS) has been applied to calculate potential soil loss at micro watershed level in the Konar basin of upper Damodar Valley Catchment of India. The main advantage of the GIS methodology is in providing quick information on the estimated value of soil loss for any part of the investigated area. The rainfall erosivity R-factor of USLE was found as 293.96 and the soil erodibility K-factor varies from 0.325 - 0.476. Slopes in the catchment varied between 0 and 83% having LS factor values ranging from 0 - 6.7. The C-factor values were computed from existing cropping patterns in the catchment and support practice P-factors were assigned by studying land slope. Average annual soil erosion at micro watershed level in Konar basin having 961.4 km^2 areas was estimated as 1.68 t/ha/yr. Further, micro watershed priorities have been fixed on the basis of soil erosion risk to implement management practices in micro watersheds which will reduce soil erosion in Konar basin.

Key words: Remote sensing, soil erosion, geographic information system, universal soil loss equation, priority.

INTRODUCTION

Land and water are the two basic natural resources for the survival of living systems. These two resources have been interacting with each other in various phases of there respective cycles. The future of the nation depends largely on the effective utilization, management and development of these resources in an integrated and comprehensive manner. Soil erosion in catchment areas and the subsequent deposition in rivers, lakes and reservoirs are of great concern for two reasons. Firstly, rich fertile soil is eroded from the catchment areas. Secondly, there is a reduction in reservoir capacity as well as degradation of downstream water quality. Sediment particles originating from the continuous process of erosion in the catchment area propagated along the river flow. When this flow accumulates into the reservoir the sediment that has been carried with the stream gets settled into the reservoir and reduces its capacity. Reduction of storage capacity of a reservoir beyond a certain limit hampers the purpose of the reservoir for which it was designed. Estimation of sediment deposition in a reservoir using conventional techniques like hydrographic survey is a cumbersome procedure. It involves huge time, manpower and even it is not cost effective.

Several empirical models based on the geomorphological parameters were developed in the past to quantify the sediment yield. Several other methods such as sediment yield index (SYI) method proposed by Bali and Karale (1977) and universal soil loss equation (USLE) by Wischmeier and Smith (1978) are extensively used for prioritization of the watersheds. The USLE has been widely applied at a watershed scale on the basis of lumped approach to catchment scale (Jain et al., 2001).

*Corresponding author. E-mail: vipulshinde123@gmail.com

In several other studies, watershed has been sub-divided either into cells or of regular grid or into units where a unique run off direction exists (Onyando et al., 2005; Wu et al., 2005). Renschler et al. (1997) used USLE and RUSLE to predict the magnitude and spatial distribution of erosion within a GIS environment using ILWIS software in catchment of 211 km2 at grid resolution ranging from 200 to 250 m to be more reasonable. Dabral et al. (2008) divided Dikrong river basin into 200 × 200 m grid cells. He found the average annual soil loss of the Dikrong river basin is 51 t/ha/yr. About 25.61% of the watershed area is found out to be under slight erosion class. The USLE model applications in the grid environment with GIS would allow us to analyze soil erosion in much more detail. It is more reasonable to use the USLE on physical basis than to apply it to an entire watershed as a lumped model. Although, GIS permits more effective and accurate application of the USLE model for small watershed, most GIS-model applications are subject to data limitations (Fistikoglu and Harmancioglu, 2002).

An W (2008) used GIS-Based hydrological model for highway environmental assessment study. He developed highway watershed model (HWM) using the watershed modeling system (WMS) to simulate the hydrology and hydraulic behavior along the stream system draining selected watersheds near I-99 highway construction site. With 15% deviation as accepted criterion, the modeling results of WMS show all total run off volumes are satisfactory. The technology of remote sensing and GIS is gaining importance as a powerful tool in the management of information in agriculture, natural resources assessment, environmental protection and conservation (Javed et al., 2009). Pandey et al. (2007) divided Karso watershed of Hazaribagh, Jharkhand State, India into 200 × 200 m grid cells and average annual sediment yields were estimated for each cell of the watershed to identify the critically prone areas of watershed. Recent studies (Pandey et al., 2007; Yoshino and Ishioka, 2005; Chowdary et al., 2004; Sharma et al., 2001; Khan et al., 2001; Sidhu et al., 1998) revealed that RS and GIS techniques are of great use in characterization and prioritization of watershed areas.

DESCRIPTION OF THE STUDY AREA

Damodar Valley Catchment lies between 23°34' to 24°9' N latitude and 85°00' to 87°00' E Longitude. The total valley area covered by DVC is approximately 24,235 sq. km. Mean annual rainfall in the basin is of the order of 1,300 mm and about 80% of rain precipitates during the monsoon (June to September). The lower valley known as Damodar catchment (Drainage area - 10966.10 km^2) has three reservoirs, namely, Tenughat, Konar and Panchet comprising of drainage area of 4395.15, 997.15 and 5573.8 km^2 respectively, which lies between 23°34' to 24°9' N latitude and 84°42' to 86°46' E longitude. Konar basin having drainage area 997.15 km^2 and 39

micro watersheds is taken for this study.

MATERIALS AND METHODS

Data used

The rainfall, run off and sediment yield data were collected from the soil conservation department of DVC, India. Daily rainfall data for the period from 1993 to 2001 in the study area was used to compute rainfall erosivity - R factor. The soil maps of the study area in the scale of 1:250,000 were traced, scanned and exported to Erdas imagine 8.5.

The scanned maps were loaded in ERDAS and georeferenced. Boundaries of different soil textures were digitized and the polygons representing various soil categories were assigned with different colours for identification. Required data like soil texture, bulk density etc. were extracted for each micro-watershed of Konar basin. Toposheet of study area was taken from DVC. SRTM Digital elevation model (DEM) was used to prepare LS factor map. The LANDSAT ETM images for the study area were downloaded from http://glcf.umiacs.umd.edu. These images were used to prepare land use/ land cover map. The micro-watershed treatment map and boundary map of 1": 4 miles scale for the upper Damodar valley were also collected from DVC Hazaribagh.

Soil erosion model- universal soil loss equation

The universal soil loss equation was used to determine the average annual soil loss and its spatial distribution on the watershed. The USLE predicts soil loss for a given site as a product of six major erosion factors (Equation 1), whose values at a particular location can be expressed numerically. The limitation of this model is that it does not estimate deposition, sediment yield, channel erosion, or gulley erosion. Thus, the USLE is suitable for predicting long-term averages and the soil erosion is estimated as follows:

$$A = R \times K \times L \times S \times C \times P \qquad (1)$$

Where; A is average annual soil loss rate (t/ha/yr), R is rainfall erosivity factor (MJ-mm/ha/h/yr), K is soil erodibility factor (t-ha-h/ha/MJ/mm), LS is topographic factor, C is crop management factor and P is conservation supporting practice factor. The data used for calculating these USLE factors is shown in Table 1.

Development of model database for universal soil loss equation

Rainfall erosivity factor (R)

The rainfall erosivity factor (R) map is prepared using daily rainfall data from Nagwan station located in the Konar basin. This map is based on 9 year-average rainfall data which is used to calculate annual average R factor values. All the storms do not produce run off and hence storms more than 12.5 mm were only used in computation as suggested by Wischmeier (1959).

Panigrahi et al. (1996) developed a model for estimation of R factor (Equation 2) from daily rainfall amount (P) for 31 years for Bhubaneshwar. They reported 12.2 average percentage deviations between the observed and calculated R factor and concluded that their model given below could be well used for computation of R factor using the daily rainfall amount.

$$R = P^2 (0.00364 \log_{10} P - 0.000062) \qquad (2)$$

Table 1. Universal soil loss equation factor – data used.

USLE factor	Data
R	Daily rainfall data
K	Soil sample analysis data
LS	SOI topographic maps
C	Digital land use/land cover map
P	Field survey data

Table 2. Soil structure code.

Code	Structure	Size, mm
1	Very fine granular	<1
2	fine granular	1 - 2
3	Medium or coarse granular	2 -10
4	Blocky, platy or massive	>10

Table 3. Soil permeability code.

Code	Description	Rate, mm/h
1	Rapid	> 130
2	Moderate to rapid	60 - 130
3	Moderate	20 - 60
4	Slow to moderate	5 - 20
5	Slow	1 - 5
6	Very slow	< 1

Soil erodibility factor (K)

The soil erodibility factor (K) was computed using field and laboratory estimated physical-chemical properties of the surface soils. The laboratory soil analysis was carried out to determine soil texture, structure, permeability and organic matter content for various soil group of the area. Wischmeier et al. (1971) developed the procedure for determination of soil erodibility factor by developing an equation based on five soil parameters, which is used in the present study.

$$100K = 2.1M^{1.14} (10^{-4}) (12 - a) + 3.25 (b -2) + 2.5 (c-3) \qquad (3)$$

Where, K = soil erodibility factor, M = percentage silt, very fine sand and sand > 0.10 mm, a = organic matter content, b = structure of the soil, c = permeability of the soil.

Soil structure code was assigned on the basis of particle size of soil using values given in Table 2. Permeability code for soil type was assigned on the basis of permeability rate using values given in Table 3. Soil erodibility factor (K) is a measure of the total effect of a particular combination of soil properties. Some of these properties influence the soil's capacity to infiltrate rain and therefore, help to determine the amount of rate of run off; some influence its capacity to resist detachment by the erosive forces of falling raindrops and flowing water and thereby determine soil content of the run off. The inter-relation of these variables is highly complex.

Topographic factor (LS)

Derivations of topographic factors (L and S) were performed by computing slope length and gradient respectively, using SOI topographical maps at a scale of 1:25,000. Combined (LS) factor for all the micro-watersheds was computed using the slope map generated from the DEM of study area. LS is the expected ratio of soil loss per unit area from a field slope to that from a 22.13 m length of uniform 9 percent slope under otherwise identical conditions. Although, L and S factors can be determined separately, the procedure has been further simplified by combining the L and S factors together and considering the two as a single topographic factor (LS) (Wischmeier and Smith, 1965). Combined LS factor layer was generated as:

(I) For slopes up till 21%, the equation modified by Wischmeier and Smith (1978) was used which is:

$$LS1 = (L / 22.1) \times (65.41 \sin^2\theta + 4.56 \sin \theta + 0.065) \qquad (4)$$

Where LS1 is the slope length and gradient factor and θ is angle of the slope.

(II) For slope steepness of 21% or more, the Gaudasasmita equation was used which is:

$$LS2 = (L / 22.1)^{0.7} \times (6.432 \times \sin (\theta^{0.79}) \times \cos (\theta)) \qquad (5)$$

Where LS2 is the slope length and gradient factor and θ is angle of the slope.

$$L = 0.4 \times Sp + 40 \qquad (6)$$

Where L is slope length in meters and Sp is slope steepness in percentage.

Crop management factor (C) and conservation practice factor (P)

The crop management (C) factor reflects the combined effect of cover, crop sequence, productivity level, length of growing season, tillage practices, residue management and the expected time distribution of erosive rainstorm with respect to seeding and harvesting date in the locality. Actual loss from the cropped field is usually much less than the amount of soil loss for a field kept continuously in fallow conditions. This reduction in soil loss depends on the particular combination of cover, crop sequence and management practices. Crop management factor is the expected ratio of soil loss from a cropped land under specific condition to soil loss from clean tilled fallow on identical soil and slope under the same rainfall conditions. In this study, the land use/land cover map was derived from the satellite images and served as a guiding tool in the allocation of C and P factors for different land use classes. The study area has been classified into seven land use classes. Crop management factor was assigned to each land use class by using available C factor values in literature for that class in same agro climatic conditions. In this study, P factor values have been assigned on the basis of percent slope of the micro watershed.

RESULTS AND DISCUSSION

Development of thematic map of universal soil loss equation factors

Rainfall erosivity (R) factor

The catchment sediment yield is more sensitive to rainfall

Table 4. Rainfall erosivity factor.

Year	Average annual rainfall (mm)	Annual R factor
1993	890	135.1
1994	1693	415.0
1995	1425	426.0
1996	1265	393.0
1997	1380	285.8
1998	1263	195.9
1999	1580	414.3
2000	1047	211.3
2001	1195	169.3
Average	1304.2	293.96

amount than to either EI_{30} or the R-factor. The daily rainfall is better indicator of variation in sediment yield with the added advantage that it can be used to characterize the seasonal distribution of sediment yield. While the advantages of using annual rainfall include its ready availability, ease of computation and greater regional consistency of the exponent. Therefore, in the present analysis R factor using daily rainfall amount suggested by Wischmeier (1959) and validated by Panigrahi et al. (1996) for Indian conditions was used. The average annual precipitation in the Upper DVC is 1300 mm with a standard deviation of 161.13 mm. Using daily rainfall data from year 1993 to 2001 and Equation 2, R factor value for Konar basin was estimated and was found as 293.96 (Table 4). Using R factor value, R factor map was prepared in ArcView3.1 and shown in Figure 1(c).

Soil erodibility (K) factor

The factors like texture, structure, organic matter content and permeability are very significant in determining soil erodibility. Soil erodibility is regulated by a complex set of physical and chemical properties and is usually determined empirically. Soil analysis data was available for all soil types found in Konar basin. K factor values for each soil type were calculated using Equation 3. K factor values are assigned to respective soil types in soil map. Using K factor values, K factor map was prepared in ArcView3.1, and shown in Figure 1(d). The value of K-factor was found to be ranging between 0.325 and 0.476.

Topographic factor (LS)

DEM generated slope length are based on the assumption that each slope plane consists of a homogenous form of slope and vegetation cover, which in practice may not be the case. While deriving topographic factors, GIS

techniques tend to predict very long slope lengths on flat to very gentle slopes, which can lead to overestimation of soil loss. As a result, the LS factor fails to fully account for the hydrological processes that affect run off and erosion, its importance as a measure of the sediment transport capacity of run off from the landscape not withstanding. SRTM DEM shown in Figure 1(b) was used to derive slope map in percent and degree. Using slope map and Equations 4 and 5, LS factor map was prepared using ArcView3.1 and shown in Figure 1(e). The elevation of the study area is ranging between 140 to 844 m and the value of LS factor for study area was ranging from 0 to 6.7.

Crop management factor (C) and conservation practice factor (P)

Information on land use permits a better understanding of the land utilization aspects on cropping pattern, fallow land, forest and wasteland and surface water bodies, which are vital for development planning/erosion studies. Remote sensing and GIS technique has a potential to generate a thematic layer LU/LC of a region. The study area has been classified into seven land use classes shown in Figure 1(a). Crop management factor was assigned to different land use patterns using values given in Table 5. Using LU/LC map and C factor values, C factor map was prepared in ArcView3.1 and is shown in Figure 1(f) Crop management factor was found to be in the range of 0.002 to 1.00. Conservation practice factor for micro watersheds of Konar basin was assigned on the basis of percent slope. Soil and Water assessment Tool (SWAT) given criteria for P factor was used for this purpose. Conservation practice factor was assigned for different slope range using values given in Table 6. Using P factor values, P map was prepared in ArcView3.1 and is shown in Figure 1(g).

Average annual soil loss of Konar Basin

The annual soil loss for micro watersheds was calculated by using annual average R (based on daily rainfall data of 1993 - 2001), K, LS, C and P factors. All the layers viz. R, K, LS, C and P were generated in GIS and over layed to obtained the product, which gives annual soil erosion map (Figure 2) for the Konar basin. This soil loss map is over layed with micro watershed map of Konar basin which contains 39 micro-watersheds to get micro watershed wise soil loss. The soil erosion rate (t/ha/yr) of a micro watershed was estimated as total soil loss of ith micro watersheds (t/yr) / total geographical area of ith micro watersheds (ha). The classification of erosion rate has given rise to five categories of soil loss intensity (Figure 2). The observed sediment yield of Nagwan catchment having 92.46 km^2 area was 2.79t/ha. The

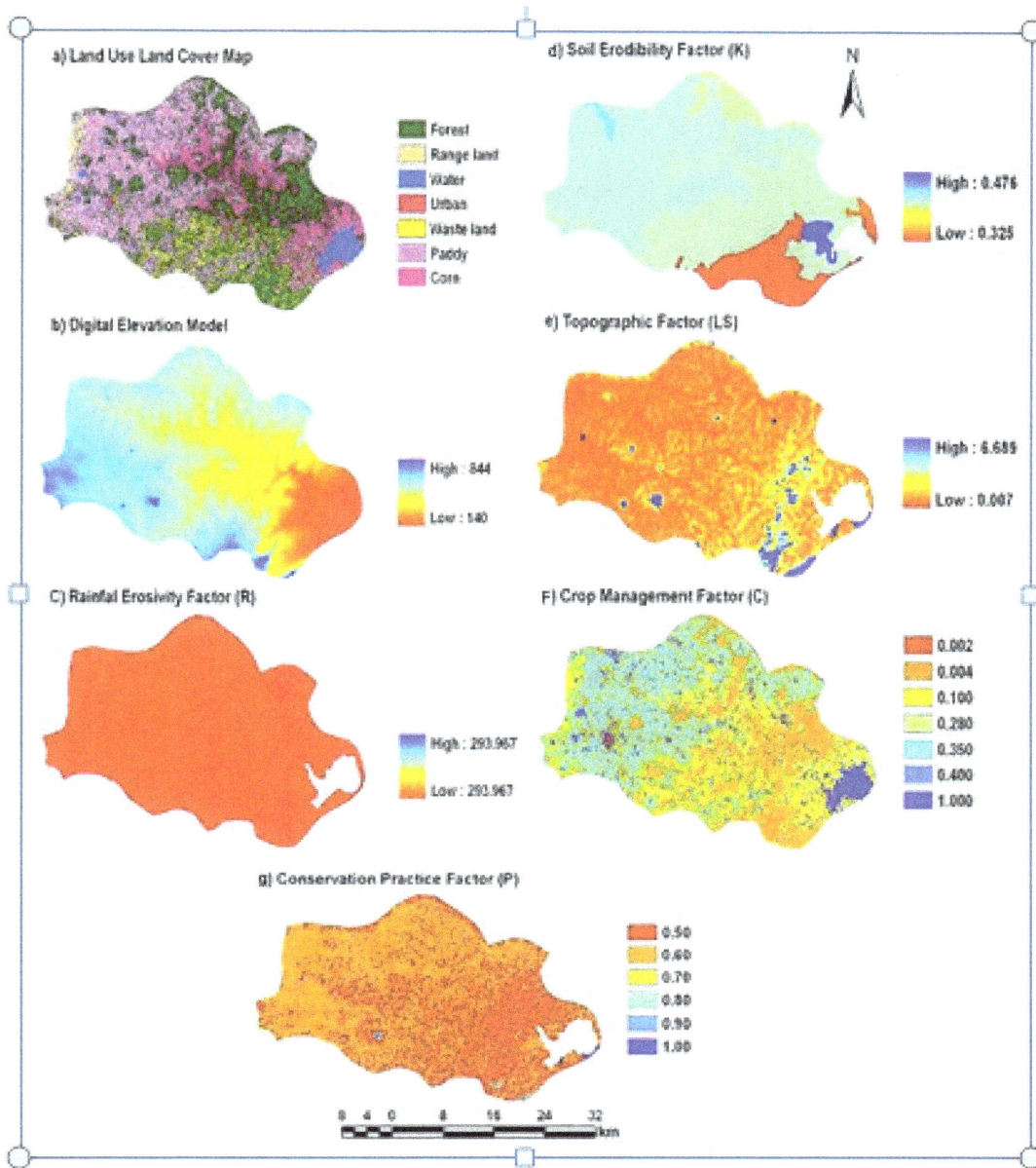

Figure 1. Thematic layers of USLE factors.

Table 5. Crop management factor values.

Land use/land cover	C value
Forest	0.004
Range	0.1
Water body	1
Urban	0.002
Wetland	0.4
Corn	0.35
Paddy	0.28

Table 6. Conservation practice factor values.

Slope, %	P value
0 - 2	0.6
2.1 - 5	0.5
5.1 - 8	0.5
8.1 - 12	0.6
12.1 - 16	0.7
16.1 - 20	0.8
20.1 - 25	0.9

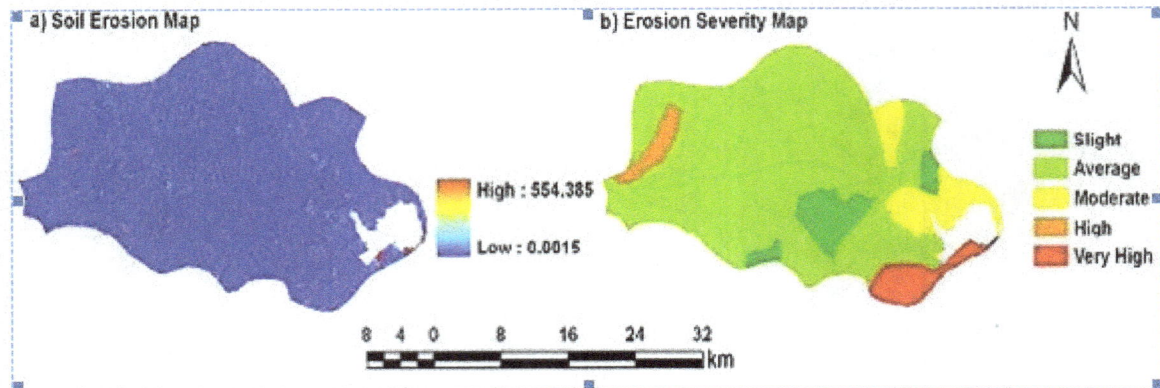

Figure 2. Soil erosion map and micro watershed wise severity map.

calculated soil loss for micro watershed having code kd1f of Konar basin which falls under Nagwan Catchment having area of 50.59 km^2 was 1.31t/ha as compared to that of 1.527t/ha as observed value. The slight difference in observed sediment yield value and calculated erosion value was due to the deposition of soil particles during erosion process, as USLE does not consider deposition of soil particles. Prioritization of micro watersheds within the basin based on soil erosion risk has been made.

Prioritization of micro-watersheds

Considering the massive investment in the watershed development programme, it is important to plan the activities on priority basis for achieving fruitful results, which also facilitate addressing the problematic areas to arrive at suitable solutions. The resources-based approach is found to be realistic for watershed prioritization since it involves an integrated approach. Prioritization of micro-watersheds was done on the basis of average annual soil loss. All the 39 micro watersheds in the study area have been prioritized by considering the results of various thematic maps derived from satellite imagery as well as rainfall and soil data. Table 7 indicates distribution of the 39 micro watersheds of Konar basin according to soil erosion intensity.

SUMMARY AND CONCLUSION

A quantitative assessment of average annual soil loss on micro-watershed basis was made using USLE with a view to know the spatial distribution in the Konar basin. The use of GIS and remote sensing data enabled the determination of the spatial distribution of the USLE parameters. Micro-watershed wise soil loss was estimated and prioritization of micro-watersheds was done on the basis of annual average erosion obtained

Table 7. MW wise average annual soil loss and priority.

MW Code	Area (km^2)	Erosion (t/ha/yr)	Priority
ka1b	44.51	10.058	1
Ka1c	18.95	1.7484	9
Ka1d	14.64	1.5926	13
Ka1f	19.6	1.067	32
Ka1g	16.51	2.1963	5
ka1h	11.42	2.3546	3
Ka1j	23.78	1.28	26
Kb1a	30.71	2.2699	4
Kb1b	11.24	0.4746	39
Kb1c	8.88	1.0978	30
Kb1d	8.35	1.6008	11
Kb1f	27.31	1.3516	22
Kb1g	21.77	2.1521	6
Kc1a	20.3	0.8972	36
Kc1b	19.4	0.9213	35
Kc1c	9.76	0.6384	38
Kc1d	25.32	1.0685	31
Kc1f	30.73	1.4888	16
Kc1g	31.62	1.5909	14
Kc1h	34.96	1.3923	20
Kc1j	8.35	1.0296	34
Kc1k	8.74	1.2296	28
Kc1m	8.6	0.8272	37
Kc2a	28.99	1.2905	25
Kc2b	37.3	1.6672	10
Kc2c	14.98	1.3338	23
Kc2d	31.22	1.5289	15
Kd1a	27.72	1.0413	33
Kd1b	39.13	1.392	21
kd1c	44	1.2667	27
Kd1f	50.59	1.3102	24
kd1g	29.47	1.4265	17
Kd1h	31.21	1.7776	8
Kd1j	28.7	1.424	19
Kd2a	26.61	1.4246	18
Kd2b	27.65	1.5949	12
Kd2c	25.19	3.9731	2
Kd2d	33.44	1.1036	29
Kd2f	29.77	1.7987	7

from 9 years daily rainfall data. Annual average soil erosion for Konar basin was 1.68 t/ha/yr at micro-watershed level. Particularly R and K are least influencing as rainfall decreases and clay proportion in soils increases downstream. The micro watershed priority-zation indicated that the micro watersheds falling under high and very high priority class requires immediate attention for soil conservation treatment. The prioritization map prepared using remote sensing and GIS technology for the present study satisfactorily matched (65%) with the priority map prepared through field based sediment yield index method of AISLUS. Hence, remote sensing and GIS technology can be used as an alternative to conventional method of soil loss estimation and subsequent prioritization of micro watershed for implementing soil conservation practices. The best management practices proposed for micro watersheds of Konar basin are; afforestation, trenching, bunding, stone wall fencing, brushwood check dams, earthen check dams, gabian structures and masonry structures.

REFERENCES

An W (2008). GIS-Based Hydrological Model - Study in Highway Environmental Assessment, ISBN-10: 3639098382, VDM Publishing House, Germany. ISBN.13: 978-3639098389,

Bali YP, Karale RL (1977). A sediment yield index for choosing priority basins. IAHS-AISH Publ. 222: 180.

Dabral PP, Baithuri N, Pandey A (2008). Soil Erosion Assessment in a Hilly Catchment of North Eastern India Using USLE, GIS and Remote Sensing, Water. Resour. Manage., 22: 1783–1798.

Deore SJ (2005), Tripathi MP (1999). Prioritization of Micro-watersheds of Upper Bhama Basin on the Basis of Soil Erosion Risk Using Remote

Sensing and GIS Technology. Ph. D. thesis, submitted to Department of Geography, University of Pune, India.

Fistikoglu O, Harmancioglu NB (2002). Integration of GIS with USLE in assessment of soil erosion, J. Water. Resour. Manag., 16: 447-467.

Jain SK, Kumar S, Varghese J (2001). Estimation of soil erosion for a Himalayan watershed using GIS technique, J. Water. Resour. Manag., 15: 41–54.

Javed A, Khanday MY, Ahmed R (2009). Prioritization of Sub-watersheds based on Morphometric and Land Use Analysis using Remote Sensing and GIS Technique,. J. Indian. Soc. Remote. Sensing. 37: 261-274.

Khan MA, Gupta VP, Moharanam PC (2001). Watershed prioritization using remote sensing and geographical information system: a case study from Guhiya, India. J Arid. Environ., 49: 465-475.

Kothyari UC, Jain SK (1997). Sediment yield estimation using GIS. Hydrol. Sci. J., 42 (6): 833–843

Onyando JO, Kisoyan P, Chemelil MC (2005). Estimation of potential soil erosion for river perkerra catchment in Kenya, J. Water. Resour. Manag., 19: 133–143.

Pandey A, Chowdary VM, Mal BC (2007). Identification of critical erosion prone areas in the small agricultural watershed using USLE, GIS and remote sensing, J. Water. Resour. Manag., 21: 729-746.

Panigrahi B, Senapati PC, Behera BP (1996). Development of erosion index model from daily rainfall data, J. Appl. Hydro.,. 9 (1,2): 17-22.

Renschler C, Diekkruger B, Mannaerts C (1997). Regionalization in surface runoff and soil erosion risk evaluation, in regionalization of hydrology. IAHS-AISH. Publ., 254: 233-241.

Sharma JC, Prasad J, Saha SK, Pande LM (2001). Watershed prioritization based on sediment yield index in eastern part of Don valley using RS and GIS, Indian. J. Soil. Conserv., 29 (1): 7-13.

Sidhu GS, Das TH, Singh RS, Sharma RK, Ravishankar T (1998). Remote sensing and GIS techniques for prioritization of watershed: a case study in upper Mackkund watershed, Andhra Pradesh, Indian. J. Soil. Conserv., 2 (3): 71-75.

Wischmeier WH, Smith DD (1978). Predicting rainfall erosion losses- a guide to conservation planning, Agricultural Handbook USDA.p. 537.

Wu SLiJ, Huang G (2005). An evaluation of grid size uncertainty in empirical soil loss modelling with digital elevation models, Environ. Monit. Assess., 10: 33-42.

Yoshino K, Ishioka Y (2005). Guidelines for soil conservation towards integrated basin management for sustainable development: a new approach based on the assessment of soil loss risk using remote sensing and GIS, Paddy. Water. Environ., 3: 235-247.

6

Atmospheric pollution from the major mobile telecommunication companies in Tanzania

M. Kasebele[1] and W. J. S. Mwegoha[2]

[1]Department of Environmental Engineering, University of Dodoma, Dodoma, Tanzania.
[2]School of Environmental Science and Technology, Ardhi University, Dar es Salaam, Tanzania.

Air pollution from five mobile telephone companies in Tanzania was investigated. Results show maximum noise level of 83 dB and a minimum of 61.4 dB, above a permissible 45 dB. Concentrations of NOx ranged from 0.135 mg/m^3(NO) to 0.18 mg/m^3(NO$_2$), above a permissible level of 250 mg/m^3. The applied Gaussian model estimated the contribution of the BS generators to the atmosphere, to be between 0.0006 µg/m^3 and 0.001 µg/m^3 at 300 m from the source to 0.35 and 0.014 µg/m^3 10 m from the source for both NO and NO$_2$ respectively while the measured values ranged from 0µg/m^3 to 10 µg/m^3 for both (NO) and (NO$_2$) at 2.5 m from the source. The levels of PM$_{2.5}$ ranged from 0.04 to 0.25 mg/m^3, compared to standard of 0.1 mg/m^3. It can be concluded that there are high levels of noise and particulate emissions at varying degrees. At least 15 m should be maintained between the base stations and the closest residence for permissible noise levels.

Key words: Atmosphere, base stations, mobile telephone, noise, particulates.

INTRODUCTION

The past decade has witnessed a number of mobile telephone companies have being operational in Tanzania (TCRA, 2009). These companies have installed Base Stations (BSs) that use diesel power generators especially during times of power cuts. Currently, eight mobile telephone companies are in place, expanding the number of their installations, consequently creating environmental problems as mentioned by Lingwala (2003) and Samuelssen et al. (2009). These include the elevated sound levels and air bone emissions coming from numerous generators that power the industry, as well as potential for groundwater pollution due to spills of oil used to fuel generators. There is also lack of personnel (for instance Sasatel, tIGO and Zantel) specifically charged to deal with environmental issues. This study therefore aims at assessing the performance of these BSs in meeting local and international atmospheric discharge standards.

MATERIALS AND METHODS

Site description

The study was conducted in Dar es Salaam city, the largest commercial city in Tanzania. Dar es Salaam is located on the eastern region of Tanzania and lies in the coast of the Indian Ocean. The sampling areas that is, Tandale Manzese and Magomeni are shown in Figure 1.

Questionnaires to local community

The sampling was conducted in Magomeni, Manzese and Tandale Wards, with about 170 households. The area also features a total of 15 mobile telephone BSs. The area is selected on the basis of documented complaints from the community to the environmental

LOCATION OF THE SAMPLING AREA

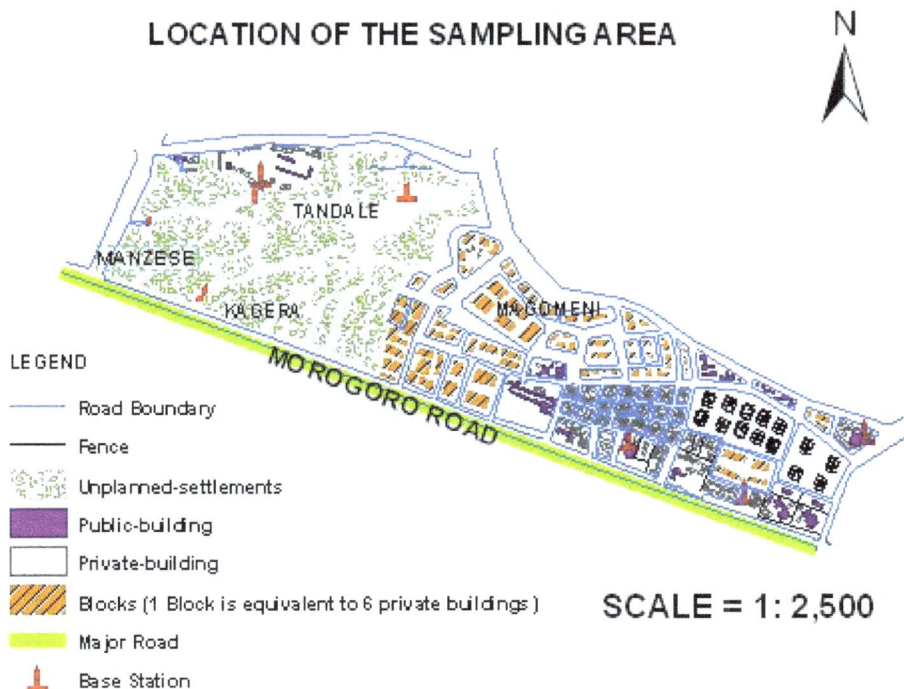

Figure 1. Location of the study area.

regulatory authorities regarding elevated levels of noise, smoke emissions from the generators and the fear of the radio frequency radiations from the transmission antenna. Questionnaire survey was conducted for chosen four closest households (about 1 to 3 m) to 14 of the base stations, making a total of 56 questionnaires.

Questionnaires to the regulatory authority

Some questionnaires were administered to the National Environment Management Council so as to crosscheck the information provided by the residents and the mobile telecommunication companies including the areas where the most complaints come from and the decisions made about the complaints.

Reconnaissance survey

The measurements were done at the 10 identified BSs in the area. A survey of the surrounding area was done prior to measurements to identify the potential areas for the measurements to be taken.

Noise level measurements

The selection of the measurement sites were based on the information rich cases (the ones providing the answer to research question and accessibility to the research data. In Table 1, various sites are chosen with respect to this criterion and the explanations of the functions are given.

The noise level was measured using Integrating-averaging sound level meter, Bruel and Kjaer Type 2240. This device used records the maximum noise level reached during the measurement procedure (the L_{max}) equivalent continuous sound level in seconds

(L_{eq}) and the peak sound levels(L_{peak}). The meter was calibrated by placing a portable acoustic calibrator, which in this case was a sound level calibrator, directly over the microphone so that it can calibrate the meter. The sound level meter calibration was done before and after each measurement session of 3 h. The sound levels were recorded in a 2 (two) meter interval from the boundary wall of the BSs to the nearest household. A total distance of 10 m was used to record the noise from the source of noise in this case the generators. Therefore a total of 5 readings were taken for each noise measurement practice.

Measurement of stack exhausts gas concentration

The instrument used for these measurements was the manual dragger pump Accuro® Pump. A manual pump using dragger tubes was used. The Dräger-Tubes® 2/c used can measure up to 200 ppm of the NO and NO_2 concentration in air in a single pump stroke). The measured volume of gas removed from the generator as an exhaust gas was drawn through a tube which contains chemicals which change in colour in response to the presence of target gas that is, NO and NO_2. The measurement period was 10 (ten) minutes for ten strokes and the concentrations of the NO_x gases was recorded for an hour average by recording in the first 10 min and in the last 10 min that is from the 50th to the 60th minute.

Measurement of stack exhausts gas exit velocity

The instrument used for these measurements was the Alnor® Velometer Series 6000P fitted with the metallic pitot probe that is used to detect the velocity of the exiting gases and the results being displayed in the analog display of the meter. The meter was held with the pitot probe direct and normal to the stack. The stack

Table 1. Sites Prone to noise pollution in the sampling area.

Measurement site	Function
House holds	Living, resting and socializing
Schools	Learning
Health centres	Providing and acquiring health services
Hotels	Resting, conferencing, working and socializing

velocity was then measured by the meter in the selected range that is, 0 to 50 m/s. 10 readings were taken in the half hour average every three minutes and the average value taken.

Measurement of particulate matter in the exhaust gases

The instrument used to measure particulate matter was the micro dust analyzer that is, MICRODUST pro 880 nm fitted with filters containing probe. The device takes measurements by sensing technique of forward light scattering (12∘ to 20∘) using 880 nm infrared range of 0.001 to 2500 mg/m^3. By this narrow range of scatter the instrument sensitivity to variations in the refractive index and the colour of measured particulate is reduced.

Calibration of the instrument was done by attaching the purge to the probe purge inlet and inserting the calibration filter in the filter position and the bulb of the purge below was squeezed rapidly 6 times. An allowance of few seconds (3 to 5 s) was given for auto-ranging and for the reading to stabilize. The squeezing was repeated whenever the readings did not stabilize. On entering these results the device would set the reading to zero.

Measurements from the mentioned points were done by orienting the probe and allowing the dust particles to be detected and filtered as it passes through the filter hole in the probe. The readings in the amount of the dust scattered therefore detected, was recorded and eventually retrieved using the WINDDUST pro Application software.

The Gaussian dispersion model

The Gaussian dispersion model (Cooper and Alley, 1994) was used to model the concentration of the gaseous pollutants. The general equation is provided by Equation 1.

$$C = \frac{Q}{2\pi * U * \delta y * \delta z} Exp\left(-\frac{1}{2}\frac{Y^2}{\delta y^2}\right) * \left[-\frac{1}{2}\frac{(Z-H)^2}{\delta z^2} + Exp-\frac{1}{2}\frac{(Z+H)^2}{\delta z^2}\right] \quad (1)$$

Where: U =Wind speed at stack height (m/s); Y = Horizontal distance from plume center line (m); Q = Emission rate (g/s); C = Steady state concentration at a point (x, y, z) g/m^3; H = Effective stack height (m), δz, δy = Standard deviations of concentration in respect to direction measured (m), these are functions of distance x and atmospheric stability; Z = Height above the ground level (m).

The Gaussian model for determination of downwind concentration of emitted pollutants is explained in Equation 2;

$$C(x,0,0,H) = \frac{Q}{\pi * U * \delta y * \delta z} Exp - \frac{1}{2}\frac{H^2}{\delta z^2} \quad (2)$$

Where:

H = Effective stack height (m); Q = Emission rate (g/s), δz, δy =Standard deviations of concentration in respect to

direction measured (m), these are functions of distance x and atmospheric stability, C = Downwind concentration (g/m^3).

From the stability class and the receptor position, δz and δy may not be obtained directly from the Pasquill-Gifford curves since, the lowest values indicated in the curves is 100 m. The formula used to calculate δz and δy are obtained using Equations 3 and 4:

$$\delta z = cx^d + f \quad (3)$$

$$\delta y = ax^b ... \quad (4)$$

Where: a, b, c, d and f are constant that are dependent on the stability class, x is distance in km.

The mentioned values can be read in Table 2 Where: a, b, c, d and f are constant that are dependent on the stability class, x is distance in km. The mentioned values can be read in Table 2.

Wind velocity calculations

The stack gas exit velocity data at site were collected at various positions from the ground level, this needed to be corrected to 10 m above the ground so that the stability classes could be determined. The mean wind speed is frequently represented empirically as a power law function of height and can be calculated using Equation 4.

$$\frac{u_2}{u_1} = \left(\frac{Z_2}{Z_1}\right)^P \quad (5)$$

Where;
u_2, u_1 = wind velocities at higher and lower elevation respectively (m/s), Z_2, Z_1 = higher and lower elevation (m), P = function of stability ($\cong$ 0.5 for very stable conditions and $\cong$ 0.15 for very unstable conditions).

The exponent p varies with atmospheric stability class and with surface roughness. Table 3 below shows various values for exponent p for various stability classes. The values of the exponent chosen are based on the rough surfaces since all the measurements were done in urban environment. For flat, open country and lakes and seas, there is less variation between the surface wind and geotrophic wind.

Estimation of Pasquill stability classes

Stability classification is designated by letters A to F. Stability class A, the most unstable, has the greatest dispersion potential stability, the most stable class which represents conditions that dampen turbulence, thereby reducing dispersion. Class A to C are limited to

Table 2. Values of curve-fit constants for calculating dispersion coefficients as a function of downwind distance and atmospheric stability.

Stability class	x<1 km					x> 1 km		
	a	b*	c	d	f	c	D	F
A	213		440.8	1.941	9.27	459.7	2.094	-9.6
B	156		106.6	1.149	3.3	108.2	1.098	2.0
C	104		61.0	0.911	0	61.0	0.911	0
D	68		33.2	0.725	-1.7	44.5	0.516	-13.0
E	50.5		22.8	0.678	-1.3	55.4	0.305	-34.0
F	34		14.35	0.740	-0.35	62.6	0.180	-48.6

b*=0.894 for all stability classes and values of x (Cooper and Alley, 1976).

Table 3. Exponents for wind profile (power law) model for rough surfaces.

Stability class	Exponent (p)
A	0.15
B	0.15
C	0.20
D	0.25
E	0.40
F	0.60

Source: Cooper and Alley (1994).

daytime whereas E to F are night time conditions only. A neutral stability classification D can occur during the day or during night time periods. Table 4 gives various stability classes with respect to sky condition. Therefore the wind speed at the stack height is obtained by Equation 6 as given below;

$$4 = u_1 \left(\frac{10}{Z_1} \right)^P$$

(6)

Control of exhaust gases

Plume rise prediction

The analytical methods for predicting the concentration of stacks effluents involve the location of a virtual or equivalent origin. Elevation H of the virtual origin is obtained by adding, the plume rise Δh to the actual height of the stack (h). There are numerous methods for calculating Δh, basically three sets of parameters control the phenomenon of a gaseous plume injected into the atmosphere from a stack. These are stack characteristics, meteorological conditions and the physical and chemical nature of the effluent. In this study the plume rise was calculated using Moses and Carson for unstable condition equation provided in equation 7 (Cooper and Alley, 1994), other equations that is, equations 8 and 9 are used to determine the mass flux and the area of the stack. The equations were chosen for Wind speed between 3 to 5; the surface is rough urban and the equations involved;

$$\Delta h = 3.47 \left(\frac{Vsd}{U} \right) + 5.15 \frac{(Q_h)^{0.5}}{U}$$

(7)

$$Q_h = \dot{m} C_p (T_s - T_a)$$

(8)

$$m = \delta_a VsA$$

(9)

$$A = \frac{\pi d^2}{4}$$

Whereby;

δ_a - density of air at 32°C= 1.165 kg/m³ (Engineering Toolbox, 2005); C_p - pressure constant specific heat capacity of air = 1.016 kJ/Kg°K (Engineering toolbox, 2005); Vs -Stack exit velocity (m/s); d- diameter of the stack (m); Ts- Temperature of the exit gases (°K); Ta- Ambient temperature (°K); Q$_h$- is the heat emission rate(KJ/s); Δh- plume rise (m).

RESULTS AND DISCUSSION

Questionnaire survey

Questionnaire survey and analysis conducted in the visited companies and the sampling areas indicated that out of the total of (4170) base stations, Environmental Impact Assessment has been conducted for only 11%, equivalent to 458 stations in Dar es Salaam. The survey also indicate that 86% of the respondents experience noise and smoke emission while the rest 25% responded that they experience noise only while the rest explains that they do not experience those effects as shown in Table 5. Moreover the questionnaire survey revealed that, 100% of the interviewed residents do not have generators for power back up. This means that the noise from the generators talked about come from the operating BSs. It was also noted that, 20% of the interviewed residents although they do not have power

Table 4. Pasquill stability classes.

Surface Wind at 10 m above the ground	Day incoming solar radiation			Night cloudiness	
	Strong	**Moderate**	**Slight**	**≥ 4/8**	**≤3/8**
<2	A	A-B	B		
2-3	A-B	B	C	E	F
3-5	B	B-C	C	D	E
5-6	C	C-D	D	D	D
>6	C	D	D	D	D

Source: Seinfeld (1986) A: extremely unstable, B: moderately unstable, C: slight unstable D: neutral, E: slightly stable, F: moderately stable.

Table 5. Environmental Management related issues in the visited companies.

Company	Zain/Airtel	Tigo	Sasatel	Zantel	Vodacom
No. of B.S in Tanzania	1250	1000	20	600	1300
No. of B.S in DSM	143	400	20	98	390
No. of B.S with EIA	68	200	20	14	24
Complaints due to operations	Yes	Yes	-	Yes	Yes
Major complaints	-Noise -Smoke	-Noise -Smoke -Fear of radiation	-	-Noise -Smoke	-Noise -Smoke

generator, they live close to other sources of noises such as flour milling machines or timber cutting machines. This makes it difficult to determine whether the noise come only from the BS operating generators or these other sources. Out of the interviewed residents 7% are from the houses surrounding the BS which EIA was done prior to project commencement.

Noise measurement results from the base stations

The noise levels measured from the boundary wall of the Cell sites as proposed in the EIA revealed the margin of failure in the implementation of the levels proposed in the Environmental Impact Assessment or addressed in the Environmental Management Plan.

Zain BSs

The first and second Zain BS with tower ID 10 and 464 respectively shows that the generators produce noise levels with the deviation from the noise levels value of 45 dB (in residential houses) and 70 dB (in industrial areas). The magnitude of this deviation has been of average of about 26 dB. This case is shown in Figure 2. According to these findings, a minimum distance of about 22 m is required above which the proposed noise level of 45 dB

can be attained.

Tigo BSs

The two TIGO BS with tower ID Dar 147 and Dar 035 respectively also show the deviation from the noise levels value of 45 dB (in residential houses) as provided in the EMP as shown in Figures 3 and 4. The magnitude of this deviation has been about 27 and 37.8 dB respectively. The two stations have different types of generators of different ages and therefore the results show the difference in the noise levels. According to this an estimated minimum distance of 15 and 20 m for each generator respectively is required to achieve the recommended noise level in the EMP.

Sasatel BSs

Results from the Sasatel BS show elevated levels of noise compared to that proposed in the EMP. The noise levels measured from two BSs with ID number Dar 18 and Dar 17 and similar making showed the levels of noises to have increased by about 25 dB from the proposed 45 dB, as shown by Figure 5. Again a distance of 20 m is required to achieve the proposed noise level in the EMP.

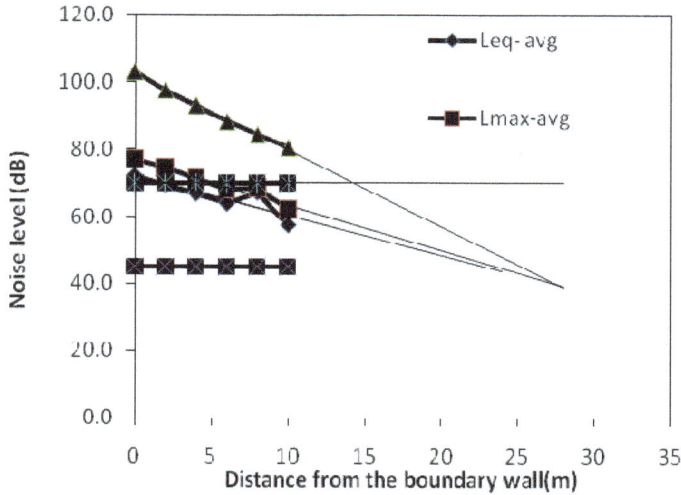

Figure 2. Variation of noise level with distance for Zain BS.

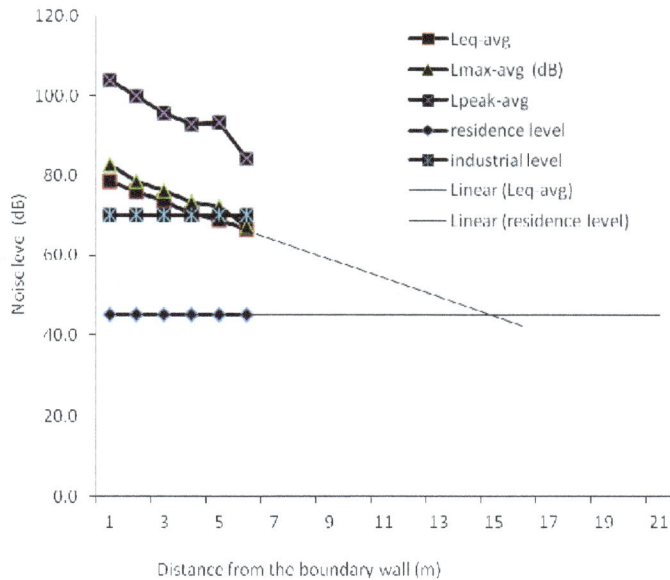

Figure 3. Variation of noise level with distance for TIGO BS 1.

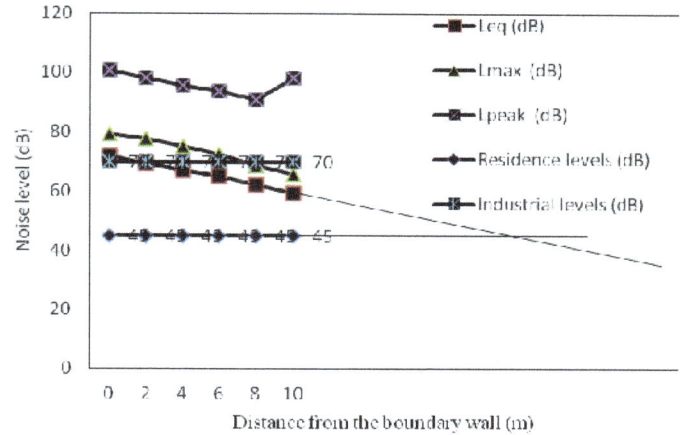

Figure 4. Variation of noise level with distance for TIGO BS 2.

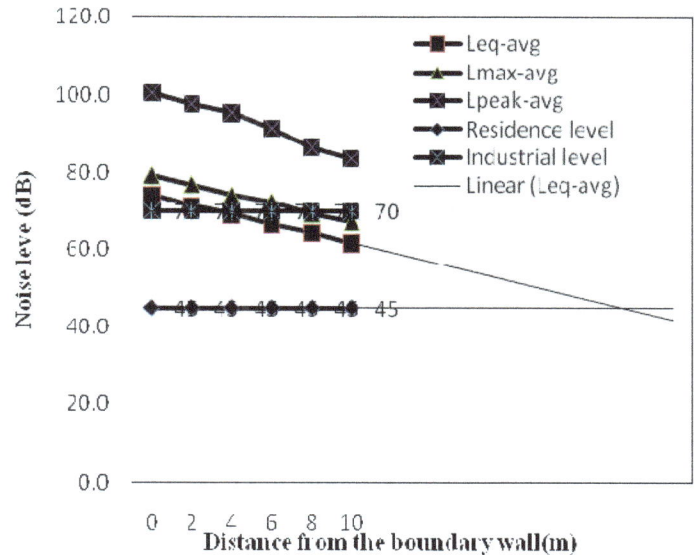

Figure 5. Variation of Noise level with distance from the Sasatel BS boundary wall.

Vodacom BSs

Results from the Vodacom BSs generators show elevated levels of noise compared to that proposed in the EMP. Even though the deviations from the proposed EMP noise level in residential levels is a bit higher in average that is by 16.4 dB compared to the other companies BSs sampled, the noise level proposed in the Noise level Management Plan in the company (that is 65 dB) is slightly lower than that in the EMP for industrial level. As shown in Figure 6, this is ok for the industrial areas even though is still higher than that for the residential areas especially in the case study areas where the distance from the boundary wall to the nearest households is hardly 2 m, in other critical cases the boundary wall of the BSs is in close contact about 0.5 m to the wall of the households' bedroom. In this case a minimum distance of 18 m from the boundary wall is required so that the proposed noise level can be attained.

Zantel BSs

Results from the Zantel BSs, as well shows the high levels of noise compared to that proposed in the EMP. The noise levels measured from two BSs generator with

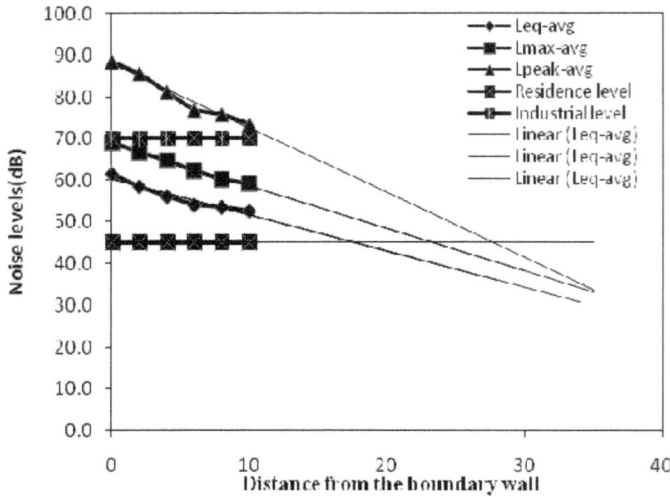

Figure 6. Variation of Noise level with distance from the Vodacom BS boundary wall.

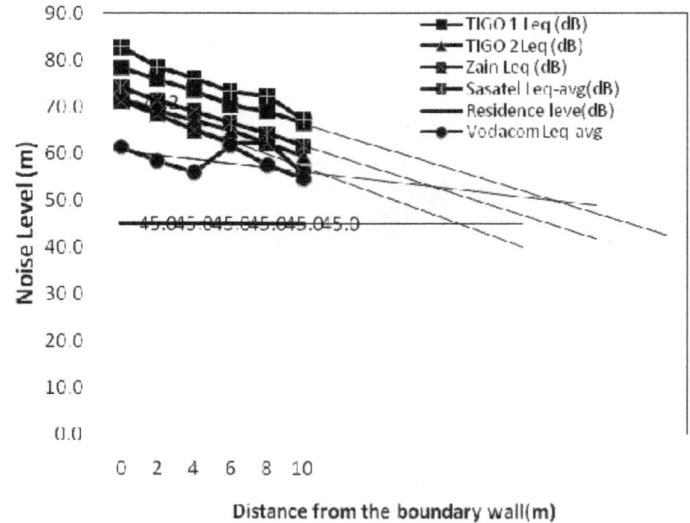

Figure 7. Variation of Noise level with distance from the Zantel BS boundary wall.

similar making showed an increment in the noise levels, by an average of about 37.8 dB from the proposed 45 dB, shown in Figure 7. From the analysis it is shown that a distance of about 26 m from the boundary wall is required from which the residents wouldn't be affected by the elevated noise, in which case the standard for the noise levels proposed in the EMP would be achieved.

Summary of the noise level measurement

A minimum of 15 m from the BSs from the site wall is

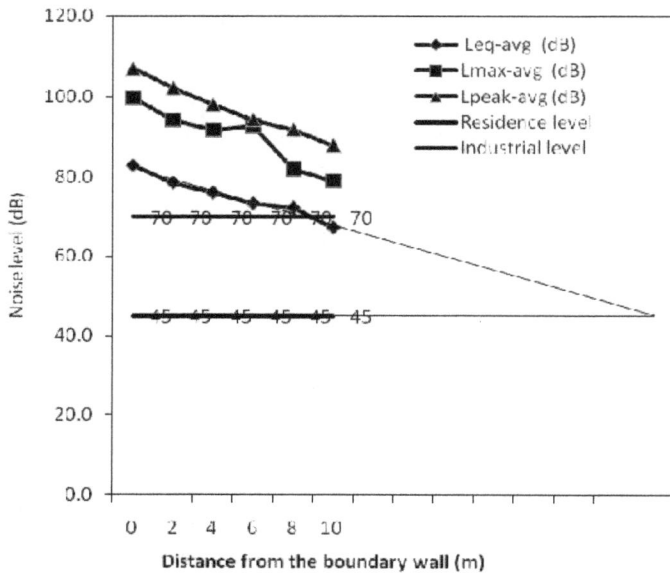

Figure 8. Noise level with distance from the boundary wall for the visited companies BSs.

required above which the effects of noise to the residents would be negligible as indicated in Figure 8.

Concentration results from the base stations

The results from the measurements of the gaseous pollutant concentration show that the gases of interests in this project were below the permissible standards for the emissions from the stacks.

NOx

Levels of NOx in base stations were low with the maximum hourly mean of 0.13 mg/m^3 for NO (Zantel BS) and 0.08 mg/m^3 for NO$_2$ (Zantel BS) compared to the permissible stack emission standard of 250 mg/m^3 (Figures 9a, 9b, 10a and 10b). It was also noted that older generator emitted more the noxious gas emission with an exception of the Zantel generators.

Orientation of the generator's stack

60% of the surveyed generators complied with the requirements of EMP while the rest did not comply with this mitigation measure as the proposed measure in the EMP states that the stacks should be vertical and with the appropriate height to enhance effective dispersion of contaminants (particularly noxious gases and particulate matter). The Gaussian Model indicates that a sufficiently tall and vertical stack has better dispersion so that the concentration of the emitted pollutant reaching the receptor is very low.

(a)

(b)

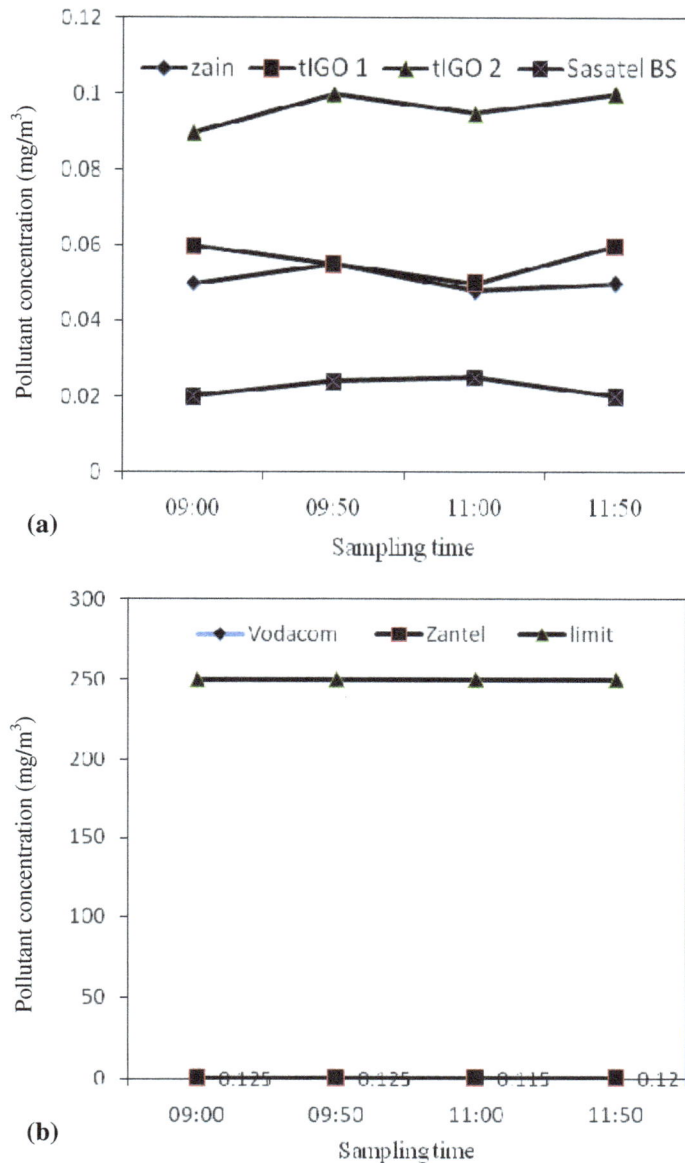

Figure 9. (a) Stack NO emissions for different generator stacks (b) NO concentration against the stack emission standard of 250 mg/m³.

The Gaussian air dispersion model results and discussion

Determination of downwind concentration of measured pollutants

Gaussian dispersion model was used to predict the average concentration of NOx downwind for hourly average based on the Equation 2 where:

H = Average effective stack height (m)
U = Average wind speed at 10 m, 4.0 m
C = Downwind concentration mg/m³

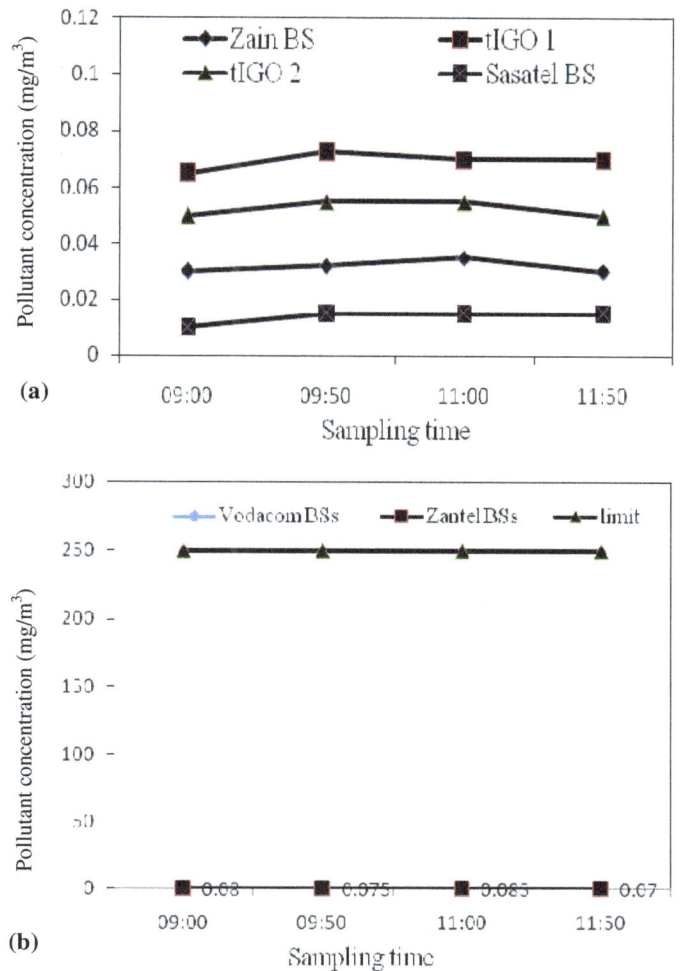

(a)

(b)

Figure 10. (a) Stack NO_2 emissions for different generator (b) NO_2 concentration against the stack emission standard 250 mg/m³.

Model results Tigo

The model results from the TIGO BSs showed peak NOx concentration of 0.35 µg/m³(NO) and 0.014 µg/m³(NO_2) at 10 m from the source. Also the measured concentration was found to be 5 and 2.5 µg/m³ for both NO and NO_2 respectively at 2.5 m from the source. The results both measured and modeled are shown in Figure 11.

Model results for the Sasatel

The model results from the Sasatel showed the peak concentration of NOx of to be 0.0033 µg/m³(NO) and 0.0056 µg/m³(NO_2) at 10 m from the source. However the measured concentration did not show any detection possibly because of the instrument detection limit. The results of modeled concentration are shown in Figure 12.

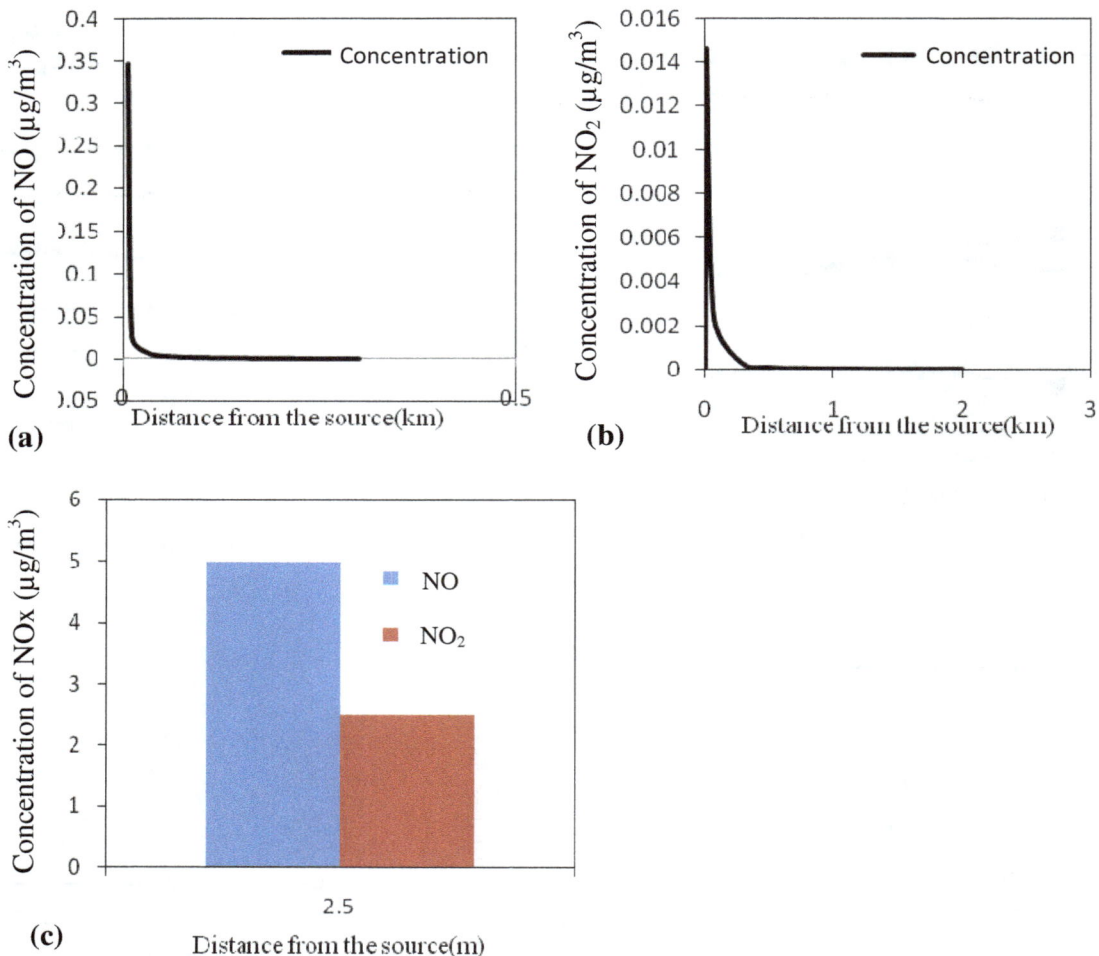

Figure 11. (a) The modeled NO conc. with distance (b) the modeled NO_2 conc. with distance (c) the measured NOx concentration with distance.

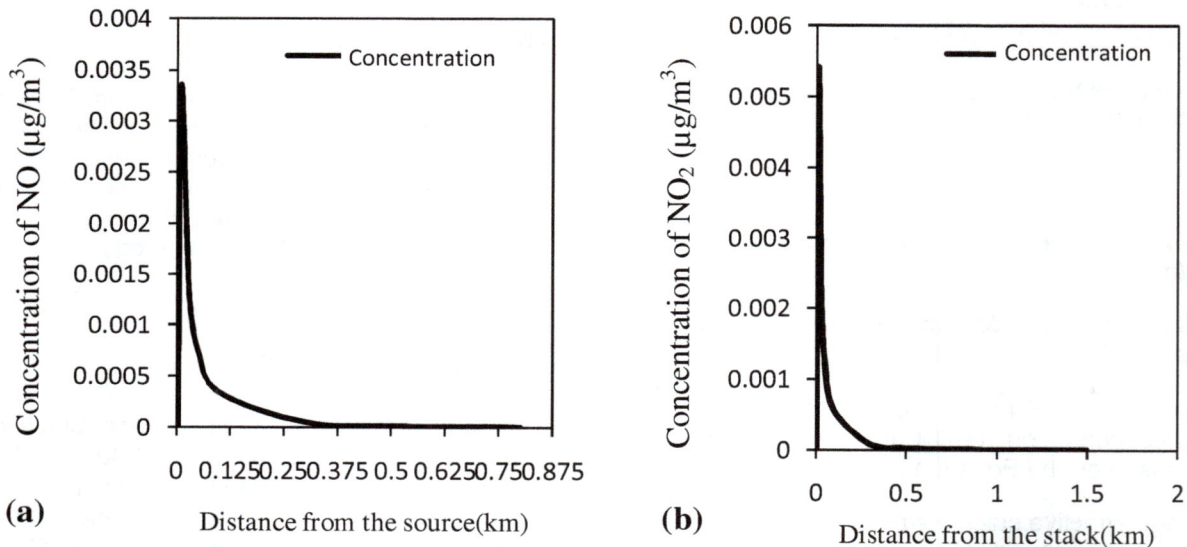

Figure 12. (a) Modeled NO concentration for Sasatel tower (b) Modeled NO_2 conc. for Sasatel tower.

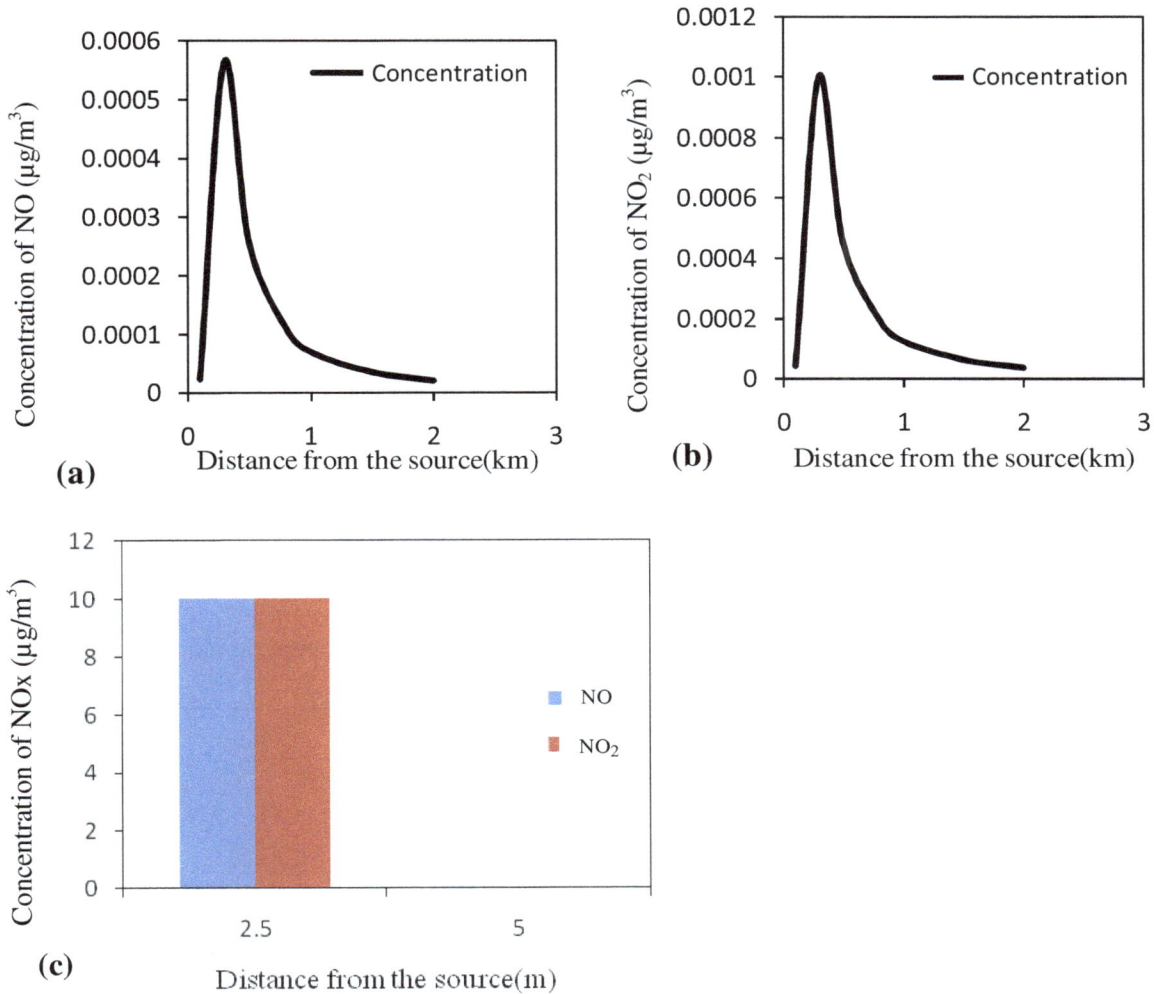

Figure 13. (a) Modeled concentration of NO for Vodacom (b) Modeled concentration of NO_2 for Vodacom BS (c) The measured NOx concentration with distance.

Model results for Vodacom

The model results from the Vodacom BSs showed peak NOx concentration of 0.0006 $\mu g/m^3$(NO) and 0.001 $\mu g/m^3$(NO_2) at 300 m from the source. However the measured concentration showed a peak concentration of 10 $\mu g/m^3$ at 2.5 m. The measured concentration was higher possibly due to contribution of other sources such as motor vehicles. Both measured and modeled concentration results are shown in Figure 13.

Model results for Zantel

The model results from the Zantel BSs showed peak NOx concentration of of 0.017 $\mu g/m^3$ (NO) and 0.012 $\mu g/m^3$(NO_2) at 300 m from the source. The measured concentration was below the instrument's detection limit.

The model output is shown in Figure 14.

Model results for Zain

The model results from the Zain BSs showed peak NOx concentration of of 0.013 $\mu g/m^3$(NO) and 0.008 $\mu g/m^3$(NO_2) at 10 m from the source. The measured concentration showed a peak concentration of 10 and 7 $\mu g/m^3$ for NO and NO_2 respectively, at 2.5 m. The measured concentration was higher compared to the modeled, first, due to the difference in positions of the measured and modeled concentration given the method of measurement, but also possibly due to contribution of other sources such as motor vehicles, since these sites are located close to the street or main roads. The results of both measured and modeled peak concentrations are shown in Figure 15.

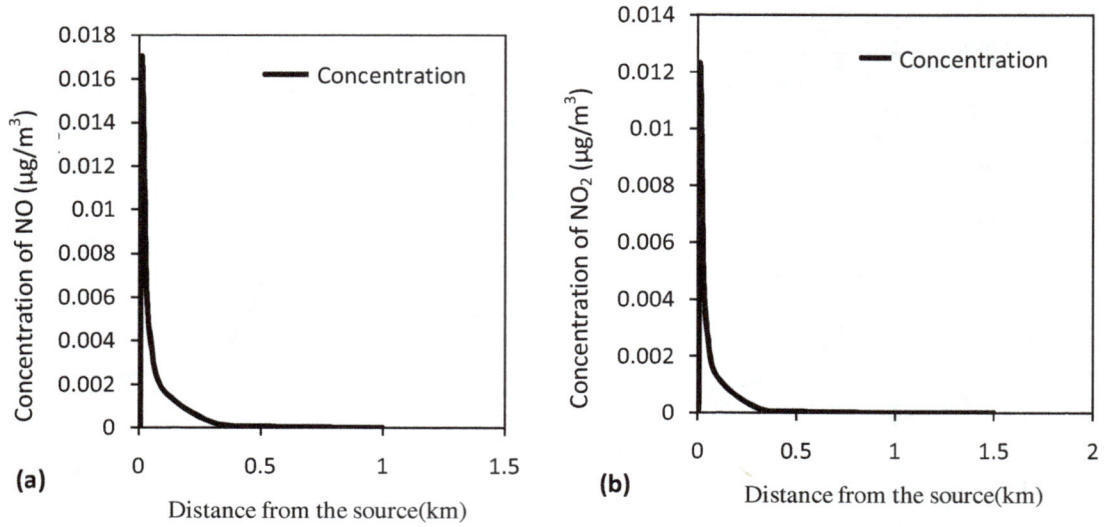

Figure 14. (a) Modeled conc. of NO for Zantel (b) Modeled conc. of NO$_2$ for Zantel.

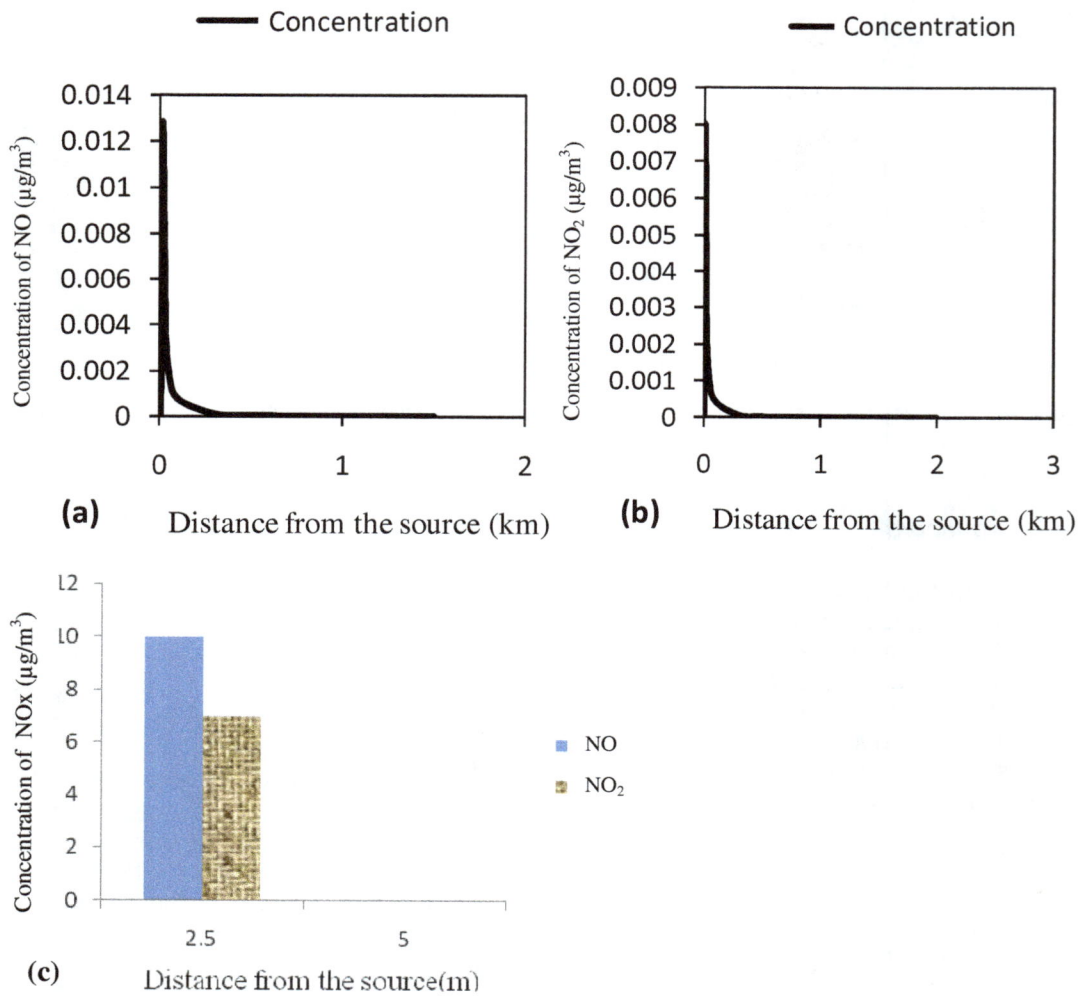

Figure 15. (a) Modeled concentration of NO for Zain tower (b) Modeled concentration of NO$_2$ for Zain BS (c) The measured NOx concentration with distance.

Particulate matter (PM) results

Tigo

The results of the stack particulate emissions for TIGO 1 and TIGO 2 BSs are as shown in Figure 16 and 17, while the ratios of the indoor to outdoor are given in Tables 1 and 2. According to the results PM concentration was found to be higher after the generators had been running for both the outside and the indoor environments. This might also be caused by the orientation of the generator stack being horizontal which caused unidirectional flow of stack smoke. The ratio of the indoor to outdoor concentration was 0.5 and 0.8, which shows that the indoor concentration was higher than the outdoor concentration for the same reason that the horizontal stacks may have influenced the concentrations in the indoor environment.

Zantel

The results of the stack particulate emissions for Zantel BSs are as shown in Figure 18, while the ratios of the indoor to outdoor are given in Table 3. These results suggest that PM concentration was higher after the generators had been running for both the outside and the indoor environment. However the PM levels in the outside ambient air was greater than in the indoor which may have been caused by other PM sources such as the moving vehicles to and from the area, given that the place is close to the street road. The results and the ratio of the indoor to outdoor concentration was found to range between 0.53 to 0.98 for before and after sampling respectively, the closeness of the ratio to 1 during sampling might have been due to proper ventilation in the classroom.

Zain

The results of the stack particulate emissions in the sampled points for Zain BSs are shown in Figure 19. Similar to Zantel, PM concentration was found to be higher after the generators had been running for both the outside and the indoor household. The ratios of the indoor to outdoor PM concentrations were such that before the generator was on Household (HH) $PM_{2.5}$ were higher than ones after the generator running.

Vodacom

The results of the stack particulate emissions in the sampled points for Vodacom BSs are as shown in Figure 20, while the ratios of the indoor to outdoor are given in Table 5. PM concentration was found to be higher after the generators had been running for both the outside and

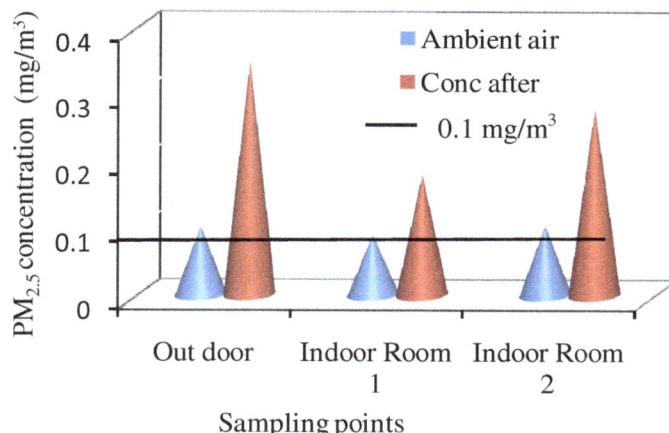

Figure 16. The PM concentration at Room 1 and 2 and outdoor.

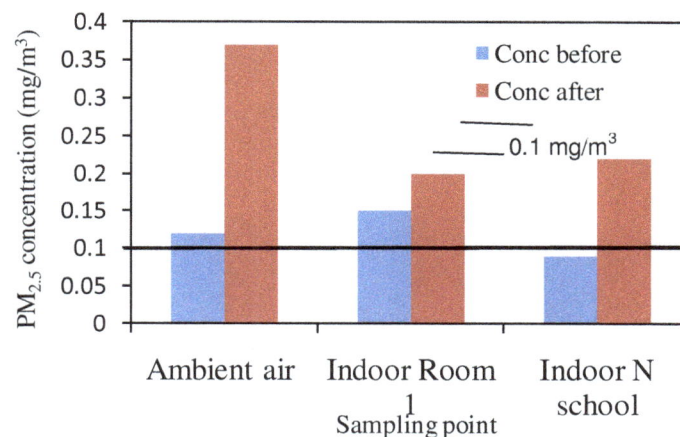

Figure 17. The PM concentration at Room 1& Nursery school class and outdoor.

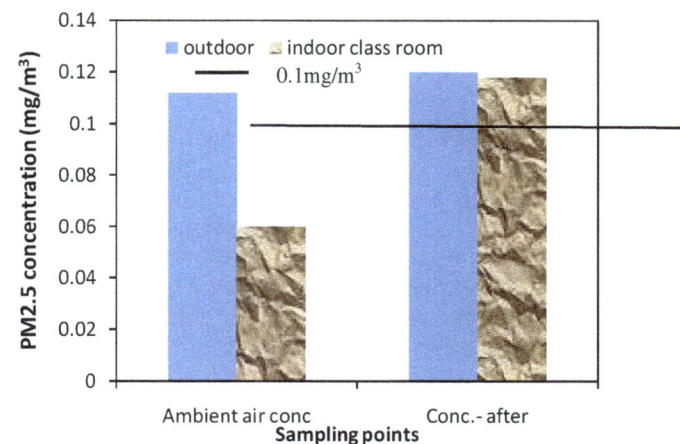

Figure 18. The PM concentration at outdoor and indoor in class room.

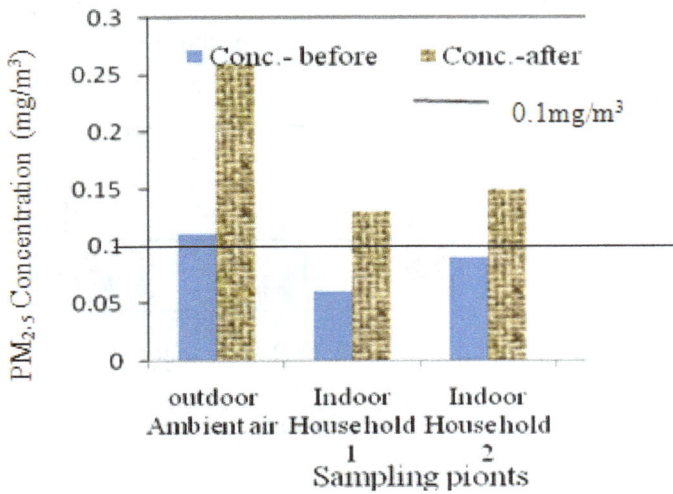

Figure 19. PM concentration at outdoor and indoor inside 2 households.

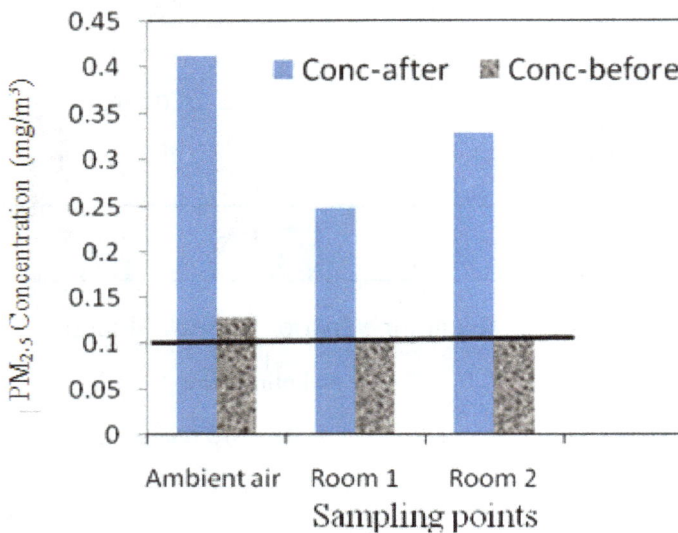

Figure 20. The PM concentration at outdoor and indoor inside class room.

the indoor environment. The ratio of the indoor to outdoor concentration was found to be 0.75 to 0.85 and 0.6 to 0.8 before and after running the generator for the two rooms, this shows that the outdoor concentration was higher than the indoor concentration in all cases which may have been contributed by the orientation of the generator stack being vertical so as to allow dispersion and the location of the rooms being a bit far from the source about 3 m.

Sasatel

The results of the stack particulate emissions at different

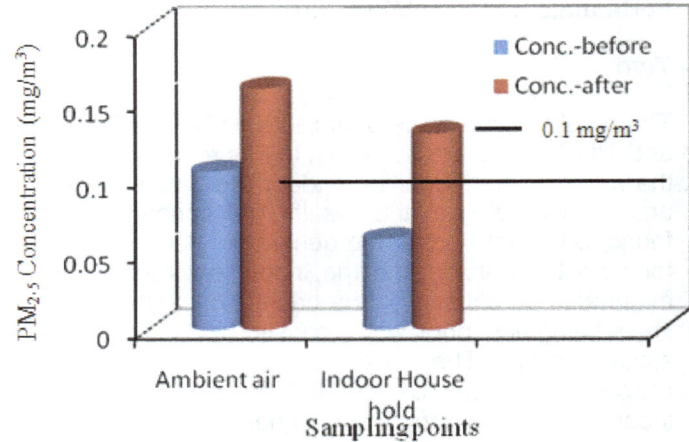

Figure 21. The PM concentration at outdoor and indoor inside class room.

sampling points for Sasatel BSs are as shown in Figure 21 while the ratios of the indoor to outdoor are given in Table 6. In this case the results of PM concentration were also found to be higher after the generators had been running in both the outside and the indoor environment. The ratio of the indoor to outdoor concentration showed that the outdoor concentration was higher than the indoor concentration which may also be contributed by the distance the rooms are located from the source and the stack being vertical so that dispersion was enhanced.

Comparison of the ambient PM$_{2.5}$ concentration between horizontal and vertical stacks

The results show that PM$_{2.5}$ concentration from the horizontal stacks cause a greater contribution in the increased ambient PM$_{2.5}$ level than those coming from vertical stack. This is shown in Figure 22.

Conclusions

The mobile telecommunication sector has played a pivotal role in the country's economic development, especially enhancing instant communication. This research was set to determine potentially adverse impacts to the environment by investigating how effective the EMPs and MPs are implemented with respect to the noise, particulate and the NOx gases pollution. Noise level was found to on the higher side compared to permissible levels. In addition, the released levels of the PM$_{2.5}$ caused a significant raise in PM level in the ambient air PM$_{2.5}$ concentration of the surrounding (indoor and outdoor) environments with the hourly average increase of about 0.25 mg/m^3 (Tigo BSs), 0.045 mg/m^3 (Zantel BSs), 0.08 mg/m^3 (Zain), 0.23 mg/m^3 (Vodacom BSs) and

Table 6. Ratios of PM concentrations (indoor and Outdoor) before and after the generator running.

Sampling point	Before		After	
	R1	R2	R1	R2
Indoor	0.09	0.104	0.18	0.28
Outdoor	0.102	0.102	0.35	0.35
Ratio In/Out	0.88	1.01	0.51	0.8

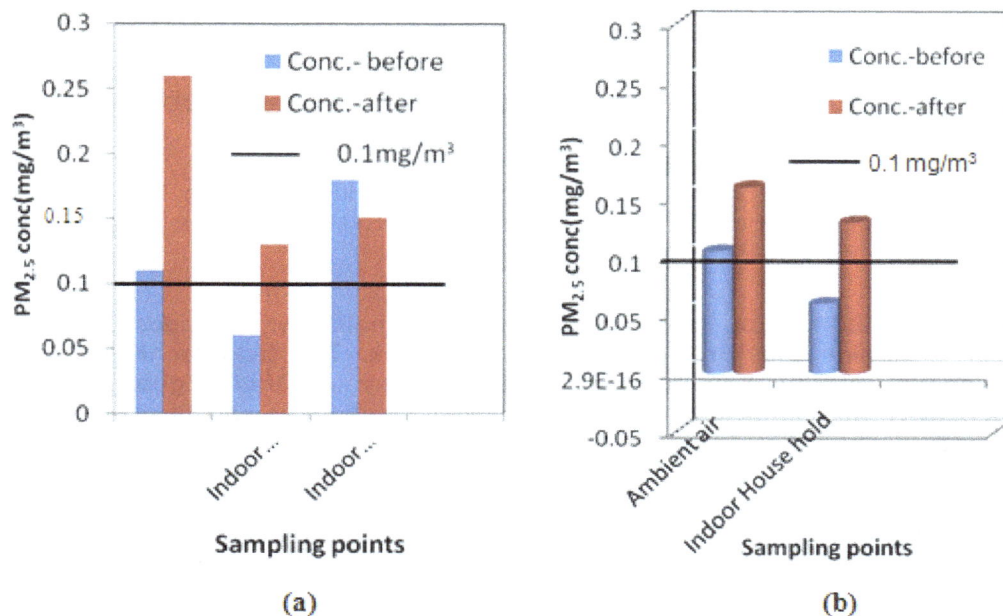

Figure 22. A comparison of the ambient $PM_{2.5}$ from the horizontal (a) and vertical (b) stacks.

0.04 mg/m^3 (Sasatel BSs) above the standard of 0.1 mg/m^3(TBS, 2007).

The EMPs do not propose the cut of values for gaseous NOx emissions however the levels of concentration at the stack exits are low with maximum hourly average of 0.18 mg/m^3(NO) and 0.135 mg/m^3(NO$_2$) compared with the permissible standard provided by the of <250 mg/m^3.

The Gaussian air pollutant plume model provided an approximation of the contribution of the BS generators to the atmosphere of maximum hourly concentration of NO and NO$_2$ respectively of about 0.35 μg/m^3 0.014 μg/m^{3}at 10 m from the source for Tigo BS, 0.0033 μg/m^3 and 0.0056 μg/m^3 at 10 m for Sasatel BSs, 0.0006 μg/m^3 and 0.001 μg/m^3 for Vodacom BSs at 300 m from the source, 0.017 μg/m^3 and 0.012 μg/m^3 for Zantel BSs at 10 m from the source and 0.013 μg/m^3 and 0.008 μg/m^3 for Zain BSs at 10 m from the source. The measured NOx values had hourly peak averages of NO and NO$_2$ respectively of about 5 μg/m^3 and 2.5μg/m^3 (Tigo BS), 10 μg/m^3 (Vodacom BS) and 10 μg/m^3 and 7 μg/m^3(Zain BS) at 2.5 m from the source.

REFERENCES

City Mayors (2006). "World's Fastest Growing Cities and Urban Areas. Available at www. citymayors.com.

Cooper CD, Alley FC (1994). Air Pollution Control: A Design Approach. Waveland Press, Illinois.

DEP Division of Compliance and assistance (2009). Environmental Management Plans. Available at http://www.dca.ky.gov/kyexcel/Environmental+Management+Plans.

Environmental Management Bureau (TBS) (2007). Engine Emissions– Health and Medical Effects. Available at http://www.emb.gov.ph.

Ling'wala S (2003). Radiation Exposure From Cellular Phone Base Station Antenna in Tanzania". Dep. Environ. Eng. UCLAS, Dar es Salaam.

Samuelsen S, McDonell V, Hack RL, Phi V, Couch P, Bolszo C, Hernandez S (2009). Fuel injection and emissions characteristics of a commercial microturbine generator". Paper GT-2004-54039. ASME Turbo Expo. Viena. Austria.

Seinfeld JH (1986). Atmospheric Chemistry and Physics of Air Pollution. Willy Inter Science, Publication, NY.

TCRA (2009). Licensing information. Available at www.tcra.go.tz.

The Engineering ToolBox (2005). Air properties. Available at www.EngineeringToolbox.com.

7

Geoelectric study of major landfills in the Lagos Metropolitan Area, Southwestern Nigeria

Oladapo, M. I.[1] , Adeoye-Oladapo, O. O. [2] and Adebobuyi, F. S.[3]

[1]Department of Applied Geophysics, Federal University of Technology, Akure, Nigeria.
[2]Department of Physics, Adeyemi College of Education, Ondo, Nigeria.
[3]Groundwater and Geophysical Services (GGS) Limited, Lagos, Nigeria.

Geoelectric study of Abule Egba, Igando and Olushosun landfills within Lagos municipality has been undertaken to determine their hydrogeologic implications. The field technique involved vertical electrical soundings utilizing Schlumberger electrode array. At Abule Egba, landfill materials are defined by resistivity varying between 1.6 Ω-m at decomposed stage and 144 Ω-m within fresh dump. The Igando landfill is defined by resistivity varying between 2.5 Ω-m at decomposed stage and 26.1 Ω-m at fresh dump. Olushosun landfill is defined by resistivity varying between 2.4 Ω-m at decomposed stage and 51.5 Ω-m at fresh dump. Interpretation of sounding curves showed that Abule Egba is underlain by fairly thick column of clayey sand indicating an unconfined aquifer. Igando landfill is underlain by thick clay column indicating a confined aquifer. The northern flank of Olushosun landfill is overlain by thin refuse dump (4.4 m) while thick refuse dump (22.9 m) overlies the central area. The hydrogeologic system at Abule Egba is vulnerable to contamination. The impermeable geoelectric characteristics of materials underlying Igando landfill offer the hydrogeologic system some form of protection. At Olushosun landfill, materials of impermeable geoelectric characteristics in the northern flank offer hydrogeologic protection while fairly permeable materials in the south offer limited hydrogeologic protection.

Key words: Geoelectric, landfill, resistivity, contamination, hydrogeologic system, aquifer.

INTRODUCTION

Land filling remains the most important technology for municipal solid waste management (Petruzzelli et al., 2007). However, groundwater contamination remains a major concern in the operation of landfills. This is due to the pollution effects of landfill leachates and its potential health risks (Lee and Jones-Lee, 1993a, b; Christensen et al., 2001; Stollenwerk and Colman, 2003; Longe and Enekwechi, 2007). Landfill is the cheapest way of disposing municipal solid wastes, but all efforts to get rid of waste also pollute the environment to some extent. For instance, landfills may pose serious threat to the quality of the environment if incorrectly secured and improperly operated while the threat to surface and ground waters could be deleterious (Longe and Balogun, 2010).

Many landfills in Nigeria have not been designed to protect the environment from pollution. As rain washes through dumped waste, some of the solids are dissolved while liquids are mixed. The water can become acidic and act on the waste in containers thus producing leachate. Leachate formation therefore is the main environmental impact during land-filling operation. The greatest contamination threat to groundwater therefore comes from the leachate generated from the fill materials

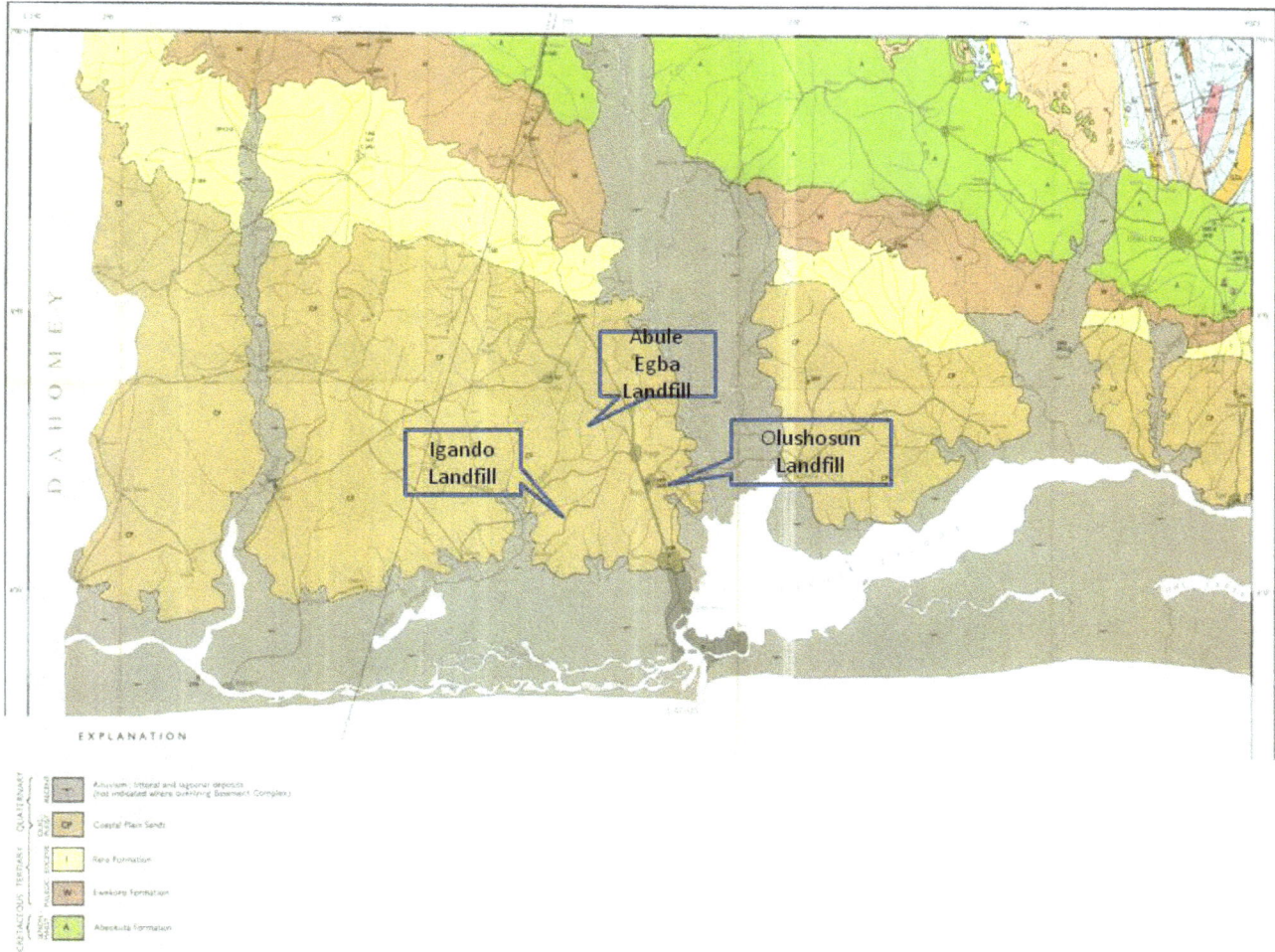

Figure 1. Geological map of Lagos showing location of the three major landfills of study.

which most often contain toxic substances especially when wastes of industrial origin are landfilled (Longe and Enekwechi, 2007; Ogundiran and Afolabi, 2008; Longe and Balogun, 2009). Composition of leachates is variable depending on type, origin and composition of the wastes, the structure, management and the age of the landfill (Ehrig, 1989; Eckenfelder, 2000).

In the Lagos area (Figures 1 and 2), the cheap mode of disposal of huge refuse matters being generated domestically and industrially has over the years, been reducing the potential sources of utilizable water for the growing population (Asiwaju-Bello and Akande, 2001). Information on sub-surface geology and aquifer characteristics below the metropolis indicate a complex lithology of alternating sand and clay deposits with three aquifer horizons (Longe et al., 1987). The water table aquifer (average thickness of 8 m) which is mostly exploited (Longe et al., 1987) is exposed to the danger of pollution by leachate effluent from indiscriminately sited refuse dump grounds (Asiwaju-Bello and Akande, 2001).

Location description

The Lagos metropolitan area is situated within latitudes N06° 23' and N06° 40' and longitudes E03° 13' and E03° 27'. There are 3 major landfills sites serving the Lagos metropolitan area. They are Abule Egba, Igando and Olushosun (which is the largest and still remains very active) (Figure 2).

Abule Egba landfill (Figures 3 and 4) is situated on the NNW flank of Lagos along the Lagos – Abeokuta highway. The site occupies a land of about 10.2 ha in the Western part of Lagos in the Alimosho Local Government and receives waste from the densely populated area. The residual life span is approximately 8 years. The Abule Egba landfill has been inactive since 2008. Four Vertical Electrical Sounding (VES) points were occupied within the refuse dump area while one VES point was occupied along Balogun Crescent as control. VES 1 and VES 2 data were acquired within the old refuse dump where landfill activities commenced in 1992. VES 3 data were

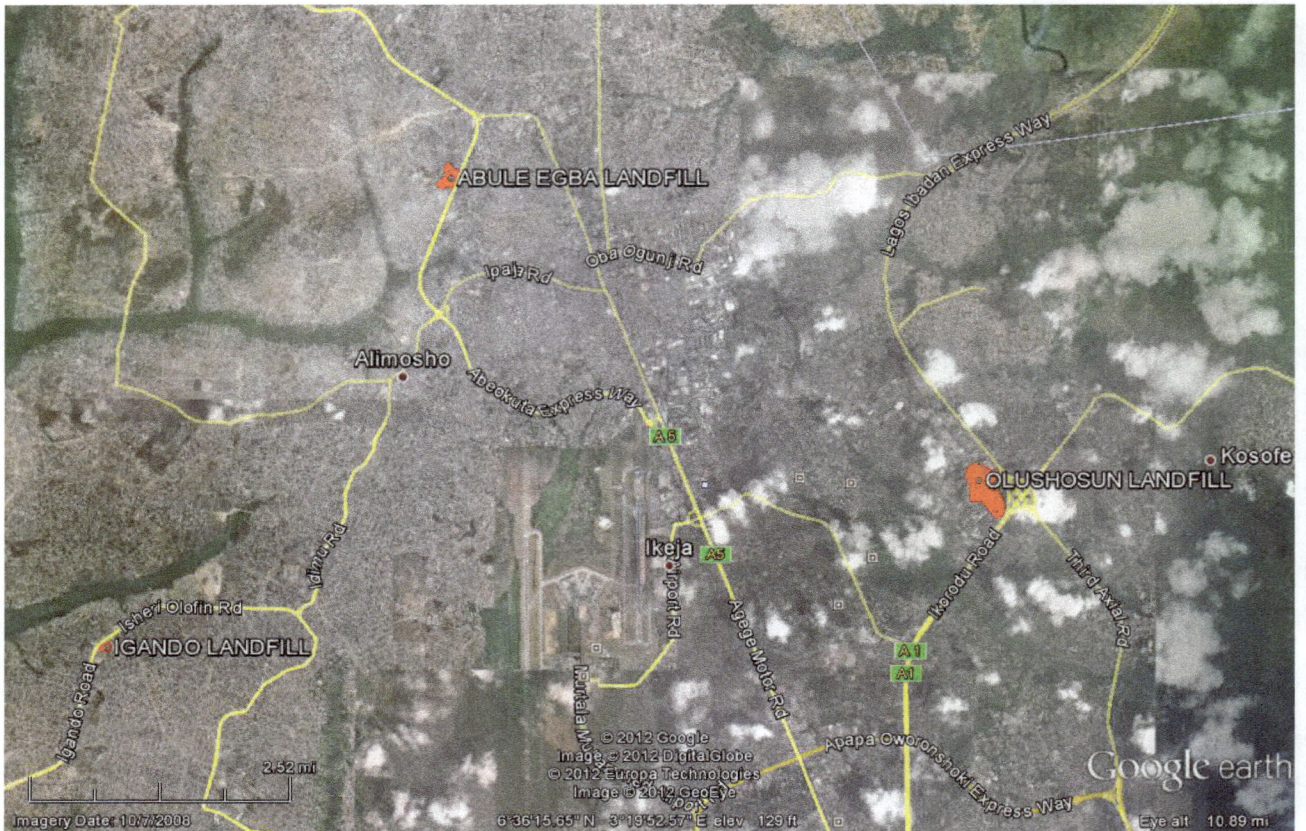

Figure 2. Satellite image of Lagos Mainland showing the location of three major landfills.

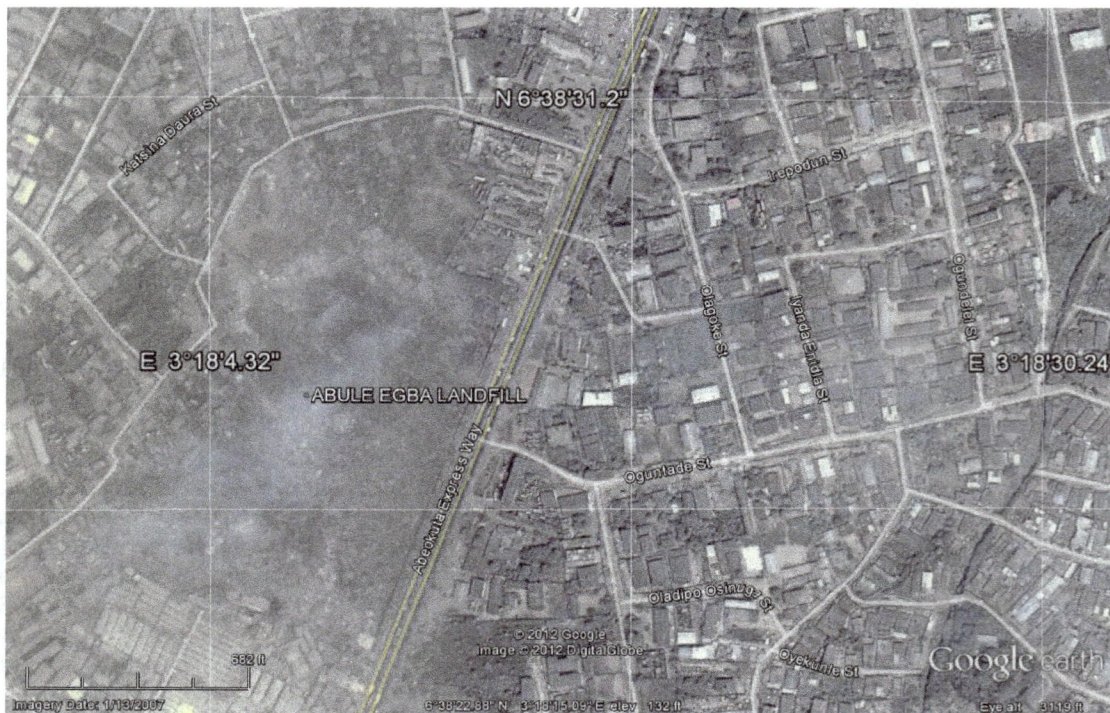

Figure 3. Satellite image showing the location of Abule Egba Landfill.

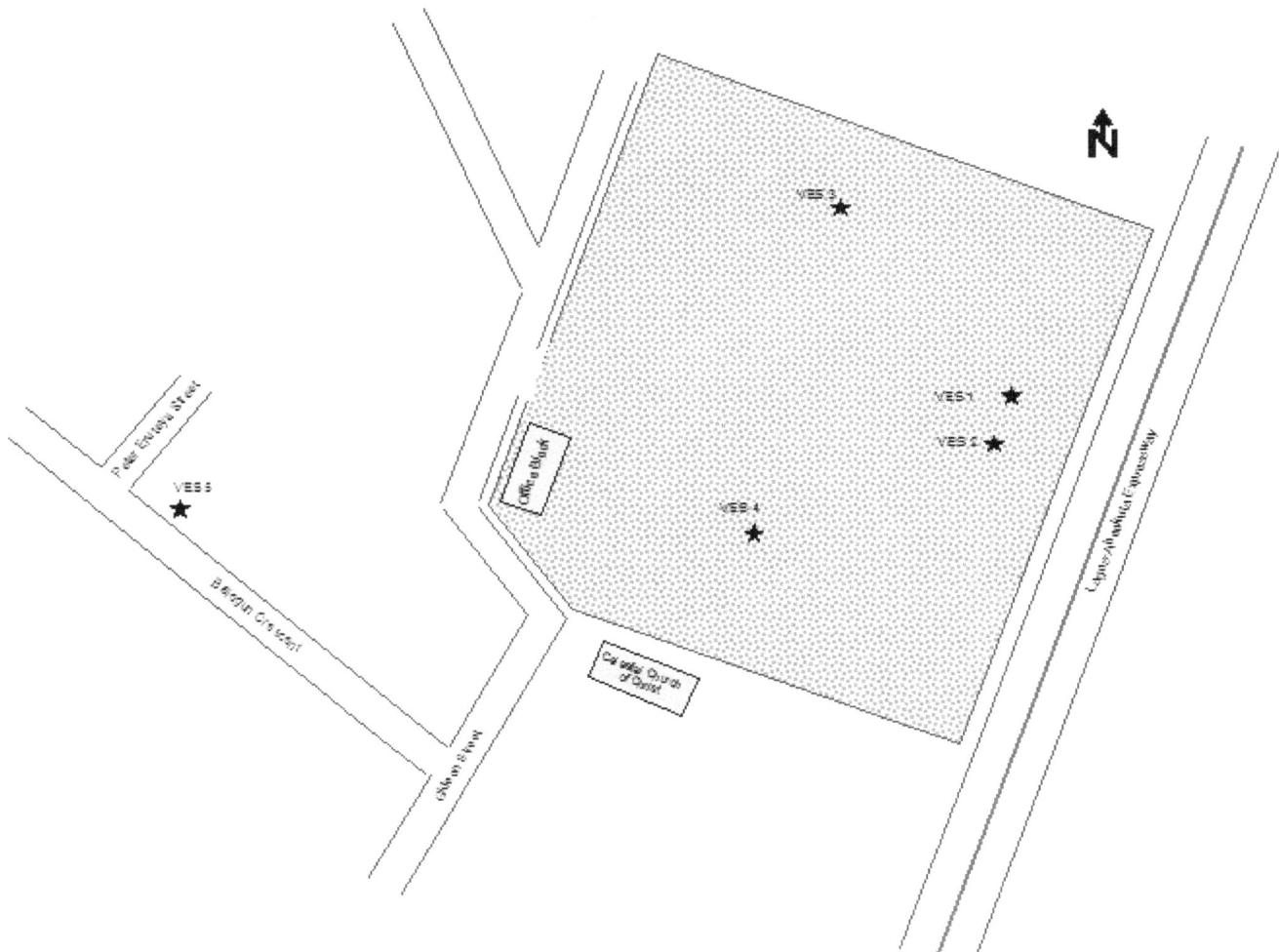

Figure 4. Layout map of Abule Egba landfill showing data points.

acquired in area of fairly recent refuse dump (2006 to 2007 phase). VES 4 data were acquired within area of long abandoned dump.

Igando landfill is situated along Lagos State University – Iba Road on 7.8 ha of land (Figures 5 and 6) with average life span of 5 years. The landfill has not been active since 5th October 2006. Four VES points were occupied (Figure 6). VES 1 was located in the central area of the landfill. VES 2 was located on the western limit of the landfill. The data were acquired on cut segment of the burrow pit. VES 3 was located on area of recent dump with fairly unconsolidated refuse. VES 4 was located close to the Lagos State University/Ojoo – Idimu roadway and served as control to enable the understanding of hydrogeologic setting in the Igando area. Olushosun landfill (Figures 7 and 8) is situated in the northern part of Lagos within Ikeja Local Government and receives approximately 40% of the total waste deposits from Lagos. The size is 42.7 ha and a residual life span

of 20 years. It is the largest in Africa, and one of the largest in the world. The landfill has been active since Friday 19th Novermber 1992. The site was initially a burrow pit where laterite was obtained for construction of roads around Lagos. The landfill was designed to receive 7,365.000 tonnes of solid waste during its operational lifespan of 10 years. This figure represents a yearly average tipping volume of 736,500 tonnes (Longe and Enekwechi, 2007). Operational design of waste to cover ratio of 9:1 was chosen for the ten years duration (Lavalin, 1992) Five VES points were occupied (Figure 8) within the landfill. VES 1 was located on active refuse dump area in the middle of the landfill. VES 2 is located on the fairly consolidated refuse column on the northern flank of the landfill. VES 3 was located on the area which was free of refuse dump at time of investigation on the northern flank. VES 4 was located within leachate drain area north of the landfill. VES 5 was located in the deep burrow pit south of the landfill.

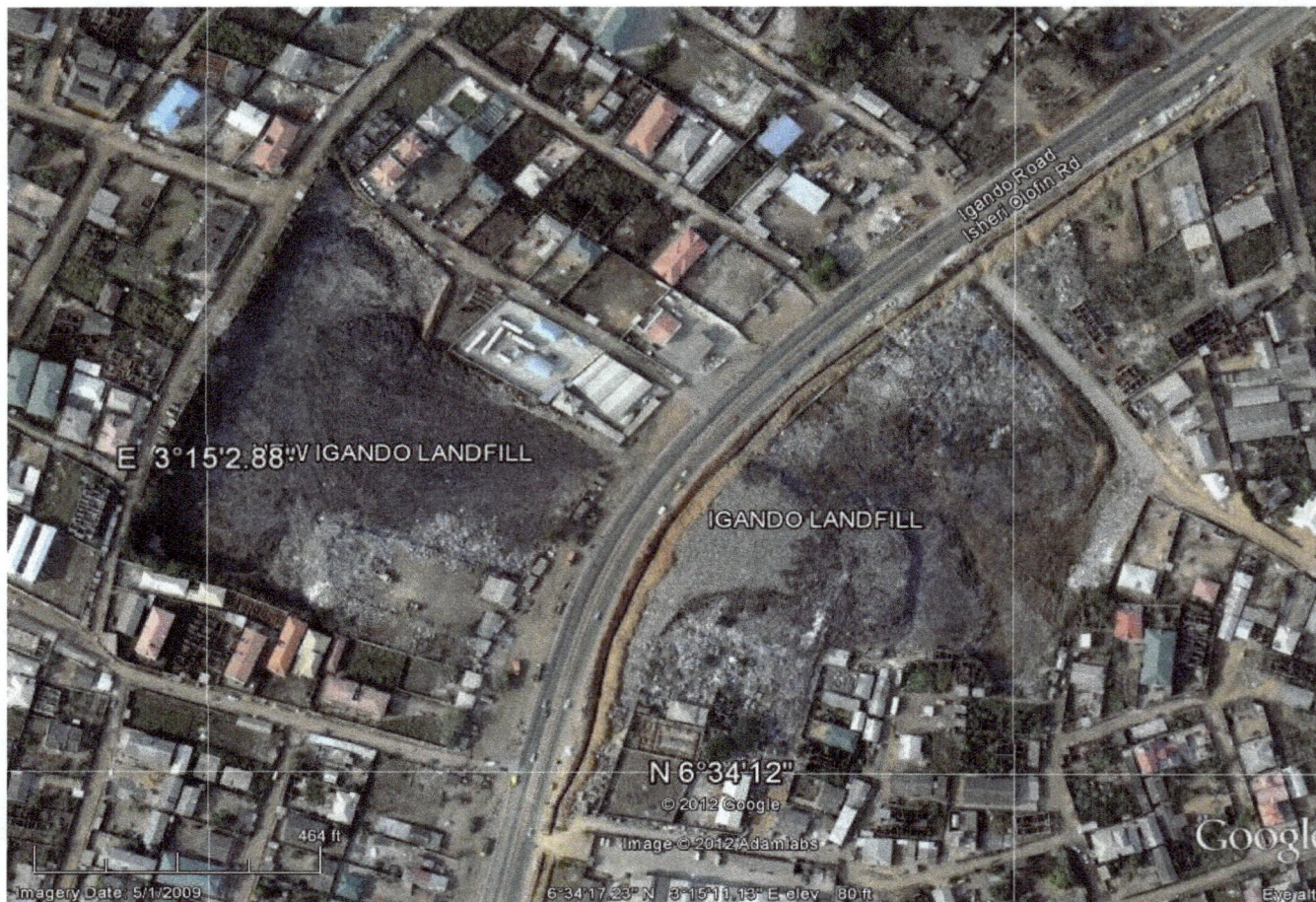

Figure 5. Satellite image showing the location of Igando Landfill (landfill east of motorway is under reference).

Local Geology/Geomorphology

Lagos metropolis is located within the Western Nigeria Coastal Zone, a zone of coastal creeks and lagoons (Pugh, 1954; Longe et al., 1987) developed by barrier beaches associated with sand deposition (Hill and Webb, 1958). Adegoke et al. (1980) recognized five geomorphologic sub-units in the coastal landscape. These are the abandoned beach ridge complex, coastal creeks and lagoons, swamp flats, forested river floodplain and active barrier beach complex.

Lagos is underlain by the Dahomey Basin (Figure 1) with lithologic constituents that are mainly sands, clays and limestones (Jones and Hockey, 1964; Longe et al., 1987; Nwankwoala, 2011). The basement complex which forms the basement rocks in the basin is overlain in succession by the Cretaceous Abeokuta Formation which is sandy with inter-bedded shales and limestone formation. Following it is the Tertiary Ewekoro Formation comprising of limestone, clays and shales and the Ilaro Formation consisting of clays and shales followed by the poorly sorted Coastal Plain Sands and Recent Alluvial

Deposits. The latter which consists of lithoral and lagoonal sediments of the coastal belt is characterized by Mangrove (saltwater) and freshwater swamps where aquifers, are readily recharged by copious rainfall thus making them vulnerable to leachate contamination in areas proximal to landfills. The lithological disposition of the aquifers gives rise to artesian and sub-artesian conditions in places. In the Lagos metropolitan area, the Coastal Plain Sands are the major aquifers generally exploited by low and medium income earners for water supply.

MATERIALS AND METHODS

Electrical resistivity method commonly used for hydrogeological, mining, geotechnical investigations, and environmental surveys (Koefoed, 1979; Loke, 1999) was adopted for the study. Subsurface electrical resistivity is related to buried materials and various geological and hydrogeological parameters such as the mineral and fluid content, porosity and water saturation (Stanton and Schrader, 2001).

The current electrode separation (AB) was varied from a minimum of 2.0 m to a maximum ranging from 80 to 450 m at the VES locations. PASI model 16-GL resistivity meter complete with

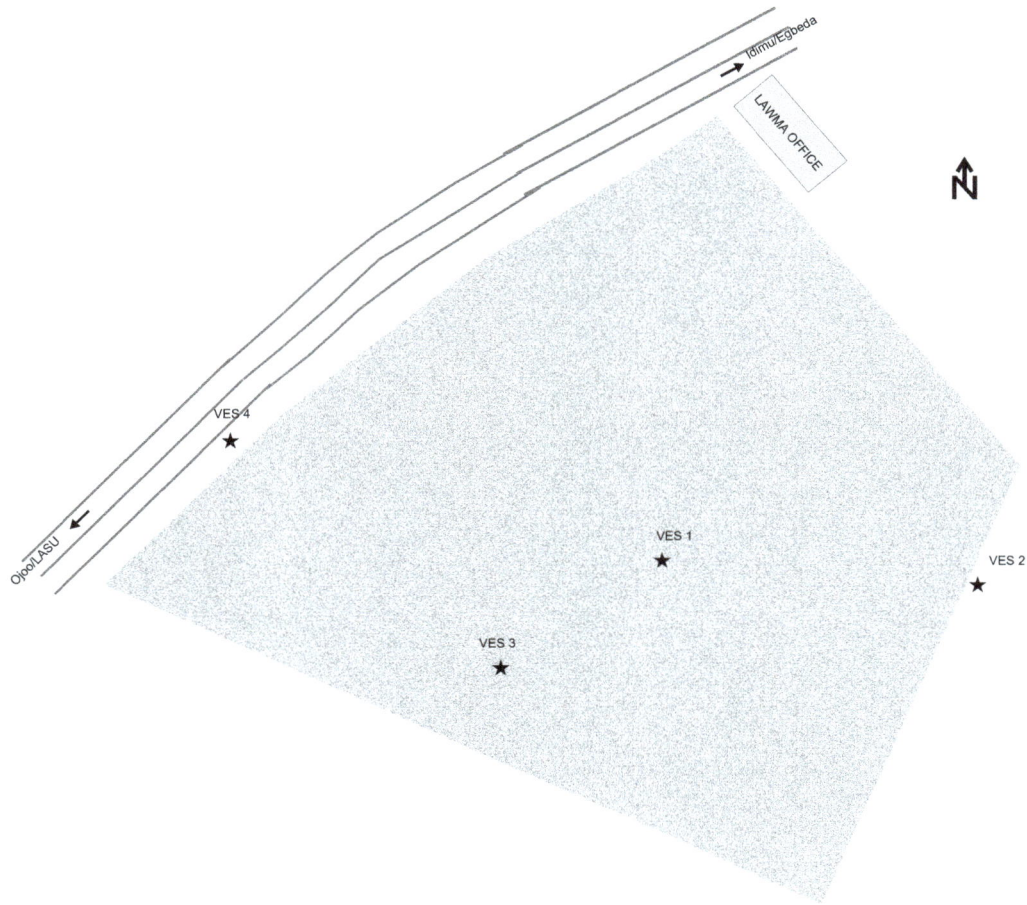

Figure 6. Layout map of Igando Landfill showing data points.

Figure 7. Satellite image showing the location of Olushosun Landfill.

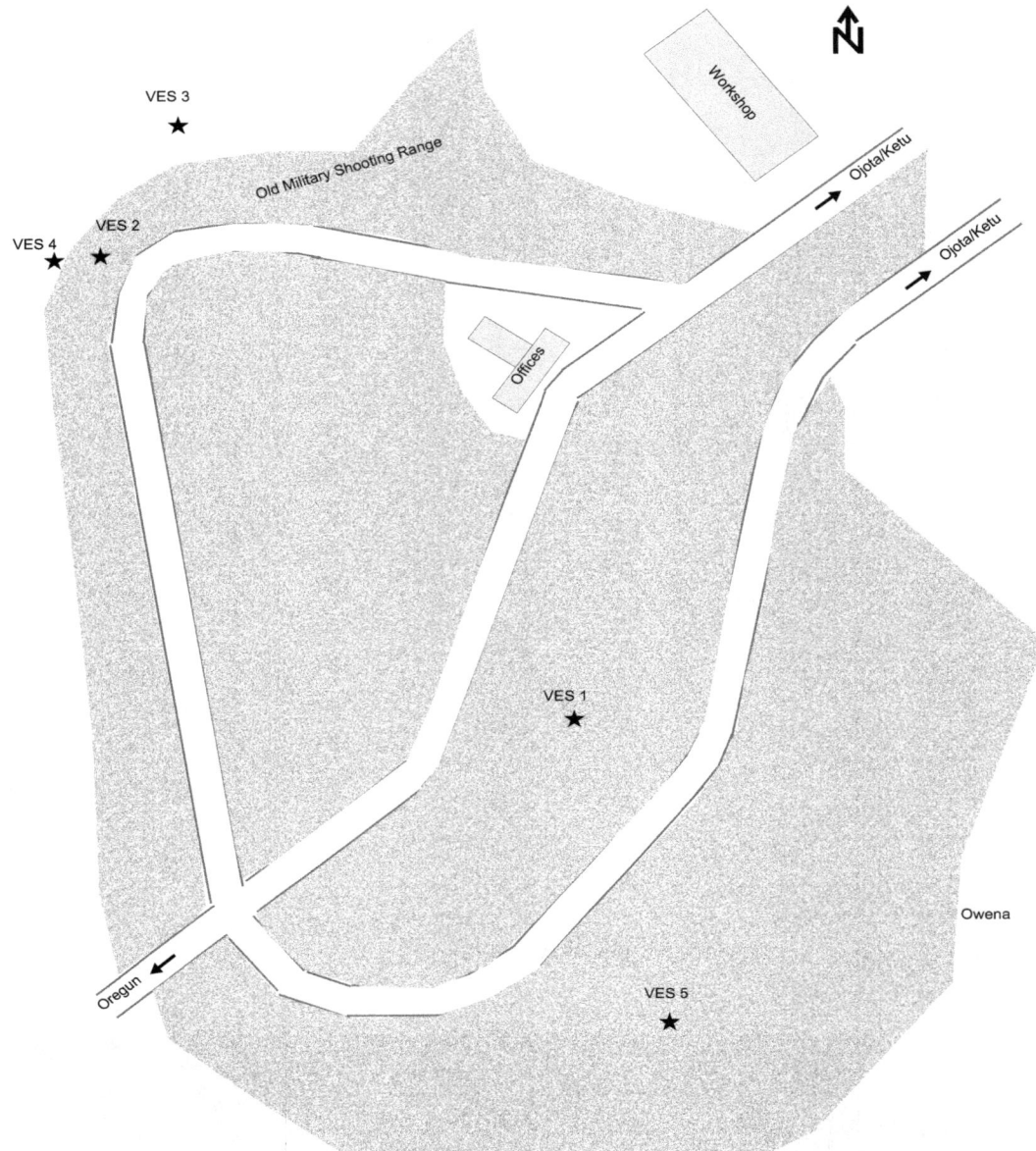

Figure 8. Layout map of Olushosun Landfill showing data points.

peripherals, (www.pasigeophysics.com) were used for field data acquisition in form of apparent resistivity measurements.

The field data acquired are presented as sounding curves. Qualitative interpretation of the curves involved evaluation of curves for geoelectric characterization of the landfills. Quantitative interpretation of the curves involved partial curve matching using two-layer Schlumberger master curves and the auxiliary curves. Outputs were modelled using computer iterations.

RESULTS

Sounding curve types obtained at Abule Egba are HKH, QH, QQH, KQH and HK (typical curves in Figure 9) while

the curve types obtained from Igando are H, K and KH (typical curves in Figure 10). Olushosun landfill is typified by H, HKQ, KH and HKH curve types (typical curves in Figure 11). The interpretation results are presented in Table 1.

DISCUSSION

The results of the geophysical studies conducted on three landfills in Lagos metropolitan area show that the refuse columns are generally characterized by low resistivity values. On the contrary, the superficial soil

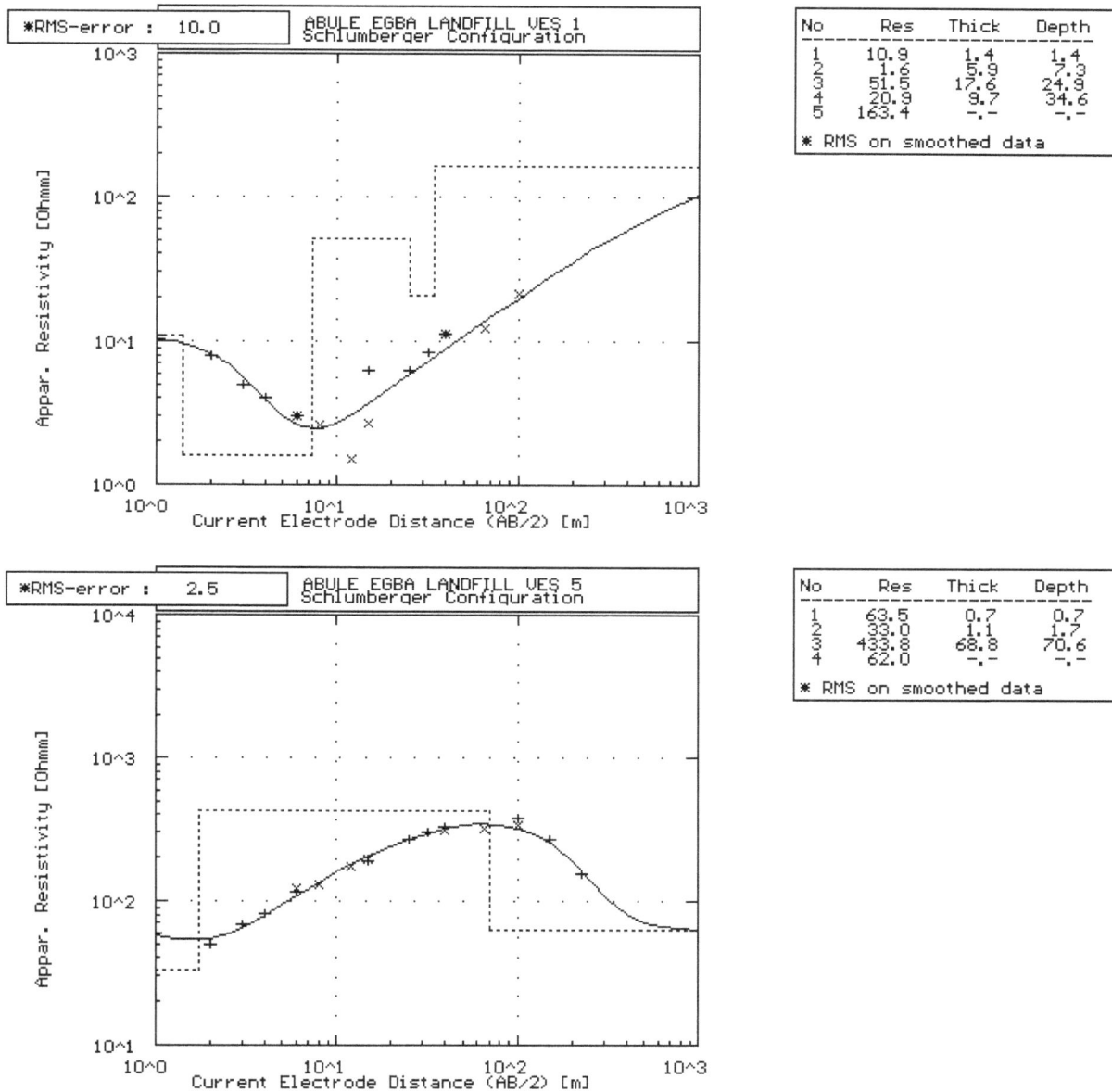

Figure 9. Typical Geoelectric Sounding Curves obtained at Abule Egba Landfill.

materials hosting the refuse are characterized by fairly high resistivity values. The contrast in the electrical resistivity characteristics of the dumps and the surrounding host rocks enabled the delineation of the refuse dump column. However, fresh domestic type refuse in the landfills are characterized by high resistivity values.

The refuse columns within Abule Egba landfill are defined by low resistivity values at decomposed stage. The decomposed resistivity values vary between 1.6 and 8.5 Ω-m. Within recent dump areas, the resistivity values vary between 14.7 and 144 Ω-m. Hydrogeologic

evaluation of VES 5 data which is located within residential area devoid of refuse presented high resistivity upper strata sequence indicative of unconfined aquifer setting. Thus, from hydrogeological point of view Abule Egba landfill is situated in an unsuitable location due to considerable groundwater recharge capability of the resistive upper sequence. Geoelectric section of Figure 12 has also shown that Abule Egba landfill environment is underlain by thick column (70.6 m) of clayey sand and sandy clay.

At Igando the refuse is defined by resistivity values varying between 2.5 Ω-m at the decomposed stage and

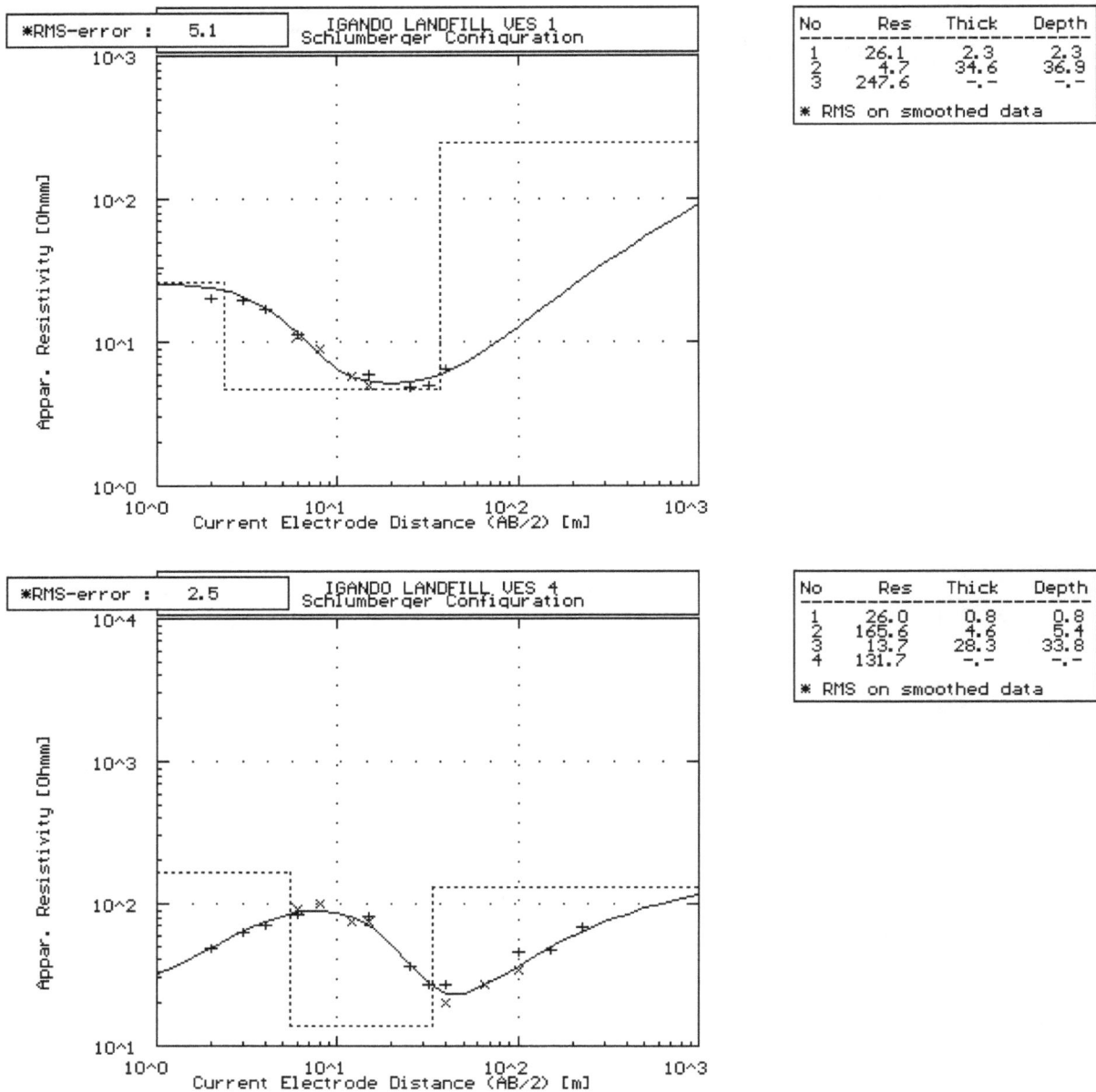

No	Res	Thick	Depth
1	26.1	2.3	2.3
2	4.7	34.6	36.9
3	247.6	-.-	-.-

* RMS on smoothed data

No	Res	Thick	Depth
1	26.0	0.8	0.8
2	165.6	4.6	5.4
3	13.7	28.3	33.8
4	131.7	-.-	-.-

* RMS on smoothed data

Figure 10. Typical Geoelectric Sounding curves obtained at Igando Landfill.

26.1 Ω-m at fairly decomposed/dry stage. The surrounding lithologic units are defined by low resistivity values that are indicative of clay. An average refuse thickness of about 9 m was obtained in the Igando landfill.

Most of the Igando landfill refuse consist essentially of domestic type trashes with the waste stream consisting of domestic, market, commercial, industrial and institutional origins. The geoelectric section at Igando (Figure 13) shows that the landfill is underlain by thick impermeable laterite/clay column. This geoelectric sequence aptly agrees with the description of Longe and Balogun (2010)

where they described the soil stratigraphy of Solous (Igando) landfill to consist of clay intercalated with lateritic clay. This lithology is capable of protecting the underlying confined aquifer from leachate contamination. The aquifer at Igando is confined with limited susceptibility to contamination as already shown by groundwater quality assessment undertaken within the water table zone of the landfill by Longe and Balogun (2010).

The refuse at Olushosun is defined by resistivity values varying between 2.4 Ω-m at decomposed stage and 51.5 Ω-m at fresh dump. The geoelectric section at Olushosun (Figure 14) shows that the northern flank of the landfill is

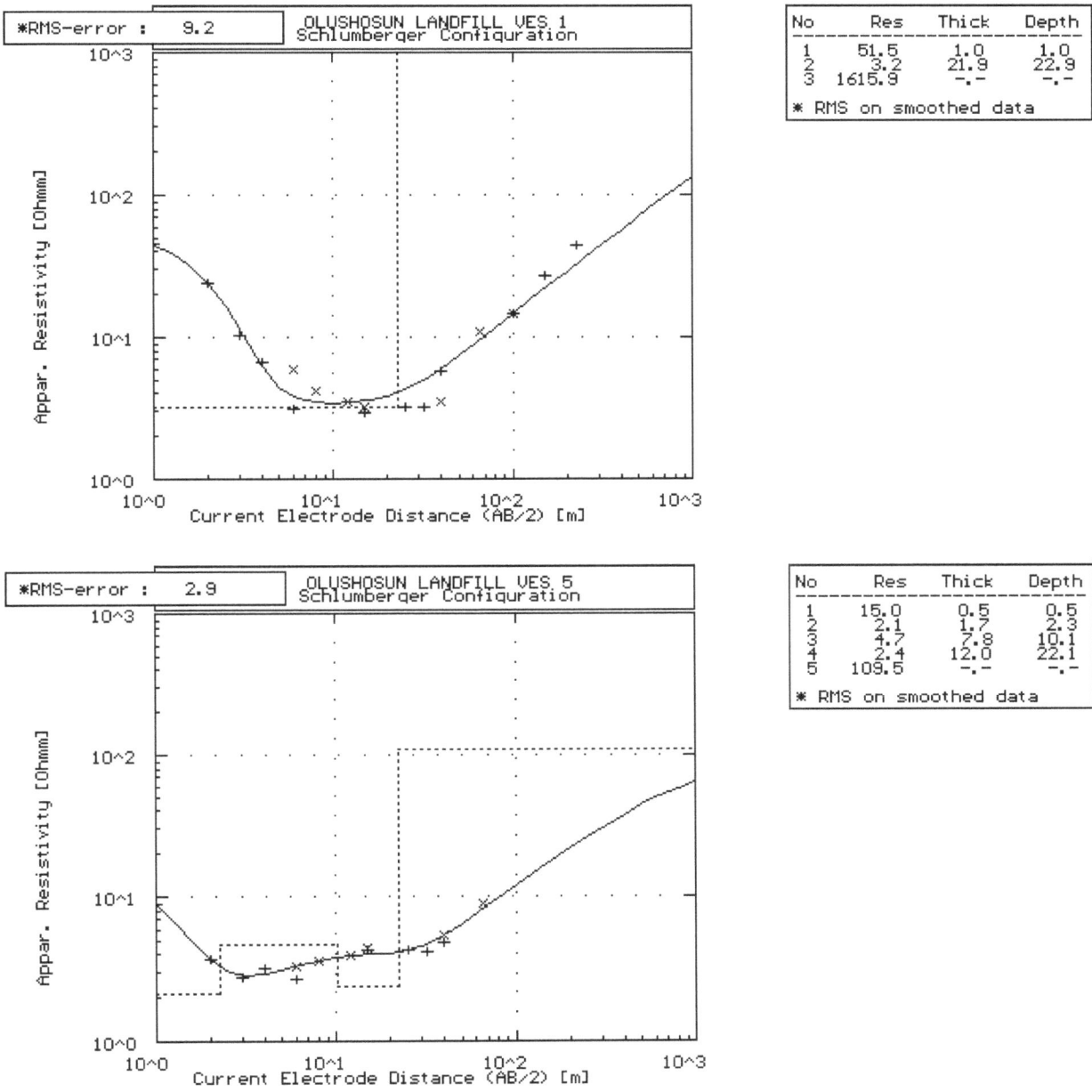

Figure 11. Typical Geoelectric Sounding Curves obtained at Olushosun Landfill.

overlain by relatively thin refuse section (4.4 m) while thick refuse section (22.9 m) overlie the central areas. The Olushosun landfill was excavated to various depth levels thus the underlying soil strata vary in composition from clay/laterite on the northern flank to clay within the deep burrow pit area of the south. The geoelectric outlay as obtained in this study is consistent with the hydrogeological setting of Olushosun earlier described by Longe and Enekwechi (2007). The sub-surface geology of the landfill which is described to be underlain by a water-bearing zone consisting of loose, medium to coarse sand with an average thickness of 10.4 m (Longe

and Enekwechi, 2007) poses a threat to aquifers in the environment.

Conclusion

The results of the geophysical investigation conducted at Abule Egba showed that the lithologic units beneath the dump area have permeable geoelectric characteristics. Thus, the hydrogeologic system at Abule Egba is vulnerable to leachate contamination. The landfill at Abule Egba has the potential to contaminate groundwater.

Table 1. Interpretation results.

Location	VES No.	Depths (m)	Resistivity (Ω-m)
		$d_1/d_2/........./d_{n-1}$	$\rho_1/\rho_2/........./\rho_n$
Abule Egba	1.	1.4/7.3/24.9/34.6	10.9/1.6/51.5/20.9/163.4
	2.	1.0/4.3/9.8	58.3/4.4/1.7/120.8
	3.	1.9/4.8/9.5/21.2	60.2/14.7/8.5/5.6/271.9
	4.	0.4/1.8/5.1/17.4	6.4/144.1/5.9/2.0/191.4
	5.	0.7/1.7/70.6	63.5/33.0/433.8/62.0
Igando	1.	2.3/36.9	26.1/4.7/247.6
	2.	0.6/4.9	255/544/152
	3.	1.0/5.7/9.1	3.8/7.7/2.5/249
	4.	0.8/5.4/33.8	26/165.6/13.7/131.7
Olushosun	1.	1.0/22.9	51.5/3.2/1615.9
	2.	0.6/4.4	8.0/2.6/660.7
	3.	0.9/3.7/13.3/25.3	1176/494/6346/874/86
	4.	0.4/2.6/12.4	1.8/56.0/4.7/61.6
	5.	0.5/2.3/10.1/22.1	15.0/2.1/4.7/2.4/109.5

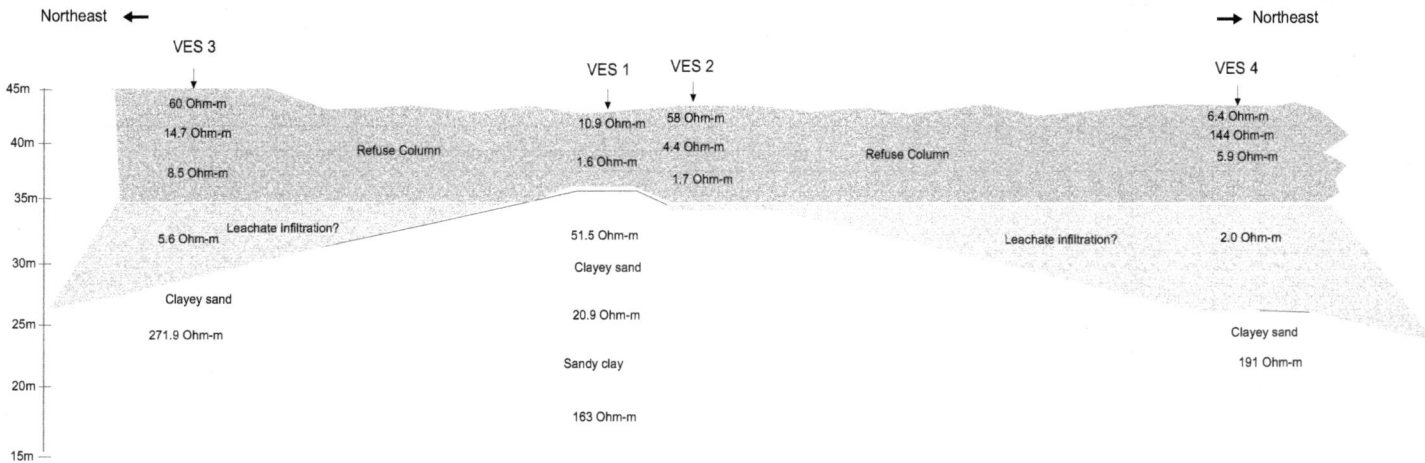

Figure 12. 2-D Geoelectric section along NE-SW section of Abule Egba Landfill.

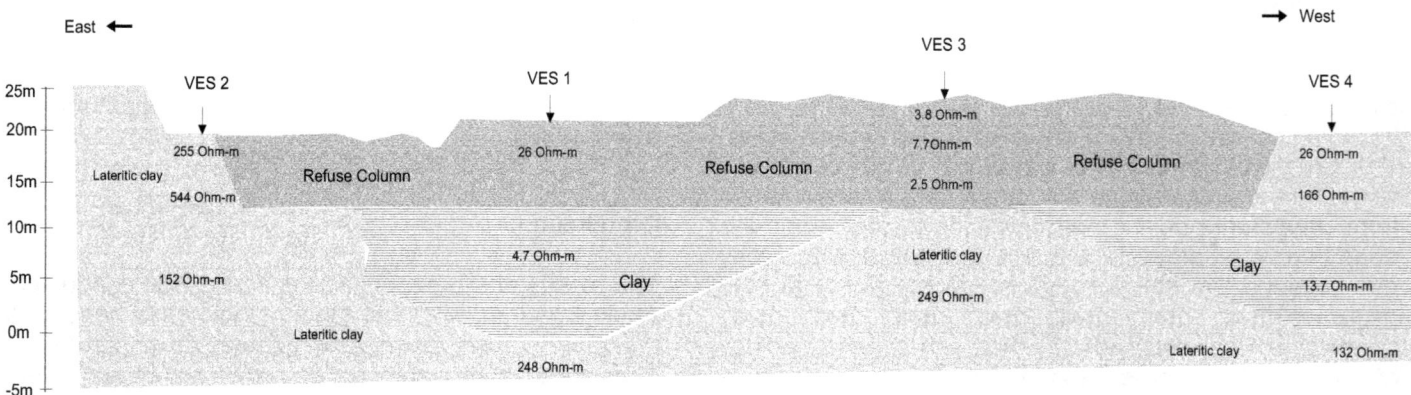

Figure 13. 2-D Geoelectric Section along E-W section of Igando Landfill.

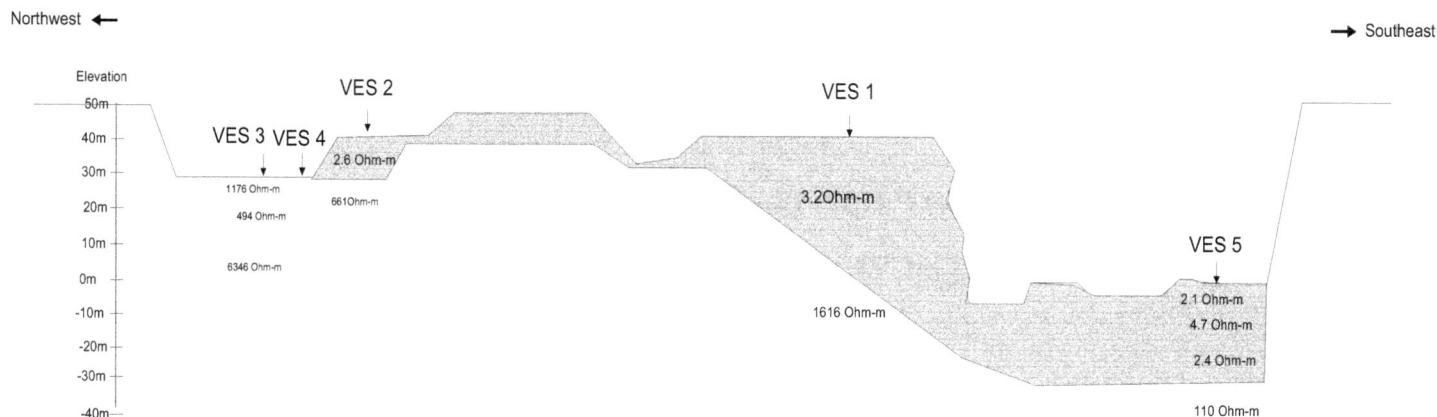

Figure 14. 2-D Geoelectric Section along NW-SE section of Olushosun Landfill.

The Abule Egba landfill is an open dump situated within a high density area with the need to evolve leachate collection system in the dump. Abule Egba landfill is situated in an unsuitable location.

At Igando, the underlying lithologic unit has impermeable geoelectric characteristics thus the hydrogeologic system is presumably protected. The existence of clay/laterite column around Igando landfill provides a means for retardation of leachate movement into the underlying aquifer. The low permeability clay and laterite medium has potential to prevent groundwater contamination at depth.

The Olushosun landfill is underlain by materials of impermeable geoelectric characteristics in the northern flank and fairly permeable geoelectric characteristics in the south around Owena phase. However, the Olushosun landfill which is the largest in the Lagos metropolitan area has no liners and leachate collection system and are therefore likely to cause groundwater contamination especially on the southern flank.

REFERENCES

Adegoke OS, Jeje LK, Durotoye B, Adeleye DR, Ebukanson EE (1980). The Geomorphology and Aspects of Sedimentology of Coastal Region of Western Nigeria. J. Mining Geol. 17:217-223.

Asiwaju-Bello YA, Akande OO (2001). Urban groundwater pollution: Case study of a Disposal site in Lagos metropolis. J. Water Res. 12:22-26.

Christensen TH, Kjeldsen P, Bjerg PL, Jensen DL, Christensen JB, Baun A, Albrechtsen HJ, Heron G (2001). Biogeochemistry of landfill leachate plumes. Appl. Geochem. 16:659-718.

Eckenfelder WW (2000). Industrial Water Pollution Control, 3rd Ed. Mc-Graw-Hill, London, UK.

Ehrig HJ (1989) Leachate treatment overview, in T.H. Christensen et.al. eds. Sanitary landfilling - Process, Technol. Environ. Impact. Acad. Press London, pp. 285-297.

Hill MB, Webb JE (1958). The ecology of Lagos lagoon II. The topography and physical features of the Lagos harbor and Lagos lagoon. Philosophic. Trans. Royal Soc. London 241:307-417.

Jones HA. Hockey RD (1964). Geology of parts of South-Western Nigeria. Bull. Geol. Surv. Nig. No. 31

Koefoed O (1979). Geosounding principles, Resistivity sounding measurements: Elsevier, Amsterdam.

Lavalin (1992). Organizational development and waste management system project. Design and operations report for the Oregun landfill site submitted to the Lagos State Waste Disposal Board.

Lee GF, Jones-Lee A (1993a). Revisions of state of MSW landfill regulations: Issues in groundwater quality. J. Environ. Manag. Rev. 29:32-54.

Lee GF, Jones-Lee A (1993b). Groundwater pollution by municipal landfills: Leachate composition, detection and water quality significance," Proc. Sardinia '93 IV International Landfill Symposium, Sardinia, Italy, 10931103.

Loke MH (1999). Electrical imaging surveys for environmental and engineering studies -- A practical guide to 2-d and 3-d surveys: Austin, Texas, Adv. Geosci. Inc., p. 57.

Longe EO, Balogun MR (2010). Groundwater Quality Assessment near a Municipal Landfill, Lagos, Nigeria. Res. J. Appl. Sci. Eng. Technol. 2(1):39-44.

Longe EO Enekwechi LO (2007). Investigation on potential groundwater impacts and influence of local hydrogeology on natural attenuation of leachate at a municipal landfill. Int. J. Environ. Sci. Technol. 4(1):133-140.

Longe EO, Malomo S, Olorunniwo MA (1987). Hydrogeology of Lagos metropolis. J. Afr. Earth Sci. 6(3):163-174.

Nwankwoala HO (2011). Coastal Aquifers of Nigeria: An Overview of Its Management and Sustainability Considerations. J. Appl. Technol. Environ. Sanit. 1(4):371-380.

Ogundiran OO, Afolabi TA (2008). Assessment of the physicochemical parameters and heavy metals' toxicity of leachates from municipal solid waste open dumpsite. Int. J. Environ. Sci. Technol. 5(2):243-250.

Petruzzelli D, Boghetich G, Petrella M, Dell'Erba A, L'Abbate P, Sanarica S, Miraglia M (2007) Pre-treatment of industrial landfill leachate by Fenton's oxidation. Global NEST J. 9:51-56.

Pugh JC (1954). A Classification of the Nigeria Coastline. J. W. Afr. Sci. 1:3-12.

Stanton GP, Schrader TP (2001). Surface Geophysical Investigation of a Chemical Waste Landfill in Northwestern Arkansas. In Eve L. Kuniansky, editor, 2001, U.S. Geological Survey Karst Interest Group Proceedings, Water-Resources Investigations Report 01-4011:107-115

Stollenwerk KG, Colman JA (2003). Natural remediation potential of arsenic-contaminated ground water, in Welch, A.H., and Stollenwerk, K.G., Eds., Arsenic in groundwater Geochemistry and Occurrence, Boston, Kluwer, pp. 351-379.

8

Regional model for peak discharge estimation in ungauged drainage basin using GIUH, Snyder, SCS and triangular models

Majid Dabbaghian Amiry[1]* and Mohammadi A. A.[2]

[1]Urban Planning, Geographic Department, Piam Noor University, Sari, Iran.
[2]Department of Watershed Management, Science and Research Branch, Islamic Azad University, Tehran, Iran.

Due to the importance of instantaneous peak discharge estimation for watershed management study in countries like Iran, the search is ongoing to correlate geomorphologic and hydrologic parameters to present models. This paper describes the use of synthetic unit hydrograph at drainage basin of Mehran (Joestan River). The obtained results were compared with recorded peak discharge in outlet of watershed. The models of Relative Mean Error (RME) and Root of Mean Square Error (RMSE) in drainage basin in central Alborz watershed were compared with each other. The results indicate the RMEs of 20.43, 40.06, 133.082, and 135.72, and RMSEs of 16.08, 14.65, 25.37 and 25.82 for GIUH, Snyder, SCS and Triangular models respectively. The daily peak discharge model was derived by 177 recorded events of daily peak discharge.

Key words: Peak discharge, parameter, model, Mehran Basin.

INTRODUCTION

Considering the world average annual precipitation (860 mm), Iran is classified as a semi-arid area with an amount of 240 mm precipitation. But his amount of precipitation doesn't cover spatial agricultural needs (Hojjati and Boustani, 2010).To address the issue, it seems that the utilization of water should be modified according to daily rate of precipitation. One of the reasonable ways to get harmony with drought is useful application of available water resources (such as surface and ground water). This strategy can not be practiced without identification of district hydrology object. Because of the lack of identification and application of hydrology science in country, we experience dangerous floods and droughts in some sites. However, in recent years, more attentions have been given to water crisis, but still there are not any recorded data in this regard. It is clear that

without studying of geomorphology and hydrology of drainage basin, execution of scientific plans for flood disaster can not be done. Drainage basin studies with attention to geomorphologic characteristics affect discharge characteristics of main rivers, their upstream and sediment generation (Nazari Samani et al., 2009). While, there is not any instrumentation for recording essential data and subsequent natural unit hydrograph, some methods can be used for determination of unit hydrograph.

Sherman (1932) concluded that the hydrograph shape must be the same for storms with same attributes. Snyder (1938) proposed a method in according with some of unit hydrograph attributes. This method is a result of researches in some cases of drainage basins in Appalachian Mountain. Some measurements were done by United States Soil Conservation Service (SCS) in different drainage basins and dimensionless hydrograph was presented (Mockus, 1957). These researches showed that if derived flood hydrograph axes in different conditions dimensioned, all of them will almost have a

*Corresponding author. E-mail: Majid_arsh@yahoo.com.

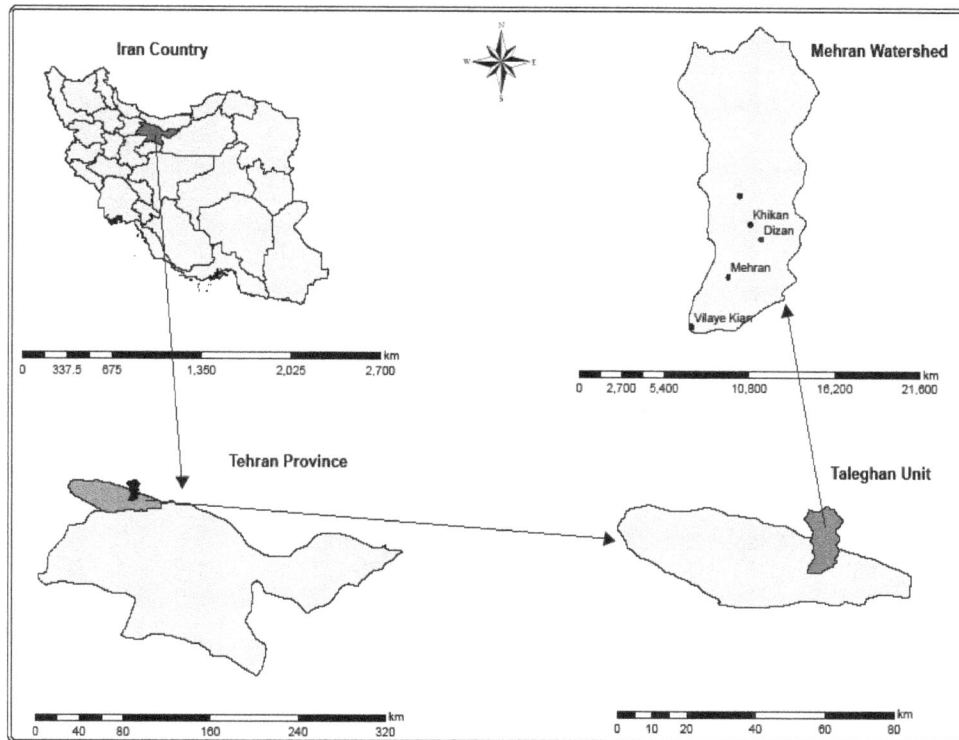

Figure 1. Location of Mehran basin.

same shape. The problems of geomorphologic instantaneous unit hydrograph (GIUH) were demonstrated in 1979 by Rodriguez- Iturbe. Recent progress in finding run off topographic was made by aid of GIUH. In previous two decades, many hydrologists (Gupta et al., 1980; Rodriguez-Iturbe et al., 1982; Krishen and Bars, 1983; Troutman and Karlinger, 1985; Agnese et al., 1988; Chutha and Dooge, 1990; Yen and Lee, 1997; Olivera and Maidment, 1999; Berod et al., 1999) were interested in run off simulations using drainage basin attribute geomorphology. The primary idea describing the engineering of stream network and results of geomorphologic responses was derived and named geomorphologic instantaneous unit hydrograph (Karvonen et al., 1999). A mathematical method and its efficiency were proposed by (Lee and Chang, 2005) as a result of studying the northern Taiwan. The results shows since the run off primarily occurs in low portions of a watershed near streams of a precipitation run off model, only the surface run off is recognized as being inadequate. And as a result, by correction of GIUH the better results can be derived. The surface flow IUH of this study could adequately reflect the variation of surface roughness conditions, and the subsurface flow IUH could reveal different soil conditions. The concept of GIUH is utilized in calculating the influence of the channel network on the delay and the shape of the hydrograph (Karvonen et al., 1999). The quantitative analysis of drainage

networks has gone through dramatics advances since 1690, mainly after Shreve's (1966) classical paper which led the way for a theoretical foundation of Horton's empirical laws. This has provided a new perspective for many other problems in fluvial geomorphology (Rodrigues-Iturb and Valdes, 1979). This article describes the most optimized model of instantaneous peak discharge estimation. To achieve this, four models including GIUH, SCS, Snyder and Triangular have been taken into account. With regards to recorded peak discharge in Mehran basin and measured data in above mentioned models, regional model for discharge estimation is earned.

MATERIALS AND METHODS

Study area

Mehran drainage basin is one of the sub basins of the Central Alborz basin. It is located in Tehran province and its Taleghan unit situated 50° 53' 24.0" to 50° 59' 19.0" East longitudinal and 34° 10' 48.0"to 36° 20' 21.0" North latitude and covers 99.71 km^2 (Figure 1).

There were rain gauge station and hydrometric station in outlet for extraction of discharge and rain statistics in the study area. In this basin, one rain gauge station: Joestan and also one hydrometric station named Mehran- Joestan were considered (Figure 2).

Main precipitation in the study area is related to Mediterranean circulation that influences the area from west in autumn through

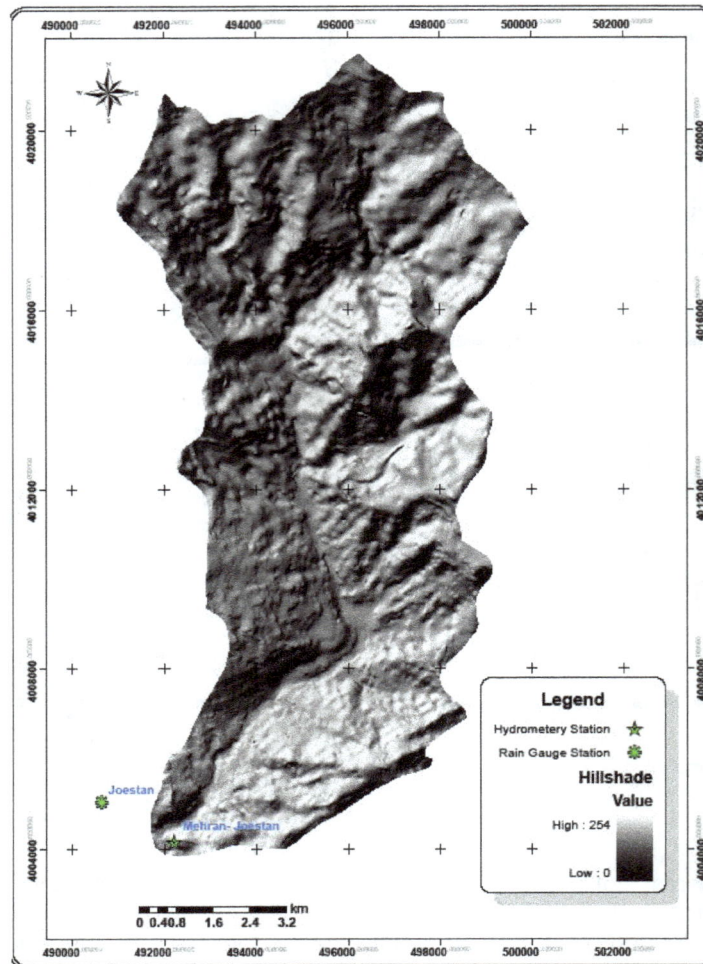

Figure 2. Location of rain gauge station and hydrometric station In Mehran basin.

spring. Since the watershed is located on the southern slopes of central Alborz, semiarid climate is predominating. Different drainage patterns can be observed, the main of which is dendritic (Figure 2). The length of its main river is about 22 km. The maximum and minimum elevations are 4390 and 1940 m respectively. Mehran drainage basin contains poor range lands and farming terrains and a small part of the watershed is garden. The total precipitation changes from 635 to 768 mm in different places of the watershed.

Extraction of rain and discharge data coincide with flood

We used flood discharge statistics and recorded rain in the station of local water institute of Tehran province and organization of water resources research. For Mehran drainage basin 15 coincidence events of rain and discharge were extracted for these events (15), 7 of which was recognized to be good (Figure 3).

Digital topographic map

Supply digital topographic map was extracted from National Cartographic Center (N.C.C.). We also extracted the followings:

stream map study drainage basin, mean slope of drainage basin area, mean weighted slope of main stream in outlet of drainage basin, main stream length from centroid to outlet of drainage basin (Figure 4)

Slope of highest stream order, stream number in each order (for determination of bifurcation ratio, R_b), stream lengths in order (for determination of length ratio, L_u) and drainage basin area in each order (for determination of area ratio, A_u). The estimation of these parameters can be handled easily and more accurately using GIS.

It is observed that the design flood is more sensitive to the design storm pattern and its time distribution (Jain et al., 2000). Lin and Oghochi (2006) have obtained the most common method implemented in major commercial GIS software, assuming minimum contributing area to determine channel head locations. However, minimum contributing areas should vary even within a small watershed according to local factors such as topography and litho logy. The infinite form and variety of drainage basins respond to the known basic geomorphologic laws exists in nature. It is expected that in the structure of the hydrologic response of a basin, a basic order should also be present which reflects the deep symmetry in formal relations between the parts involved in Horton's geomorphologic laws (Rodriguez- Iturbe et al. 1982).

It must be mentioned that bifurcation ratio ($R_B=N_u/N_{u+1}$), Length

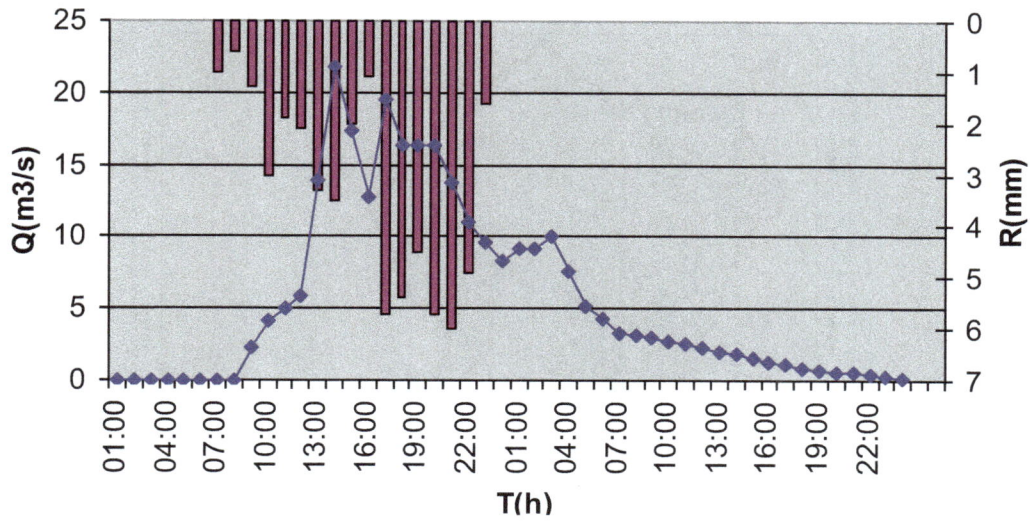

Figure 3. Observed Hydrograph at 7, 8 Nov. 2006 .

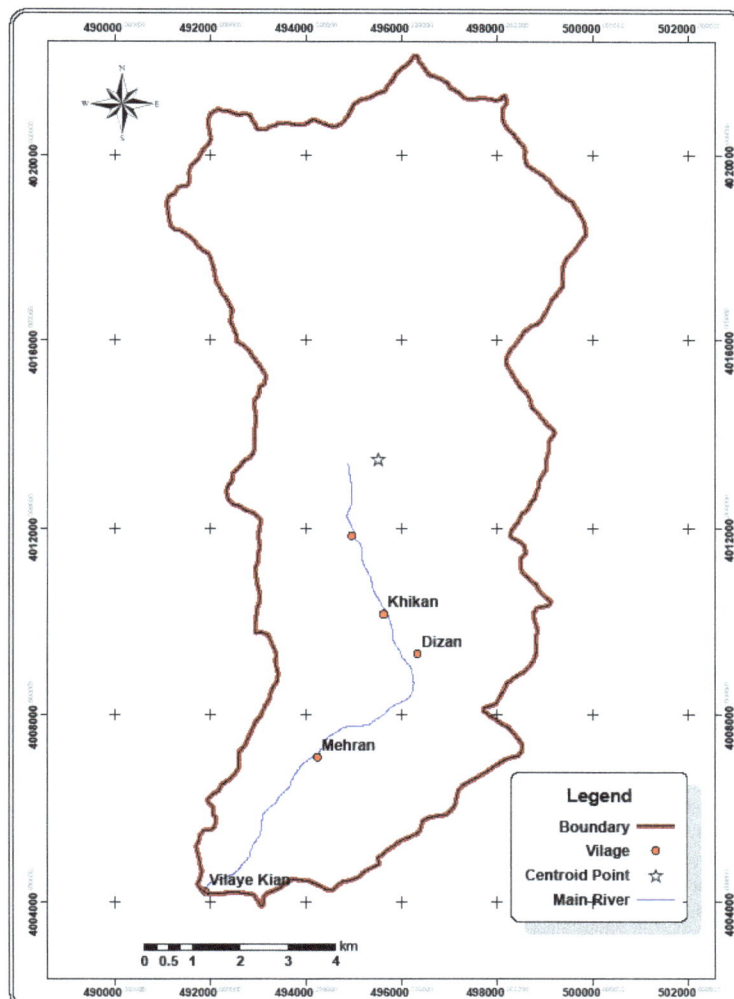

Figure 4. The main stream from centroid to outlet of Mehran basin.

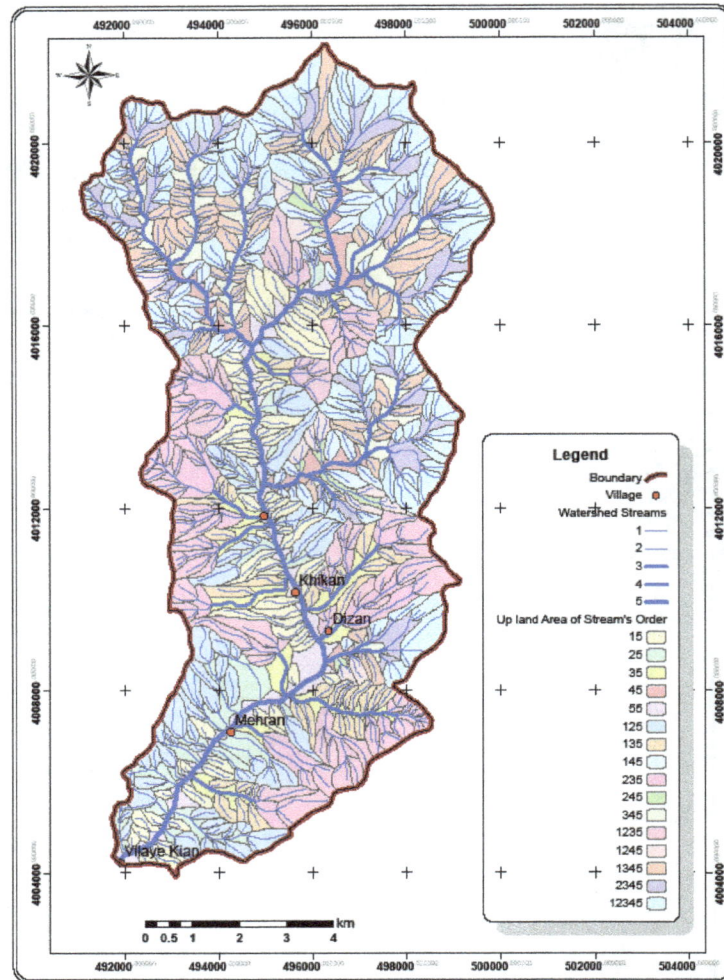

Figure 5. Up land area for each order of stream.

ratio (R_l=L_u/L_{u-1}) and area ratio (R_A=A_u/A_{u-1}) from this relation were calculated. N_u, N_{u+1}: are the numbers of stream of orders U and U+1; L_u, L_{u-1}: are the mean length of streams of orders U and U-1; A_u, A_{u-1}: are the mean area of the basins of orders U and U-1.

A river basin is made up of two interrelated systems: the channel network and the hill slopes. The hill slopes control the production of storm water runoff which, in turn, is transported through the channel network towards the basin outlet. The runoff- contributing areas of the hill slopes are both a cause and an effect of the drainage network growth and development. This cause-and-effect relationship may be visualized through the following consideration taken from Gupta et al. (1980) and Rodriguez- Iturbe (1993). It must be attended to each water ways including stream area with spatial order that entered to stream and with stream by upper order, and then it needs to be reached to outlet. For instance water way 245 in Mehran drainage basin consists of a stream area of order two that is entered to streams with order four and then entered to stream of order five. Therefore for each basin, there exist at most $2^{\Omega-1}$ water ways (Ω is the biggest stream order in each of basin) (Zhang and Govindaraju, 2003). In Mehran drainage basin, there are 16 water ways. For earning each order's area, tables of different water way gained and for earning each water way, the upper entered water way area must be attended (Figure 5).

Flow velocity

Following formula was introduced by Rodriguez- Iturb et al. (1979) to calculate the flow velocity for one special storm:

$$V_\Omega = 0.665\alpha_\Omega^{0.6}(i_rA)^{0.4}$$
$$\alpha_\Omega = S_\Omega^{0.5}/nB^{2/3}$$
(1)

Where V_Ω is the flow velocity (m/s), i_r is rain intensity (cm/h), A is drainage basin area (km^2), S_Ω is slope of the main river in drainage basin outlet (%), n is Mannig's roughness coefficient and B is the mean flow width in outlet of drainage basin (m).

Instantaneous peak discharge estimation

The classical theory of the instantaneous unit hydrograph (IUH) relates the rainfall excess over catchments to the direct runoff at the catchments' outlet rests based on three basic assumptions: lumped system, linearity, and time invariance (Rooso, 1984). GIUH model and relations presented by Rodriguez- Iturb et al. (1979) (Formula

2).

$$q_p = \frac{1.31}{L_\Omega} [R_L^{0.43} V]$$ (2)

Where L_Ω is the biggest length of Main River (km), V is flow velocity (m/s), q_p is peak discharge in (hr^{-1}) (Formula 3).

$$Q_P/Q_e = t_r \cdot q_p \left(1 - t_r * \frac{q_p}{4}\right) \qquad Q_e = i_r \cdot A$$

$$\rightarrow \quad t_b \geq t_r$$ (3)

Q_p: is the exited peak discharge (m^3/s), Q_e is the effective discharge (m^3/s), q_p is the peak discharge of geomorphologic instantaneous unit hydrograph (hr^{-1}), t_r is the time of effective precipitation (h), i_r is rain intensity (cm/h) and A is the drainage basin area (km^2).

Peak discharge estimation

Other models such as Snyder, SCS and Triangular have also been studied using relations presented in references such as Snyder (1938) and SCS Engineer- ing Handbook. Washington. D. C. (1968).

SCS model

The method of peak discharge estimation employed by the Soil Conversion Service (SCS), U. S. Department of Agriculture, uses an average number of natural UHs for watersheds varying widely in size and geographical location. In the SCS model, the lag time, Tl, shall be determined using watershed physical properties such as the area, main river length, average slope and CN (Curve Number). The synthetic unit hydrograph can then be computed. The SCS model permits computing the peak discharge for a watershed that has insufficient observed rainfall–runoff data.

A Unit Hydrograph (UH) is defined as the direct runoff hydrograph (DRH) produced by 1 unit (inch) of effective rain (runoff) uniformly distributed over a basin. Unit hydrographs can be combined with precipitation data and basin data to determine the DRH for a particular basin. The curve number was determined with respect to land use and soil hydrological group maps in different antecedent moisture conditions (dry, average and moist) and hydrological conditions. The losses estimation is the sum of the interception, infiltration, and transmission of the soil and surface (in mm). The runoff calculation is given below:

$$S = \frac{25400}{CN} - 254$$ (4)

Runoff was calculated using the following formula (5):

$$Q = \frac{(P - 0.25)^2}{P + 0.85}$$ (5)

Q: Run off (mm), S: Losses (mm) and P: Maximum precipitation in 24 h (mm).

After calculating runoff caused by a storm event, the maximum flood discharge was calculated using the following formula (6):

$$Q_{max} = \frac{2.083 AQ}{t_p}$$ (6)

Q_{max}: maximum flood discharge (m^3/s), A: Basin area (km^2), Q: Run off (mm) and tp: time of flood crest which is evaluated by time of concentration (tc) in minute.

Snyder model

Snyder (1938) was the first to propose a unit hydrograph technique that could be used on un gauged basins. His method was based on a number of watersheds in the Appalachian Highlands ranging in size from 10 mi^2 to 10,000 mi^2. Snyder's equations are:

$$T_p = C_t (L.L_c)^{0.3}$$ (7)

Where t_p is basin lag, L is length of the main stream from the outlet to the divide, L_c is Length along the main stream to a point nearest the watershed centroid and Ct is coefficient usually ranging from 1.8 – 2.2 (Ct has been found to vary from 0.4 in mountainous areas to 8.0 along the Gulf of Mexico).

$$Q_P = 640 \left(\frac{C_p A}{t_p}\right)$$ (8)

Where Q_p: peak discharge, A: drainage area and C_p: storage coefficient ranging from 0.4 to 0.8 where larger values of Cp are associated with smaller values of Ct.

$$T_b = 3 + \frac{t_p}{8}$$ (9)

Where T_b: the time base of the hydrograph. For small watersheds, Eq 9 should be replaced by multiplying tp by a value of from 3 to 5 as a better estimate of Tb. Eqns. 7,8 and 9 define points for a unit hydrograph produced t_p an excess rainfall of duration D = tp/5.5. For other rainfall excess durations D', an adjusted formula for tp becomes:

$$T_{p'} = t_p + 0.25(D' - D)$$ (10)

Where tp' is the adjusted lag time for duration D'. Once the three quantities tp, Qp, and tb are known, unit hydrograph can be sketched so that the area under the curve represents 1,0 in of direct runoff from the watershed.

In this application the two items of data are: C_p is Storage coefficient plus and T_p is catchments' lag times.

Triangular model

Since the height and base-width of the triangular are constrained to be simple functions of its time to peak, Tp, the triangular UH is a parameter model. Employing a statistical relationship between Tp and catchments' characteristics, the triangular model can be applied in regionalization mode for catchments un gauged for flow.

Table 1. Numbers and dates of events studied in drainage basin.

Date of events	Intensity i_r (cm/h)	Events Num.
20, 21 Apr 2003	2.55	7
29, May, 2003	3.8	
24, 25 Apr 2004	2.225	
26, 27 Apr 2005	3.55	
19, 20 May 2005	0.66	
7, 8 Nov 2006	0.66	
27, 28 Apr 2007	3.28	

The triangular model is widely employed for flood hydrology. For hydrological analysis of lower flows, or for characterizing whole flow regimes, the parameter triangular model is, not surprisingly, limited by its conceptual simplicity (e.g. hydrograph recessions are not characteristically linear)

Models calibration

Relative Mean Error (RME)

Relative Mean Error relation for calculated peak discharge from observed peak discharge is presented in Formula (11, 12):

$$RME = {}^1\!/_n \sum RE_i \quad \text{(11) and}$$

$$RE_i = {[(Q_{op} - Q_{cp}) * 100]}\big/{Q_{op}} \quad \text{(12)}$$

Where RE_i is the relative error percentage for each of events, Q_{op} is the observed peak discharge and Q_{cp} is the calculated peak discharge.

Root of Mean Square Error (RMSE)

Root of Mean Square Error of peak discharge is presented in Formula (13, 14):

$$RMSE = \left[{}^1\!/_{n(\sum_{i=1}^{n} SE_i)} \right]^{1/2} \quad \text{(13)}$$

and

$$SE_i = \left(Q_{op} - Q_{cp} \right)^2 \quad \text{(14)}$$

In which SE_i is the relative error for each event, Q_{op} is the observed peak discharge and Q_{cp} is the calculated peak discharge.

Models presentation for daily and instantaneous peak discharge

Geomorphologic parameters can be derived from digital model

easily. Geomorphologic parameters are also used in rainfall- run off modeling (Fleurant and Ronald, 2006). In this section, with regards to factors in studied models and recorded daily and instantaneous peak discharge data in Mehran- Joestan hydrometer station, attempted to present regional model for peak discharge estimation.

RESULTS

As mentioned before, from 15 coincident events of rain and discharge, 7 events were recognized to be good. The results of rainfall and discharge coincidence extraction were presented in Table 1.

With regard to geomorphologic factors for each of these models, using Geographic Information Systems (GIS), Digital Elevation Model (DEM) production and stream nets for Mehran basin were earned (Table 2).

Two important preferences of study models (especially GIUH model) beside geomorphologic factors are flow and rainfall factors. GIUH model's factors such as Mannig's roughness coefficient, slope of the main river in basin's outlet, mean flow width in outlet of basin for flow velocity calculation in Formula (1) were considered. (Table 3).

After earning factors of each study model we can apply them. Beside the results of each model, events date and observed discharge for accidental comparison are also presented (Table 4 and Figure 6).

To check the validity of each model, error functions are determined. The results of Relative Mean Error (RME) and Root of Mean Square Error (RMSE) investigations are presented in Table 5. The results indicate that GIUH model with the RME of 20.43 and RMSE of 16.089 has the lowest error among other study models.

The results of this research show that it is not possible to create the regression model for instantaneous peak discharge, because there are not enough recorded events. Therefore we have attempted modeling to present daily peak discharge. For Mehran drainage basin, 177 daily flood events with regarding to harmony between rain hyetograph and flood hydrograph were recognized to be good and applicable. With calculated factors in studied models in this research and other measured parameter, the regression equation is calculated (Formula 15). Within the last two decades or

Table 2. Geomorphologic calculated parameters in Mehran drainage basin.

Streams order	Number of streams	Length of streams (km)	Mean Length of streams (km)	Upstream drainage basin area (km²)	Mean Upstream drainage basin area (km²)	Mean stream length from upstream to outlet (km)	Main stream distance from outlet to centroid of drainage basin (km)	Mean slope of drainage basin (m/m)	Mean slope of main stream in outlet of drainage basin (m/m)
1	598	286.21	0.4786	67.67	0.11				
2	120	72.330	0.6027	59.28	0.49				
3	27	36.998	1.3703	65.99	2.44	22.07	11.749	0.244	0.01955
4	5	9.352	1.8704	48.53	9.70				
5	1	16.548	16.548	99.71	99.71				

Table 3. The required parameters for measurement flow velocity from kinematic wave parameters.

Drainage basin	Mannig's roughness coefficient (n)	slope of main river in drainage basin outlet S_Ω (%)	drainage basin area (km²)	Rain intensity I_r (cm/h)	mean flow width in Outlet of drainage basin B (m)
Mehran	0.0382	1.95	99.71	It's different for any events in drainage basin	7.089

Table 4. Date of events and peak discharge estimation (m³/s) from using models in Mehran drainage basin.

Events Date	Mehran drainage basin				
	Qp (o.)	Qp (Tri.)	Qp (SCS)	Qp (Sny.)	Qp (GIUH)
20,21 April 2003	55.46	48.483	47.893	21.39	18.73
29 May, 2003	23.51	54.83	54.145	22.09	23.366
24, 25 April 2004	8.97	58.67	57.92	22.45	9.052
26, 27 April 2005	18.54	33.66	34.108	22.45	16.94
19, 20 May 2005	34.054	32.77	32.35	22.1	12.59
7, 8 Nov. 2006	22.57	31.136	31.54	21.73	22.83
27, 28 April 2007	22.35	51.46	50.827	21.73	21.803

so, one of the simplest approaches to the problems of rainfall-runoff modeling has been through the application of linear theories (Dooge 1973).

$$Q_P = 2.416V + 1.336A - 0.039B - 2.591 \quad (15)$$

$$R^2 = 0.952 \qquad R = 0.97$$

Where V is the flow velocity, A is discharge area and B is wetted perimeter in discharge area.

Discussion and conclusions

With regards to (Table 5), it can be concluded that the GIUH, Snyder, SCS and Triangular models could provide better estimation respectively. With regards to Tables 4 and 5, the GIUH and Snyder models have the same results to some extent.

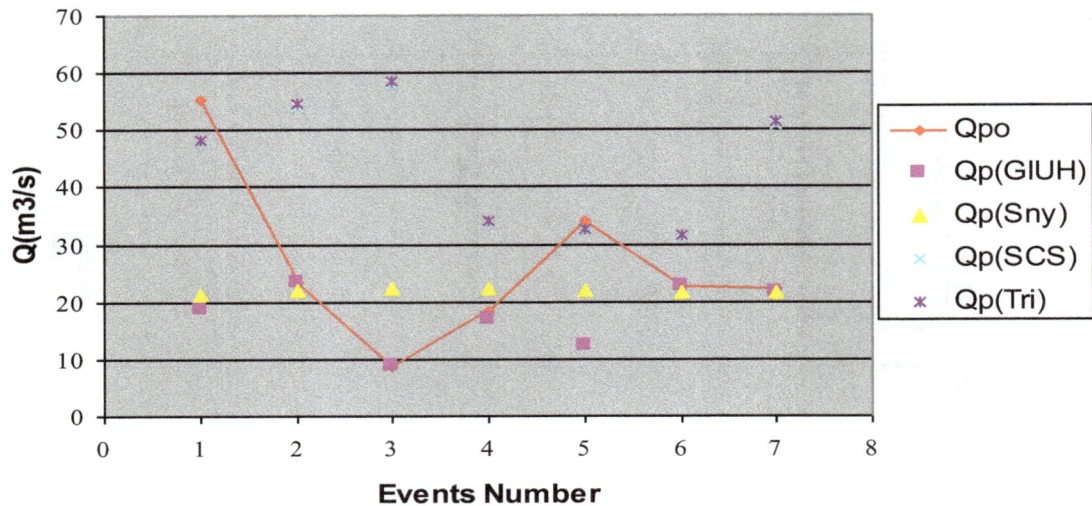

Figure 6. Observed discharge and peak discharge estimation from using models.

Table 5. Comparison of study models in drainage basin with index of Relative Mean Error (RME) and root of mean square error (RMSE).

Study models	Mehran drainage basin	
	RMSE	RME
GIUH	16.089	20.43
Snyder	14.65	40.062
SCS	25.371	133.082
Triangular	25.828	135.722

Kumar et al. (2002) have modeled the unit hydrograph and correlated GIUH model parameters with Clark model parameters. But in our study, besides GIUH model parameters, we also used SCS, Snyder and Triangular models parameters to model peak discharge estimation. Mossa (2008), studied hydrologic characteristics such as stream and slope nets, flow hydraulic and spatial rainfall distribution of GIUH model in seven basins of south west of France. Results shows that splitting the basin to sub basins in two points of stream nets for GIUH model determination is sufficient but we did split the basin to 16 sub basins on the upper stream's order ($2^{\Omega-1}=2^{5-1}=16$). Also the analysis shows that the GIUH model has more sensitivity to stream topology, spatial rainfall distribution and characteristics of flow hydraulic. Thus, with regards to Fig (5), characteristics of flow hydraulic have been improved by our study.

In the study done in Paskohak drainage basin by Rahimian and Zare (1995) to compare the results of GIUH with SCS, Snyder and Triangular methods, it has been concluded that GIUH has a better coincidence with observed hydrograph. That was the reason that our study utilizes this result. Jain and Sinha (2003) studied Horton

laws with their applications in GIUH model on UN gauged basin with fifth order in Himalayan Mountains. The results shows discharge with 50 years return period have good accomplishments with observed data. Their study confirms our results. Kumar et al. (2007) have used GIUH model for extraction run off hydrograph in Ajar basin of India. Results comparisons from error functions (such as root mean of standard error) in six events have the best results. Their study also confirms our results.

Ghiassi (2004) has estimated the hydrograph by GIUH and GCIUH methods, and by other synthetic methods such as Snyder, SCS and Triangular, he has compared the results. This study has been done in reprehensive basins of Kassilian in northern Iran and Lighvan in northwest of Iran. It is mentioned that the GIUH by ROSSO method also acquired. After then these methods compared with observed hydrograph, the results were acceptable and they have no significant differences. The other results of this research project show that for peak discharge estimation, hydrographs of GIUH, Triangular, SCS and Snyder methods have the best estimation respectively. Therefore Ghiassi's results for GIUH are matched with our results. Montazeri et al (2004) showed

Table 6. Events analysis of Geomorphology model in Mehran drainage basin

Model	Drainage Basin	Date of events	Problem	Reason
GIUH	Mehran	20, 21 April 2003	Qpo>Qpe	Snow melt and rainfall with up continuous

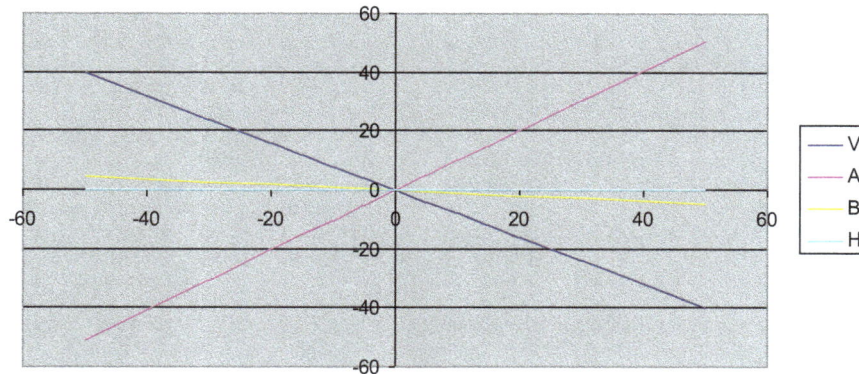

Figure 7. Sensitivity analysis of daily peak discharge model's factors in Mehran basin.

that using GIS technique for extraction of required parameters for Clarck synthetic hydrograph and comparing them with observed hydrograph in outlet will result in a good harmony between them. Therefore for this study also used this technique.

Based on the results obtained the GIUH model is the best model for the estimation of instantaneous peak discharge. Data obtained for each of the events in the Mehran drainage basin demonstrate that in one of the events in 20 and 21 April, 2003, the observed discharge (Qp_o) was greater than the estimated discharge by model (Qp_e). The reason is presented Table 6.

Sensitivity analysis of daily peak discharge model's factors (Formula 15) in Mehran drainage basin in Figure 7 demonstrates that the factors like flow velocity and discharge area have the major effects on model's sensitivity. So, accurate measurement of these parameters enhances the efficiency and will provide a more accurate model.

For many drainage basin in the world, that do not have hydrometric station or have an incomplete data, it is recommended that if there is a gauge rain station, the Geomorphologic model for peak discharge estimation to be used. Is it does not exist, the Snyder model is recommended.

For the same kinematical conditions, size and scale in the GIUH model are not reflected through the area on the basin but through the length of the storms (L_Ω). Two basins may be considered hydrologicaly similar when they have identical $R_L^{0.43}/ L_\Omega$ which controls q_p. Due to the values of R_L existed in nature, we may assume that $R_L^{0.43} \approx R_L^{0.38}$, two basins will be similar when they have equal values of ($R_L^{0.43}/ L_\Omega$) and (R_B/R_A), (where L_Ω is in

Kilometers when comparing different values of $R_L^{0.43}/ L_\Omega$ Rodriguez- Iturbe and Valdes, 1979). With regards to above mentioned problems for further confidence of GIUH model, it is recommended that this model is used in other drainage basins in the world and the results analyzed.

In the end, with regard to the need for these models in the world, it is recommended that models with these characteristics are presented.

ACKNOWLEDGEMENTS

We would like to thank TAMAB (Research Organization of Water Resources) and Surveying Organization for providing the data of this study and for helping us with data pre-processing.

REFERENCES

Agnese C, D'Asaro D, Giordano G (1988). Estimation of the time scale of the geomorphologic instantaneous unit hydrograph from effective stream flow velocity. Water Resour. Res., 24(7): 969-978.

Berod DD, Singh VP, Musy A (1999). A geomorphologic kinematic-wave (GKW) model for estimation of flood from small alpine watersheds. Hydrol. Process. 13: 1391-1416.

Chutha P, Dooge JCI (1990). The shape parameters of the geomorphologic unit hydrograph. J. Hydrol., 117: 81-97.

Dooge JCI (1973). Linear theory of hydrologic systems, Tech. Bu. U.S. Dep of Agr., Washington, D. C., 436-448.

Fleurant C, Kartina RB (2006). Analytical model for a GIUH, Hydrol. Process. 2: 3879-3895.

Ghiassi N (2004). Application of geomorphologic instantaneous unit hydrograph in Kasilianand Lighvan basins, Final report of research plan, Soil conservation and watershed management institute. pp. 161-165. (abstract in English).

Gupta V, Waymire E, Wang C (1980). A representation of an instantaneous unit hydrograph from geomorphology. Water Resour. Res. 16(5): 855-862.

Hojjati MH, Boustani F (2010). An Assessment of Groundwater Crisis in Iran, Case Study: Fars Province. World Academy Sci. Eng. Technol., 70: 476-480.

Jain SK, Singh RD, Seth SM (2000). Design Flood Estimation using GIS Supported GIUH Approach, Water Resour. Manag., 14: 369-376.

Jain V, Sinha R (2003). Derivation of unit hydrograph from GIUH analysis for a Himalayan river. Water Resour. Manag., 17: 355-375.

Karvonen T, Koivusalo H, Jauhiainen M, Palko J, Weppling K (1999). A hydrological model for predicting runoff from different land use areas. J. Hydrol., 217: 253-265.

Krishen D, Bras R (1983). The linear channel and its effect on the geomorphologic, IUH. J. Hydrol., 65: 175-208.

Kumar RC, Chatterjec C, Lohani AK, Kumar S, Sing RD (2002). Sensitivity Analysis of the GIUH based Clark Model for a catchment. Water Resour. Manag., 16: 263-278.

Kumar RC, Chatterjee C, Lohani AK, Sing RD, Kumar S (2007). Run off estimation for an UN gauged catchment using Geomorphologic Instantaneous Unit Hydrograph (GIUH) Models, Hydrol. Process. 21: 1829-1840.

Lee KT, Chang CH (2005). Incorporating subsurface-flow mechanism into geomorphology-based IUH modeling, J. Hydrol., 91-105.

Lin Z, Oghochi T (2006). Drainage density and slope angle in Japanese bare lands from high-resolution DEMs, CSIS Discussion Paper≠ 56:17.

Mockus V (1957). "Use of Storm and Watershed. Characteristics in Synthetic Hydrograph Analysis and. Application,". AGU, Pacific Southwest Region Mtg., Sacramento, Calif.

Montazeri S, Rahnama M, Akbarpour A (2004). Instantaneous unit hydrograph determination with using Clarck method and GIS technique in Karde dam drainage basin, international conference of watershed management, water resource and soil management. Keraman, Iran: pp198-207. (abstract in English).

Mossa R (2008). Distribution on the Geomorphologic Instantaneous Unit Hydrograph transfer function. Hydrol. Process. 22: 395-419.

Nazari SA, Ahmadi H, Mohammadi A, Ghoddousi J, Salajeghe A, Bogg G (2009). Factors Controling Gully Advancement and Models Evaluation (Hable Rood Basin, Iran). Water Resour. Manag., Published Online, DOI 10.1007/s 11269-009-9512-4.

Olivera F, Maidment D (1999). Geographic information systems (GIS) - based spatially distributed model for runoff routing. Water Resour. Res., 35(4): 1135-1146.

Rahimian R, Zare M (1995). Application of geomorphologic instantaneous unit hydrograph for synthetic hydrograph in UN gauged drainage basin, Collection of article; third conference of hydrology, ministry of energy: pp 203-227. (abstract in English).

Rodriguez-Iturb I, Valdes JB (1979). The geomorphologic structure of hydrologic response. J. Water Resour. Res., 15(6): 1409-1420.

Rodriguez-Iturbe I (1993). The geomorphological unit hydrograph, Channel Network Hydrol., 43-68.

Rodriguez-Iturbe I, Devoto I, Valdes JB (1979). Discharge response and hydrologic similarity: the interrelation between the geomorphologic IUH and storm characteristic. Water Resour. Res., 15(6): 1435-1444.

Rodriguez-Iturbe I, Gonzales-Sanabria M, Bras R (1982). A Geomorphoclimtic theory of the instantaneous unit hydrograph. Water Resour.Res., 18(4): 877-886.

Rooso R (1984). Nash model relation of Horton order ratios. J. Water Resour. Res., 20: 914-920.

SCS (Soil Conservation Service) (1968). Hydrology, Suppl. A to Sec.4. Engineering- ing Handbook. Washington. D. C

Sherman LK (1932). Stream flow from rainfall by unit- graph method. Engineering News Record 108: 501-505.

Shreve RL (1967). Infinite topologically random channel networks. J. Geol., 75:178-186.

Snyder FF (1938). Synthetic unit- graphs. Transactions, American Geophysics Union 19: 447-454.

Troutman BM, Karlinger MR (1985). Unit hydrograph approximation assuming linear flow through topologically random channel networks. Water Resour. Res., 21(5): 743-754.

Yen BC, Lee KT (1997). Unit hydrograph derivation for ungauged watersheds by stream-order laws. J. Hydrol. Eng., 2(1): 1-9.

Zhang B, Govindaraju RS (2003). Geomorphology-based artificial neural network (GANNS) for estimation of direct runoff over watersheds. J. Hydrol., 273: 18-34 (P.16).

9

Evaluation of the water quality status of Lake Hawassa by using water quality index, Southern Ethiopia

Adimasu Woldesenbet Worako

Department of Water Resources and Irrigation Management, Dilla University College of Agriculture and Natural Resources, Ethiopia.

Lake Hawassa is one of the eight Major Ethiopian Rift Valley Lakes and the smallest among them which is situated in southern regional state; it is a closed basin system and receives water from only perennial Tikurwuha River and runoff from the catchment areas. It is an important source of water for surrounding rural communities for various uses like domestic, irrigation, livestock watering, fishing and recreation. Quality of the lake water is vital for the surrounding rural and urban communities for proper and safe use of the lake. The present study was designed to determine the water quality status of the lake for multiple designated water uses by employing the water quality index. To assess the status water samples were collected in monthly intervals for a period of three months from December to February (dry period), 2011/12. From all water quality parameters analyzed turbidity, Mn, Na^+, K^+, F^-, PO_4^{3-}, total coliform and fecal coliform were higher than the recommended limits of national and international standards for designated water uses. Based on the water quality index calculation the lake water is categorized under marginal category which reveals the water is frequently threatened and impaired and as well departs from natural condition. Accordingly the lake water is under fair category for irrigation and aquatic life; however, it needs great care on selection of crops and soil condition. The lake is under higher risk by deleterious anthropogenic activities on watershed and it needs mitigation measures to prevent it from further deterioration.

Key words: Water quality index, Lake Hawassa, water quality status, designated water use.

INTRODUCTION

Water is an indispensible and basic element which supports life and the natural environment, a prime component for industry, a consumer item for human beings and animals and a vector for domestic and industrial pollution (Colin and Quevauviller, 1998). Access to adequate water for domestic purposes, irrigation, sanitation, and solid waste disposal are the four basic needs which impact significantly on socio-economic development and the standard of life. Status of water quality is highly imperative to the sustainability of natural ecosystems and any development activities and it demands great monitoring and regulation. However, currently the meager freshwater resources are becoming more unconducive for the required uses due to different point and non-point sources of pollution from the catchment area.

Water pollution also aggravates the problem of water shortage. Due to this fact the world people living under water-stressed condition ranges from 1.4 to 2.1 billion (Vorosmarty et al., 2000; Oki et al., 2003; Arnell et al., 2004) and peoples affected by unsafe, poor sanitation and hygiene reaches 54.2 million per year with 1.7 millions death (WHO, 2005). So, assessing the status of water quality periodically is quite urgent to save the world from severe water quality initiated functional stress and scarcity.

Ethiopia is a developing country which is endowed with a number of lakes and large rivers which gives immense value to overall economic development. For instance, the country has 12 river basins, 11 fresh lakes, 9 saline lakes, 4 crater lakes and over 12 major swamps/wetlands. However, the water scarcity and inadequacy is the main feature of the country today. In addition to scarcity the quality of water is also threatened as common to all developing countries (Milda, 2009). Among freshwater resources, Lake Hawassa is one of the Major Rift Valley lakes in Ethiopia and used for various purposes by semi-urban and urban dwellers. But the lake has been subjected to many pollutants generated from neighboring industries (like Hawassa Textile factory, floury factory, sisal factory, etc), agriculture activities, service rendering centers (near the lake which release their effluent without any treatment like resorts), hospitals, urban storm water and sewage, and other activities on the catchment (Zinabu and Zerihun, 2002). Specifically, Hawassa textile factory and Hawassa Referal Hospital's discharge to the lake is seriously degrading its viability since their effluents has become over the set standards to the environment (Yosef et al., 2010; Abayneh et al., 2003; Demeke, 1989).

Research on lakes water quality status on regular basis and its impact on the lake ecosystems and on the potential of the lake water resources for multiple designated uses like drinking, irrigation, recreation and aquatic life are very limited. Therefore, this study was undertaken to avail basic information for the determination of the water quality status of the lake and the main constraining factors that limits its suitability for various designated water uses.

Separate assessment of water quality suitability for the intended uses is time consuming and does not yield appropriate systems to monitor and control the quality of water bodies. Thus, evaluation of the water quality status of the lake by using water quality index is employed. Water quality index (WQI) is one of the most effective tools to aggregate and communicate information on the quality of water to the concerned citizens and policy makers (Puri et al., 2011). It numerically summarizes the information from multiple water quality parameters into a single value that can be used to compare data from several sites and months. The use of WQI simplifies the results of analysis related to a water body as it aggregates in one index

of all parameters analyzed (Warhate and Wankar, 2012). There are a number of indices developed in many parts of the world to evaluate water quality status and pollution extents of water bodies like U.S NSFWQI (Sharifi, 1990), BCWQI (CCME, 1995), OWQI (DEQ, 2003), and Smith's Index (Smith, 1987). For this investigation an indices developed by the British Columbia Ministry of Environment, Lands and Parks and modified by Alberta Environment which is CCME WQI (1.0 model) was used. This index provides a numerical values in between 0 (worst water quality) and 100 (best water quality) with five descriptive categories such as excellent (CCME WQI value = 95-100), good (CCME WQI value = 80-94), fair (CCME WQI value = 65-79), marginal (CCME WQI value = 45-64) and poor (CCME WQI value = 0-44) (CCME, 2001). This study was designed to determine the lake water suitability for drinking, irrigation, recreation and aquatic life by employing the CCMEWQI water quality index calculation method.

METHODOLOGY

Lake Hawassa is one of the eight major Ethiopian Rift Valley lakes which cover an area of about 94km^2 (Yemane, 2004) and the smallest in comparison with other central Rift Valley natural lakes. It is situated 275 km south of the capital city Addis Ababa and west of Hawassa town. The lake is located between 06° 58'to 07° 14′ N latitudes and 38° 22' to 38° 28' E longitudes with an elevation of 1685 masl and is bounded by various mountains such as Mt. Tabor (1810 masl) and Mt. Alamura (2019 m.a.s.l) (Yemane, 2004). Hence, the surface and sub-surface drainage is towards the lake and it's the main destination for any type of contaminants generated from catchment areas. The catchment area of the lake is 1250 km^2 (Girma and Ahlgren, 2009) with closed basin feature and receives only one perennial river from eastern escarpment, Tikur Wuha River. This river is extremely affected by various industries on the basin like Hawassa Textile Factory, Hawassa Sisal factory, Hawassa Flour Factory, Tabor Ceramic Factory, etc (Yosef et al., 2010).

The area receives a mean annual rainfall of 950 mm and has a mean annual air temperature of 19.8°C (Arkady and Brook, 2008). The area is characterized by three main seasons; long rainy season (locally called kiremt) in the summer from June-September (mean annual total rainfall accounts from 50 to 70%), dry period (locally called bega) which extends between October and February and short rain season (locally called belg) during March and May, when about 20 to 30% of the annual rainfall falls. Mean monthly rainfall is above 100 mm from April to September with August showing the highest 124 mm and the lowest rainfall occur in November, December and January (Halcrow, 2010). It has maximum depth of 22 m and a mean depth of 11 m (Elias, 2000). Evaporation from the lake is estimated to be 1710 mm/year, the average annual inflow and outflow(underground flow) is 1440 and 570 mm, respectively as well as the total volume of the lake water is 1.3 km^3(Tenalem, 1998; Gugissa, 2004; Arkady and Brook, 2008).

Grab sampling was done in monthly interval for three months (December 2011 up to February 2012) at ten selected sampling sites from surface 30 cm and 1 m bottom of the lake. The sampling sites are selected based on the relative importance, location and magnitude of human influences. Sample site S1 (Inlet of Tikurwuha River to the lake), S2 (around Haile resort), S3 (around Lewi resort), S4 (Referral Hospital), S5 (at the center of the lake), S6 (direct opposite to Haile resort, rural side), S7 (direct opposite to Lewi

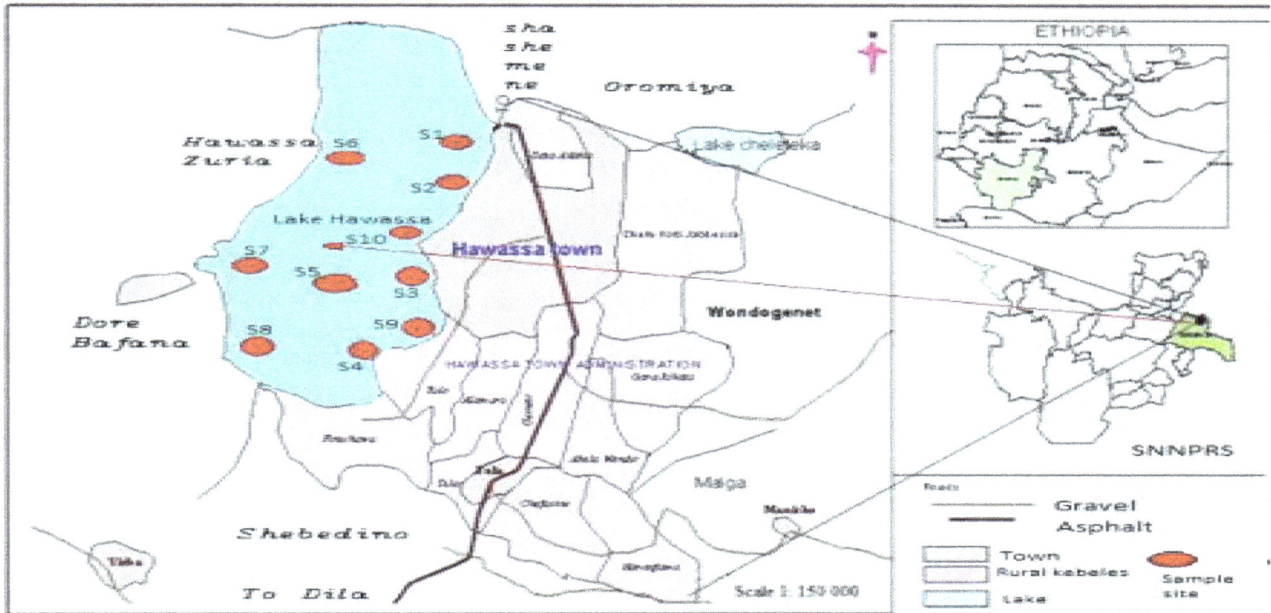

Figure 1. Location map of the study area and sampling sites on the Lake Hawassa.

resort, rural area side, Dore-Bafana), S8 (direct opposite to Referral hospital, rural side), S9 (around Amora-Gedel, town storm water and sewage entrance site) and S10 (around Fikir-Hayike, recreational center) (Figure 1). The water quality parameters analyzed in this study were illustrated on Table 1.

The water quality index was computed following CCME WQI (CCME, 2001) by using the following formula:

$$CCME\,WQI = 100 - \left(\frac{\sqrt{F_1^2 + F_2^2 + F_3^2}}{1.732} \right)$$

where F_1 (scope)-is the number of variables whose objectives are not met, F_2 (frequency)-is the frequency with which the objectives are not met and F_3 (amplitude)-is the amount by which the objectives are not met. The divisor 1.732 normalizes the resultant values to a range between 0 and 100, where 0 represents the "worst" water quality and 100 represents the "best" water quality.

The calculations of these three parameters to determine CCME WQI were described as follows:

1. F_1 (Scope) represents the percentage of variables that do not meet their objectives at least once during the time period under consideration ("failed variables"), relative to the total number of variables measured:

$$F_1 = \left(\frac{Number \cdot of \cdot failed \cdot variables}{Total \cdot number \cdot of \cdot variables} \right) X100$$

2. F_2 (Frequency) represents the percentage of individual tests that do not meet objectives ("failed tests"):

$$F_2 = \left(\frac{Number \cdot of \cdot failed \cdot tests}{Total \cdot number \cdot of \cdot tests} \right) X100$$

3. F_3 (Amplitude) represents the amount by which failed test values do not meet their objectives. F_3 is calculated in three steps:

a. excursion = $\left(\dfrac{Failed\,test.value}{Objective} \right) - 1$

b. nse = $\dfrac{\sum_{i=1}^{n} excursion}{No.of.total.tests}$

c. $F_3 = \left(\dfrac{nse}{0.01nse + 0.01} \right)$ is an asymptotic function that scales the normalized sum of the excursions from objectives (nse) to yield a range between 0 and 100.

The quality criteria of each analyzed parameters were compared to prescribed limits of various international and national standards like WHO (2004), CCME (2009), USEPA (2000), FAO (1985), EEPA(2003) and other guidelines for those designated water uses. After the CCME WQI value was determined with respect to site and month the lake water quality was ranked as per the CCME WQI ranking.

RESULTS AND DISCUSSION

Among analyzed water quality parameters reports which above the recommended limits for drinking water use were turbidity, BOD_5, Mn, fluoride, Na^+, K^+, PO_4^{3-}, total coliform and fecal coliform; for irrigation uses MAR, KR, SPP, and others common to drinking water use; for recreational uses clarity, turbidity, TC and FC as well as

Table 1. Standard water quality parameters determination methods and instruments used.

Parameters	Determination method and instrument
Temp., EC, TDS and salinity	pH and conductivity meter (HANNA pH211)
BOD_5 and DO	Modified winkler-Azide dilution technique
Turbidity	Nephelometeric (HACH, model 2100A)
Secchi depth	20 cm diameter of Secchi disk
NO_3^-, NO_2^-, PO_4^{3-}, NH_3 and NH_4.	Photometric measurements using flame photometer
Chloride	Mohr Agregetrometric titration method
Fluoride	Spectrophotometerically by Ampule method (HACH, Model 41100-21)
COD	Determined by dichromate reflux method through oxidation of the sample with potassium dichromate in sulphuric acid solution followed by titration
Mg, Na, K, Ca, Cr, Cd, Cu, Mn, Zn and Pb	Determined by atomic absorption spectrometer, AASP (Varian SP-20) using their respective standard hollow cathode lamps (APHA, 1995; APHA, AWWA and WPCF, 1998)
Iron	Determined by using UNICAM UV-300 thermo electrode.
TC and FC	Most probable number method (MPN/100 ml)
Indices (SAR, MAR, SSP, KR and TH)	Richards (1954), Raghunath, (1987), Todd(1980), and Kelly's,(1963) empirical formulas

for aquatic life sustenance Mn, Cu and Zn (Table 2) were the main constraining parameters which were above the recommended limits of WHO, EEPA, CCME,USEPA and FAO guidelines for designated water uses.

pH and turbidity

The determination of pH of the water is very important since it affects the solubility and availability of micronutrients like Zn, Mn, Fe and Cu and how they can be utilized by aquatic organisms and also reduces the performance of water treatment systems and disinfectants in water supply. The pH of the lake water ranged from 6.98 to 7.71 with an average value of 7.54. The value of pH decreased in the lake in comparison to the former research done by Alemayehu (2008), 8.5 and Elizabeth et al. (1994), 8.8. This may reveal the increment of organic matter load to the lake ecosystem as decomposition of organic matter leads to decrease in pH, acidity (WHO, 1984).

Nevertheless, with reference to pH value it is within the permissible limit (6.5-9.0) for drinking, irrigation, recreation and aquatic life (WHO, 2006; CCME, 2001; EEPA, 2003). The consumption of more turbid water may constitute a health risk as excessive turbidity can protect pathogenic microorganisms from the effect of disinfectants, and stimulate the growth of bacteria (Zvikomborero, 2005). The turbidity of the lake water was found to be higher than the prescribed limits (<5NTU) for drinking and recreation purposes (Table 2) (WHO, 1993; CCME, 1999).

BOD₅

BOD is a measure of the amount of oxygen that bacteria will consume while decomposing organic matter under

aerobic conditions (Tenagne, 2009). Unpolluted, natural waters should have a BOD_5 of 5 mg/L or less but on this study the lake water BOD_5 value is on average 117 mg/L. The elevated values of BOD_5 in the lake may show the high level of pollution and it is concentration is beyond the permissible limits of EPA guideline (<5 mg/L) for aquatic, drinking and recreation use (Table 2) (USEPA, 2000).

Na⁺ and K⁺

The concentration of Na^+ ion ranged from 300.95 to 414.11 mg/L with an average value of 331.14 mg/L which is higher than the permissible limits (200 mg/L) for drinking and irrigation water use (WHO, 1983, 2006). The consumption of eminent Na^+ ion in drinking water leads to hypertension, congenial heart diseases and kidney problems (Singh et al., 2008) where as in irrigation water it may cause crusting, plugging, soil dispersion and sealing of surface pores (FAO, 1985) which leads to infiltration problem and structural instability of the soil. The dominance of Na^+ ion over other major cations could be attributed due to weathering of acidic rocks (Alemayehu, 2008) (Table 2). In all sampled sites the value of K^+ was beyond the permissible limits for drinking and irrigation water use (WHO, 1984). An elevation of potassium in the lake indicates the effect of hospital effluents, septic system effluents, and other anthropogenic activities beside the natural sources.

Fluoride

The most prominent sources of fluoride in water are a natural weathering of mineral bed rocks (WHO, 2004) and it is a common problem mainly in the Rift Valley

Table 2. Mean physicochemical and bacteriological water quality characteristics of the Lake Hawassa in ten sampled sites.

Parameters	Site sample taken										WHO, FAO, CCME
	S1	S2	S3	S4	S5	S6	S7	S8	S9	S10	
EC	701	756	752	757	755	755	756	756	756	756	1500
TDS	420.8	454.7	450.7	454	452.2	453.2	453.7	453	455.6	453.5	1000
pH	6.98	7.71	7.58	7.73	7.66	7.54	7.58	7.54	7.53	7.59	6.5-8.5
Temp	21.33	20.98	21.33	21.25	21.17	21.23	21.05	21.25	21.33	21.32	15-30
DO	11.2	17.37	18.35	17.78	18.4	19.17	21.42	20.55	18.85	15.4	>5
Turb.	20.98	7.02	6.98	6.82	6.95	6.98	6.93	6.87	6.92	7.97	<5
Fe	0.180	0.071	0.072	0.078	0.073	0.075	0.074	0.072	0.078	0.080	0.3
BOD_5	94.5	56.17	73.33	138.2	92.17	133.5	143	144.8	157.7	136.7	<5
Cu	0.046	0.006	0.005	0.011	0.005	0.005	0.005	0.005	0.005	0.001	2
F^-	2.31	14.32	11.83	17.29	14.45	12.36	15.65	13.9	13.27	12.9	1.5
Cl^-	31.31	28.95	31.91	33.09	28.95	31.91	28.9	31.91	31.31	30.1	250
TH	124.2	106.1	107.9	126.97	122.6	125.3	130.5	124.7	113.3	137.2	500
Mn	0.489	0.056	0.039	0.043	0.036	0.034	0.056	0.052	0.043	0.040	0.05
Zn	0.32	0.31	0.19	0.16	0.23	0.17	0.12	0.12	0.16	0.16	5
Mg^{2+}	28.54	24.25	24.67	29.24	28.29	28.93	29.96	28.81	26.08	31.92	200
Ca^{2+}	2.72	2.53	2.55	2.68	2.49	2.51	2.92	2.48	2.39	2.34	100
K^+	71.82	75.18	70.80	85.04	74.87	70.54	69.46	78.71	71.39	72.76	20
Na^+	300.9	341.6	301.3	348.8	325.6	324.1	414.1	317.6	315.6	321.9	200
NO_3^-	3.02	8.87	6.29	8.46	5.26	4.47	3.84	3.30	4.64	4.54	45
PO_4^{3-}	1.36	1.11	0.98	1.07	1.15	0.99	0.97	0.85	1.28	1.42	0.02
Cr	ND	ND	ND	ND	ND	ND	ND	ND	ND	ND	0.05
Pb	ND	ND	ND	ND	ND	ND	ND	ND	ND	ND	0.01
Cd	ND	ND	ND	ND	ND	ND	ND	ND	ND	ND	0.003
Ni	ND	ND	ND	ND	ND	ND	ND	ND	ND	ND	0.02
TC	12333	15,833	12,167	9,500	12,833	6,000	8,667	10,667	10,000	20,833	<50
FC	213	130	63	78	47	67	58	85	92	163	<10
Clari	0.50	0.63	0.66	0.68	0.66	0.75	0.68	0.68	0.66	0.64	1.2*
SAR	12.19	14.22	12.81	13.91	13.22	12.98	16.01	12.63	13.17	12.56	26**
SSP	84.04	88.34	86.46	86.08	85.93	85.47	88.29	85.47	86.26	83.78	60**
KR	5.74	6.89	6.31	6.42	6.25	6.06	7.18	5.82	6.34	5.71	1**
MAR	94.23	93.84	93.94	94.53	94.55	94.80	94.40	94.92	94.57	95.37	50**

All units are in mg L^{-1} saving temperature, turbidity, clarity, EC, and pH which are expressed in °C, NTU, m, µS cm^{-1}, and non-dimensional, respectively. TC and FC units in MPN/100 ml and MAR, SAR, KR and SSP by %. *-indicates only for recreational use and **-express only for irrigation use. The bold ones indicate that analyses result is above the permissible limits except clarity which is below acceptable level.

lakes of eastern African countries (Tamiru, 2006) due to geological factor.

In the present investigation the concentration of fluoride ranged from 2.31 to 17.29 mg/L with an average value of 12.83 mg/L. Drinking water with high fluoride concentration above the permissible limit (1.5 mg/L) may causes dental fluorosis and if continuously consumed for a long period with the concentration 3 to 6 mg/L and above may lead to skeletal fluorosis and skeletal crippling (Kloos and Redda, 1999). The lake water fluoride concentration is twelve times higher than the permissible limits for drinking, irrigation and livestock watering purposes (CCME, 1999; WHO, 1998, 2006) and hence not suitable for these designated purposes.

Nutrients

The most known principal limiting nutrients in freshwater lakes of Rift Valley lakes are nitrogen and phosphorus. Nitrogen can exist in water in four forms like NH_3, NO_3^-, NO_2^- and NH_4^+ which may cause groundwater and surface water pollution in excessive quantity through leaching, stimulate algal growth in surface water that increases maintenance costs in irrigation practices, carcinogenic and blue-baby diseases in infants of human being. But currently the concentration of NO_3^- in Lake Hawassa is within the permissible limit (WHO, 2006; Ayers and Westcot, 1985) for drinking and irrigation. However, the concentration of phosphate is higher than the recommended limits (0.005-0.02 mg/L) to freshwater healthy ecosystem (USEPA, 2000) (Table 2) and hence, the lake is categorized in eutrophic state index as Carlson (1977). However, according to Chapman (1996) the nutrients levels in lake water show great impairment of the lake ecosystem by point and non-point sources of pollution. Nitrite and nitrate should be less than 0.001 and 0.1 mg/l for conducive aquatic life (Murdoch et al., 2001) but lake water has high nutrient contents which depart more from natural desirable levels.

Total coliform and fecal coliform

The concentration of total coliform and fecal coliform in the lake were higher than the recommended limits for drinking water (WHO, 2006) and EU (1998), less than 50 and 10 MPN/100 ml, respectively. Irrigation water requires safe water for production of horticultural crops like vegetables and fruits to prevent transmission of diseases causing pathogens (bacteria, viruses and protozoa). Bacterial diseases such as cholera, typhoid fever, gastroenteritis and salmonellolis may happen when the concentration of total coliform in irrigation water becomes above 1000 MPN/100 ml (WHO, 1983; CCME, 1999). Recreational water quality is highly dependent on bacteriological quantities for direct or indirect recreation. However, the lake has high total and fecal coliform which is above the permissible limits of WHO (1989) and CCME (1999), <500 MPN/100 ml and it impedes the suitability for the required intention (Table 2). In general, the lake water is not suitable for drinking, irrigation, and recreation as well as fin fishes harvesting purposes basically based on bacteriological (TC and FC) concentrations.

Heavy metals

Trace levels of dissolved metals in surface water are essential for proper biological functioning in both plants and animals (CCME, 2009). Generally, the concentration of heavy metals in the lake was relatively high at the Inlet of Tikurwuha River due to point sources of pollution from Hawassa textile factory and other factories which discharge their waste directly into this river. However, except Mn, Cu and Zn other metals such as Cd, Cr, Pb and Ni (Table 2) are within the permissible limits to all designated water uses (CCME, 2009; EU, 1998; WHO, 1998). The level of Mn concentration is higher for drinking uses in three sites and for irrigation and aquatic life in one site (S1) while Cu and Zn content were above the recommended limits to only aquatic life (CCME, 1999; WHO, 1983; USEPA, 2000).

Water quality index (WQI) calculation

The WQI was computed based on the three parameters

F_1, F_2 and F_3 for drinking water uses. The values obtained were:

$$F_1 = \left(\frac{Number\ of\ failed\ variables}{Total\ number\ of\ variables}\right)X100 = \left(\frac{9}{26}\right)*100 = 34.62$$

$$F_2 = \left(\frac{Number\ of\ failed\ tests}{Total\ number\ of\ tests}\right)X100 = \left(\frac{87}{2600}\right)*100 = 3.3\#$$

$$F_3 = \left(\frac{nse}{0.01nse + 0.01}\right) = \left(\frac{4.562}{0.01*4.562 + 0.01}\right) = 81.46$$

$$CCMEWQI = 100 - \left(\frac{\sqrt{F_1^2 + F_2^2 + F_3^2}}{1.732}\right) = 100 - \left(\frac{\sqrt{34.62^2 + 3.35^2 + 81.46^2}}{1.732}\right) = 48.86 \cong 49$$

The WQI value computed for irrigation water use was:

$$F_1 = \left(\frac{Number\ of\ failed\ variables}{Total\ number\ of\ variables}\right)X100 = \left(\frac{9}{20}\right)*100 = 45$$

$$F_2 = \left(\frac{Number\ of\ failed\ tests}{Total\ number\ of\ tests}\right)X100 = \left(\frac{74}{2000}\right)*100 = 3.7$$

$$F_3 = \left(\frac{nse}{0.01nse + 0.01}\right) = \left(\frac{0.332}{0.01*0.332 + 0.01}\right) = 24.96$$

$$CCMEWQI = 100 - \left(\frac{\sqrt{F_1^2 + F_2^2 + F_3^2}}{1.732}\right) = 100 - \left(\frac{\sqrt{45^2 + 3.7^2 + 24.96^2}}{1.732}\right) = 70.21 \approx 70$$

The WQI value computed for recreation water use was:

$$F_1 = \left(\frac{4}{6}\right)*100 = 66.67 \ ; \ F_2 = \left(\frac{33}{600}\right)*100 = 5.5 \ ; \ F_3 = \left(\frac{0.409}{0.01*0.409 + 0.01}\right) = 29.01$$

$$CCME\ WQI = 100 - \left(\frac{\sqrt{66.67^2 + 5.5^2 + 29.01^2}}{1.732}\right) = 58$$

The WQI value computed for aquatic life was:

$$F_1 = \left(\frac{3}{8}\right)*100 = 37.5 \ ; \ F_2 = \left(\frac{20}{800}\right)*100 = 2.5 \ ; \ F_3 = \left(\frac{0.088}{0.01*0.088 + 0.01}\right) = 8.1$$

$$CCME\ WQI = 100 - \left(\frac{\sqrt{37.5^2 + 2.5^2 + 8.1^2}}{1.732}\right) = 77.8 \approx 78$$

Based on the above separate water quality index computation for the designated water uses, the index for drinking (CCME WQI = 49) and recreational (CCME WQI = 58) purposes falls in marginal category, whereas the index for irrigation (CCME WQI = 70) and aquatic life

(CCME WQI = 78) uses falls within fair category. The overall or cumulative CCME WQI of the Lake Hawassa is 49 and hence it is under marginal category. According to the CCME ranking this water quality is frequently threatened or impaired and its condition often exceeds natural or desirable levels (CCME, 2001). So, mitigation measures should be developed in watershed overall activities, that is, for point and non-point sources of pollutions.

CONCLUSIONS AND RECOMMENDATION

The current study evaluated the physicochemical and bacteriological water quality characteristics of Lake Hawassa for multiple designated water uses like drinking, irrigation, recreation and aquatic life. The parameters of water quality analyzed and examined from various sampling sites in the lake show unsuitability of the water for drinking and recreational uses; but with some great care it is fair for irrigation and aquatic life. Based on the calculated cumulative water quality index it is ranked under marginal category of CCME WQI 49 which indicates the lake water is frequently threatened and impaired for those designated water uses.

Water quality of the lake was highly impaired on the town side of Hawassa that's due to inlets of various factories effluents like Hawassa textile factory, sisal factory, soft drink factory, ceramic factory and sewage as well as regional Hawassa referral hospital effluents. The lake is affected by both point and non-point sources of pollution beside the natural factors. Hence checking the effluent standards of the surrounding factories, controlling the service rendering center waste disposal system and constructing the municipal wastewater and storm water treatment plant are extremely essential to protect the lake water quality from further deterioration.

Conflict of Interest

The authors have not declared any conflict of interest.

REFERENCES

Abayneh A, Taddese W, Chandravanshi BS (2003). Trace metals in selected fish species from Lake Awassa and Ziway. SINET: Ethiopia J. Sci. 26(26):103-114.

Alemayehu T (2008). Environment resources and recent impacts in the Hawassa collapsed caldera. Main Ethiopian Rift Q. Int. 189:152-162.

APHA (American Public Health Association) (1995). Standard methods for the examination of water and waste water. 19th Ed. Washington DC, USA.

APHA, AWWA, WPCF (1998). Standard Methods for the Examination of Water and Waste-water 18th Ed. Washington, U.S.A. pp. 2-27, 4-108, 4-117, 4-131, 10-26.

Arkady MD, Brook A (2008). Water Balance and Level of regime of Ethiopian Lakes as Integral Indicators of Climate Change. Faculty of Hydrology, Russian State Hydrometeorology University, Saint Petersburg, Russia.

Arnell NW, Livermore MJL, Kovats SP, Levy E, Nicholls R, Parry ML, Gaffin SR (2004). Climate and socio-economic scenarios for global-scale climate change impacts assessments: Characterising the SRES storylines. Global Environ. Change 14:3-20.

Ayers RS, Westcot DW (1985). Water quality for agriculture. FAO Irrigation and Drainage, paper No. 29(1):1-109.

Carlson RE (1977). A Trophic State Index for Lakes. Limnol. Oceanogr. 22(2):361-369

CCME (Canadian Council of Ministers of the Environment) (1995).Protocol for the derivation of Canadian sediment quality guidelinesfor the protection of aquatic life. CCME EPC-98E. Prepared by Environment Canada, Guidelines Division, Technical Secretariat of the CCME Task Group on Water Quality Guidelines, Ottawa.[Reprinted in Canadian environmental quality guidelines, Canadian Council of Ministers of the Environment, 1999, Winnipeg.]

CCME (1999). Canadian water quality guidelines for the protection of aquatic life: Dissolved oxygen (freshwater). In: Canadian environmental quality guidelines, 1999, Canadian Council of Ministers of the Environment, Winnipeg.

CCME (2001).Canadian Water Quality Index 1.0Technical report and user's manual. Canadian Environmental Quality Guidelines Water Quality Index Technical Subcommittee, Gatineau, QC, Canada.

CCME (2009). Guidelines for Canadian recreational water quality. Prepared by the Federal-Provincial-Territorial Working Group on Recreational Water Quality of the Federal- Provincial-Territorial Committee on Health and the Environment.

Chapman D (1996). "Water Quality Assessments: A Guide to the Use of Biota, Sediments and Water." Environmental Monitoring. Second Edition. UNESCO, WHO, and UNEP. E&FN Spon, London UK.

Colin F, Quevauviller P (1998). Monitoring of water quality. The contribution of Advanced Technology. Elsever Ltd.

Demeke A (1989). Study on the age growth of adult Oreochromis niloticus in Lake Awassa, Ethiopia. M.Sc. Thesis, school of Graduate studies, Addis Ababa University.

DEQ (2003). The Oregon Department of Environmental Quality. Available from: http://www.deq.state.or.us/lab/WQM/WQI/wqi main.htm.

EEPA (Ethioian Environmental protection Authority) (2003). Guideline Ambient Environment Standards for Ethiopia. Prepared by EPA and UNIDO under ESDI project US/ETH/99/068/Ethiopia. Addis Ababa.

Elias D (2000). Reproductive biology and feeding habits of the cat fish Clarias garipinus (burchell) in Lake Awassa, Ethiopia. SINET: Ethiopia J. Science. 23:2.

Elizabeth K, Zinabu G, Ahgren H (1994). The Ethiopian Rift Valley lakes Chemical Characteristics of a Salinity Alkalinity series. Kluver Academic Publishers, Belgium. Hydrobiologica 288: 1-12

EU (European Union) (1998). Council Directive 98y83yEC of 3 November 1998 on the quality of water intended for human consumption. Official J. European Comm. p. L330y32 –L330y54.

FAO (1985). Water quality for agriculture, by R.S. Ayers & D.W. Westcot. Irrigation and Drainage Paper No. 29 (Rev. 1). Rome.

Girma T, Ahlgren G (2009). Seasonal variations in phytoplankton biomass and primary production in the Ethiopian Rift Valley Lakes Ziway, Awassa and Chamo-The basis for fish production. Elsevier Sc. Limnologica 40(2010):330-342.

Gugissa E (2004). Urban Environmental impacts in the town of Awassa Ethiopia, Addis Ababa University, Unpublished MSc Thesis Addis Ababa University, Ethiopia.

Halcrow G (2010). Rift valley lakes basin integrated resources development plan study project. Lake Hawassa sub-basin integrated watershed management feasibility study. Main report Vol.1.

Kelly WP (1963). Use of saline irrigation water. Soil Sci., 95(4):355-391

Kloos H, Redda T (1999). Distribution of fluoride and fluorosis in Ethiopia and prospects for control. Trop. Med. Int. Health 4:355 -364.

Milda L (2009). The environmental impact caused by the increasing demand for water. Water and resources management-a case study in Ethiopia. MSc thesis. TAMK University of Applied Sciences Environmental Engineering, Tampere.

Murdoch T, Cheo M, OLaughlin K (2001). Streamkeeper's Field Guide:

Watershed Inventory and Stream Monitoring Methods.

Oki T, Agata Y, Kanae S, Saruhashi T, Musiake K (2003). Global water resources assessment under climatic change in 2050 using TRIP.IAHS Publ., 280:124–133.

Puri PJ, Yenkie MKN, Songal SP, Gandhore NV, Sarote GB, Dhanorkar DB (2011). Surface water (lakes) quality assessment in Nagpur City (India) based on water quality index (WQI). RASAYAN J. Chem. 4:43-48.

Raghunath IM (1987). Groundwater. 2nd Edn., Wiley Eastern Ltd., New Delhi, India.

Richards LA (1954). Diagnosis and improvement of saline and alkali solids, Agric. Handbook 60, U.S. Dept. Agric., Washington, D.C., P. 160.

Sharifi M (1990). "Assessment of Surface Water Quality by an Index System in Anzali Basin". In The Hydrological Basis for Water Resources Management. IAHS 197:163-171.

Singh AK, Mondal GC, Suresh K, Singh TB, Tewary BK, Sinha A(2008). "Major ion chemistry, weathering processes and water quality assessment in upper catchment of Damodar River basin, India", Environ. Geol. 54:745-758.

Smith DG (1987). "Water Quality Indexes for Use in New Zealand's Rivers and Streams", Water Quality Centre Publication No. 12, Water Quality Centre, Ministry of Works and Development, Hamilton, New Zealand.

Tamiru A (2006). Groundwater occurrence in Ethiopia. Addis ababa University, Ethiopia. With the support of UNESCO.

Tenagne A (2009). The impact of urban storm water runoff and Domestic waste effluent on water quality of Lake Tana and Local ground water near the city of Bahirdar, Ethiopia.

Tenalem A (1998). The hydrological system of the Lake District basin, Central Main Ethiopian Rift. PhD thesis. International Institute for Geographic Information Science and Earth Obseration(ITC), Enschede.

Todd D (1980). Groundwater Hydrology. 2nd Edn. John Wiley and Sons. New York.

USEPA (U.S. Environmental Protection Agency) (2000). Nutrient Criteria Technical Guidance Manual: Rivers and Streams. United States Environmental Protection Agency, Office of Science and Technology. Washington, DC. EPA-822-B-00-002.

Vorosmarty CJ, Green P, Salisbury J, Lammers RB (2000). Global water resources: vulnerability from climate change and population growth. Science 289:284-288.

Warhate SR, Wankar KG (2012). The evaluation of water quality index around Welkorela-Pimperi coal mines. Sci. Rev. Chem. Commun. 2(3):197-200.

WHO (2006). Guidelines for safe recreational water environments, Swimming pools and similar environments. Geneva. Vol. 2

WHO (2005). Water, Sanitation and Hygiene Programming Guidance Water Supply and Sanitation Collaborative Council and World Health Organization, 2005 Printed in Geneva1219 Chatelaine, Geneva, Switzerland.

WHO (2004). Guidelines for Drinking Water Quality, Third Edition. Volume 1: Recommendations. World Health Organization Geneva.

WHO (1993). Guidelines for Drinking Water Quality. 1st Ed. Geneva.

WHO (1989). Health Guidelines for the Use of Waste water in Agriculture and Aquaculture. Report of a WHO Scientific Group. Geneva, Switzerland: World Health Organization Press.

WHO (1984). Guidelines for Drinking-Water Quality, Vol. 2. Health Criteria and Other Supporting Information. Geneva: WHO.

WHO (1983). Guidelines for Drinking Water Quality, Vol. 3. World Health Organization, Geneva.

Yemane G (2004). Assessment of the water balance of Lake Awassa catchment, Ethiopia, M.Sc. thesis, International institute for Geo-information Science and Earth Observation (ITC), Enscheda, the Netherlands.

Yosef HG, Bekele L, Behailu B, Kefyalew S, Wondewosen S (2010). Environmental impact assessment and policy on Lake Hawassa, SNNPRs. SOS-Sahel, Ethiopia.

Zinabu G, Zerihun D (2002). The chemical compostion of the effluent from Awassa Textile factory and its effects on aquatic biota. SINET: Ethiop. J. Sci. 25(2):263-274.

Zvikomborero H (2005). An assessment of the water quality of drinking water in rural districts in Zimbabwe. The case of Gokwe south, Nkayi, Lupane and Mweezi districts. Physics Chemistry Earth 30:859-866.

10

Study of geosynthetic clay liner layers effect on decreasing soil pollution in the bed of sanitary land fills

Rouhollah Soltani Goharrizi[1], Fazlollah Soltani[2] and Bahador Abolpour[3]

[1]Collage of Natural Resources, Department Of Environment, Islamic Azad University, Bandar Abbas, Iran.
[2]Department of Civil Engineering, Faculty of Engineering, Graduate University of Advanced Technology, Kerman, Iran.
[3]Department of Chemical Engineering, Faculty of Engineering, Shahid Bahonar University of Kerman, Jomhoori Blvd., Post Code 76175, Kerman, Iran.

According to importance of preventing pollution of soil by leachate in the landfills, the bed of landfills has to be designed in the way that it can prevent soil pollution. This research has studied the effect of geosynthetic clay liner (GCL) layer to decrease soil pollution in the bed of landfills. Two different kind of soil were transferred to laboratory for doing experiments. In the first step, the penetration experiments have been done by water and leachate on condensed soil samples without use of GCL layers. In the second step, the penetration experiments have been repeated by water and leachate on condensed soil samples with GCL layers located on the top of soil samples. The results of this research show that, the rate of permeability in soil samples with GCL layers are decreased about 98%.

Key words: Leachate, geosynthetic clay liner (GCL) layer, soil pollution, solid waste

INTRODUCTION

Geosynthetic clay liners (GCLs) represent a relatively new technology (developed in 1986) that is currently gaining acceptance as a barrier system in municipal solid waste landfill applications. GCL technology offers some unique advantages over conventional bottom liners and covers. For example, GCLs are fast and easy to install, have low hydraulic conductivity and have the ability to self-repair any rips or holes caused by the swelling properties of the bentonite from which they are made.

Many studies have been conducted to assess the performance of GCL permeated with various types of liquids containing cations. Egloffstein (2001) studied GCLs and also its application in contact with leachates or chemical solutions. Petrov and Rowe (1997) investigated GCL chemical compatibility by hydraulic conductivity testing. Petrov et al. (1997a, b) also worked on GCL hydraulic conductivity. GCLs permeated with chemical solutions and leachates are studied by Ruhl and Daniel (1997). Shackelford et al. (2000) evaluated the hydraulic conductivity of GCLs permeated with non-standard liquids. Hydraulic conductivity of GCLs to tailings impoundment solutions are investigated by Shackelford et al. (2010). Hydraulic conductivity and swelling of nonprehydrated GCLs permeated with single species salt solutions are studied by Jo et al. (2001). Some of them also worked on the long-term hydraulic conductivity of a GCL permeated with inorganic salt solutions (Jo et al., 2005). Ashmawy et al. (2002) studied the hydraulic performance of untreated and polymer-treated bentonite in inorganic landfill leachates. Shan and Lai (2002) investigated the effect of hydrating liquid on the hydraulic properties of geosynthetic clay liners. Kolstad et al. (2004a, b) worked on hydraulic conductivity and swell of nonprehydrated GCLs permeated with multispecies inorganic solutions and also dense prehydrated GCL permeated with aggressive inorganic solutions. Lee et al.

(2005) investigated the correlating index properties and hydraulic conductivity of GCLs. Guyonnet et al. (2005, 2009) presented a correlation between permeability, microstructure, and surface chemistry and also studied performance-based indicators for controlling GCLs in landfill applications. Katsumi et al. (2008) studied the long-term barrier performance of modified bentonite materials against sodium and calcium permeant solutions and also presented evaluating methods to modify the chemical resistance of GCLs. Rosin-Paumier et al. (2011) investigated the impact of a synthetic leachate on permittivity of GCLs. And also Islam and et al. (2004) were conducted Continuous flow experiments using sand-packed columns to investigate the relative significance of bacterial growth, metal precipitation, and anaerobic gas formation on biologically induced clogging of soils.

The measured hydraulic conductivity values are often highly variable at equivalent effective stresses even though there is a general trend of decreasing hydraulic conductivity with increasing static confining effective stress. The aim of this research is to study the efect of Geosynthetic Clay Liner Layers as a barrier for decreasing the rate of permeability in soil samples. Experiments have been done by water and leachate on condensed soil samples without use of GCL layers. Then the penetration experiments have been repeated by water and leachate on condensed soil samples with GCL layers.

EXPERIMENTATION

Materials

Soil samples

At this stage, two different kinds of disturbed soil samples from two different places were brought to the laboratory. Collection of Soil samples were conducted by the way of disturbed sampling in the one meter depth of soil. Then samples were put in the packet and were transferred in the lab without losing their moisture. Characteristics of mines and development project of the soil samples are:

(a) Dam project of kashman (Iran, Central Khorasan province, City of Mashhad, 10 km Road Chanaran, Ali Abad Mine).
(b) Train project of Esfahan-Tehran (Km 750+337 path, Boreholes from deep between 0.4 to1.5 m).

Leachate and GCL layers

In this study leachate testing requirements was prepared from organic fertilizer (compost) factory in Esfehan. The characteristics of this leachate are shown in Table 1.
GCL layers characteristics made in the laboratory are given in Table 2.

Primary testes

Determination of primary moisture content

Results of experiments to determine primary moisture content are

shown in Table 3.

Grading tests

Grading tests were conducted for soil samples by standard sieves. The results are shown in Figure 1. Slope and shape of the gradation curve can be described by the coefficient of uniformity and the coefficient of curvature. Table 3 presents the coefficient of uniformities and the coefficient of curvature in soil samples.

Determination of Atterberg limits, maximum dry density and optimal moisture density

Plastic and liquid limits of soil samples are presented in Table 3. Furthermore, density tests were conducted on soil samples in the way of normal proctor. Optimum moisture content and maximum dry density of the soil samples are presented in Table 3.

Preparation of samples

The permeability tests were conducted by the use of penetration cell with variable load. Photographs of experimental setup are shown in Figure 2. Preparation steps of soil samples are shown in Table 4.

RESULTS AND DISCUSSION

Summary results of permeability test have been shown in Figure 3. These diagrams can be divided to two independent areas that each area relates to sample conditions in penetrometer cell.

Area 1

This area shows the changes of permeability coefficient of composed sample in cell. In this case, sample consists of compressed soil column in cell in adding the prepared GCL insulation curtain established in the top layer of compressed soil. As diagrams show in this area, amounts of permeability coefficient have a tangible difference from other areas. This is logical, because in this case, passing fluid must cross from two different and resistant areas (compressed soil and GCL insulation curtain). In this case, numerical amounts of permeability coefficients in these ranges will be changed for different soil samples. Existence of insulation curtain in top layer will lead to creation of flow routes of the main entrance.

Permeability coefficients for soil type A: 2.46321E-10 to 8.26465E-09 m/s

Permeability coefficients for soil type B: 7.29016E-11 to 2.97735E-09 m/s

Area 2

This area indicates changes in permeability coefficient in compressed sample of soil (without establishment of GCL

Table 1. Average concentration of chemical constituent in the waste leachate produced at the organic fertilizer (compost) factory in Esfehan in 10 sampling.

I (mg/L)	Zn (mg/L)	Cu (mg/L)	Cr (mg/L)	Cd (mg/L)	TFS (mg/L)	TVS (mg/L)	TS (mg/L)	COD (mg/L)	pH
2.38	5.81	1.21	1.27	0.46	16219.6	22595.9	38815.5	38562.5	4.98

Table 2. Some properties of GCL layers.

Commercial name of layer	BENTOMAT
Material of Bentonite	Grain bentonite sodic
Mass of Bentonite (g/m^2)	5000
Material of top-mat-polymer layer	Woven polypropylene
Mass of top-mat-polymer layer (g/m^2)	110
Material of bottom-mat-polymer layer	Not woven polypropylene
Mass of bottom-mat-polymer layer (g/m^2)	220
Total mass of layer (g/m^2)	5330
Total thickness of layer in dry condition (mm)	6

Table 3. Quantities of primary moisture content, coefficient uniformity and coefficient curvature, Plastic and liquid limits, and the optimum moisture content and maximum dry density of soil samples.

Soil sample	Kashman dam project (A)	Train project Esfehan to Tehran (B)
Primary moisture (%)	0.8	1.1
Coefficient of curvature (C_C)	0.9	0.8
Coefficient of uniformity (C_U)	2.2	2.8
Plastic limits (w_P)	15.7	13
Liqued limits (w_L)	18.5	35
Optimum moisture content, w_{opt} (%)	9.8	8.9
Maximum dry density, ρd_{max} (Kg/m^3)	1889.9	1803.4

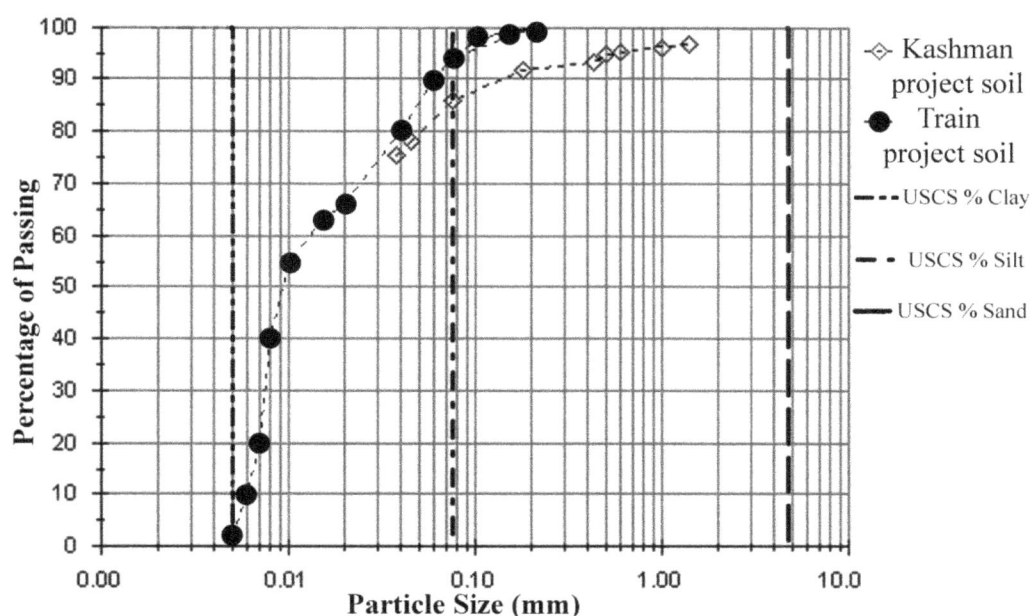

Figure 1. Gradation curve of kashman and train project soils.

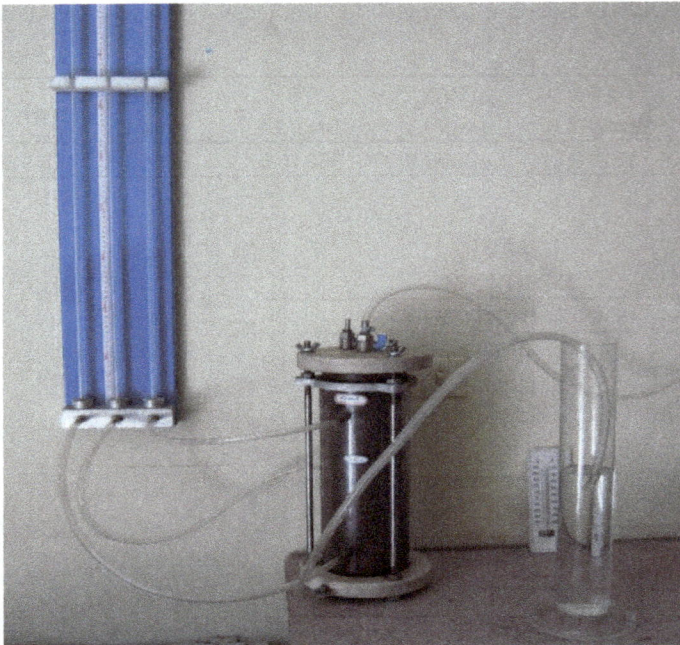

(a)

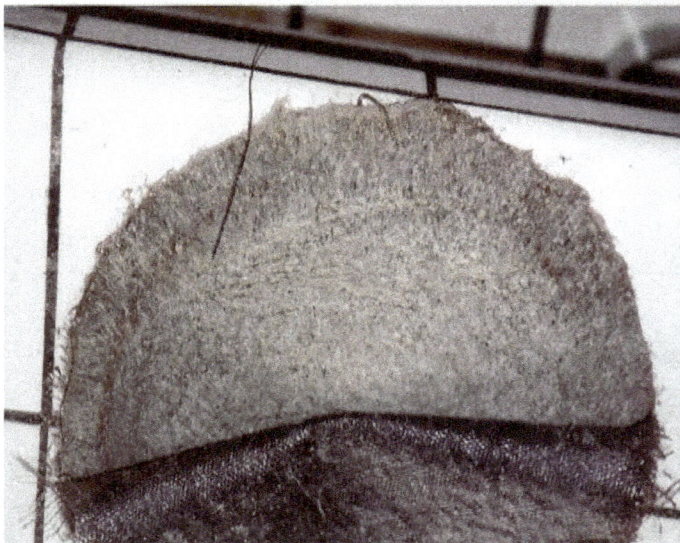

(b)

Figure 2. (a) Experimental setup and cell (b) Manufactured GCL.

entrance flow routes of fluid to the soil mass will be distributed in the whole surface of the sample section uniformly.

Permeability coefficients for soil type A: 1.78806E-08 to 3.12233E-08 m/s
Permeability coefficients for soil type B: 9.26869E-11 to 2.80485E-10 m/s

In the Figure 3 It is seen that with increase in energy of soil compaction for the composed sample based in penetrometer cell, permeability coefficient will be decreased. The reason can be found in changing the structure of based soil. In another word, increase in energy of soil compaction leads to exiting more air from soil sample. This decrease in vacant space of the soil leads to increasing in soil resistance against crossing of fluid (water or leachate). In addition, in the case of fixed energy of soil compaction it is seen that, with change in the type of the passing fluid from the composed sample, permeability coefficient will be changed. This phenomenon is explainable in the way that, change in the fluid type, leads to change in the fluid viscosity. On the other hand, change in the fluid viscosity is an important factor in changes of permeability. In other words, increase in viscosity (converting the passing fluid water to leachate) leads to reduction of permeability. It is also concluded that the method of preparation of GCL insulation curtain before establishment in penetrometer cell has an important role in the permeability of composed sample. As it can be observed, increase in overload when preparing GCL insulation curtain, causes decrease in permeability coefficient of composed sample in a same condition. It can be explained in the way that, increasing overload to GCL insulation curtain causes more compression of bentonite paste available in the curtain. This compression in paste is related to porosity reduction concept of bentonite paste.

Although there is not any experiment that is conducted in the same condition with our research but a laboratory study of land fill leachate transport in soils have been done by Islam et al. (2004). In their research an influent acetic acid concentration of 1750 mg/L which was in leachate decreased the soil's hydraulic conductivity from an initial value of 8.8×10^{-5} m/s to approximately 7×10^{-7} m/s in the 2 to 6 cm section of the column of soil.

Conclusion

The leachate of waste that deposits in the landfills can have some harmful effect on soil and underground water. GCL are materials that are manufactured from betonith, these material reduce the permeability of leachate in the landfill. The results of this research show that there are some factors such as viscosity of fluid and energy of soil compaction that influence the permeability of soil. Using the GCL in the sanitary landfills reduces the rate of

insulation curtain) based in cell. In this case, the sample consists of one compressed soil column in cell only. As the diagrams indicate, in this area, amounts of permeability coefficient have significant changes rather than other areas. This is logical, because in this case passing fluid will cross from a resistant area (compressed soil). In these conditions, numerical amounts of permeability coefficient will be located in the following ranges. In this case, lack of insulated curtain that causes

Table 4. Preparation of soil samples A and B.

Sample test conditions	A																				B																			
	1	2	3	4	5	6	7	8	9	10	11	12	13	14	15	16	17	18	19	20	1	2	3	4	5	6	7	8	9	10	11	12	13	14	15	16	17	18	19	20
Soil compaction by normal proctor	✓	✓	✓	✓	✓	✓	✓														✓	✓	✓	✓	✓	✓	✓													
Soil compaction by modified proctor								✓	✓	✓	✓	✓	✓	✓														✓	✓	✓	✓	✓	✓	✓						
Preparing GCL layer under 25 kpa	✓	✓						✓	✓												✓	✓						✓	✓											
Preparing GCL layer under 50 kpa			✓	✓						✓	✓												✓	✓						✓	✓									
Preparing GCL layer under 75 kpa					✓	✓						✓	✓												✓	✓						✓	✓							
Preparing GCL layer under 100 kpa							✓							✓													✓							✓						
Water as a fluid	✓	✓	✓	✓	✓	✓	✓	✓	✓	✓	✓	✓	✓	✓							✓	✓	✓	✓	✓	✓	✓	✓	✓	✓	✓	✓	✓	✓						
Leachate as a fluid															✓	✓	✓	✓	✓	✓															✓	✓	✓	✓	✓	✓

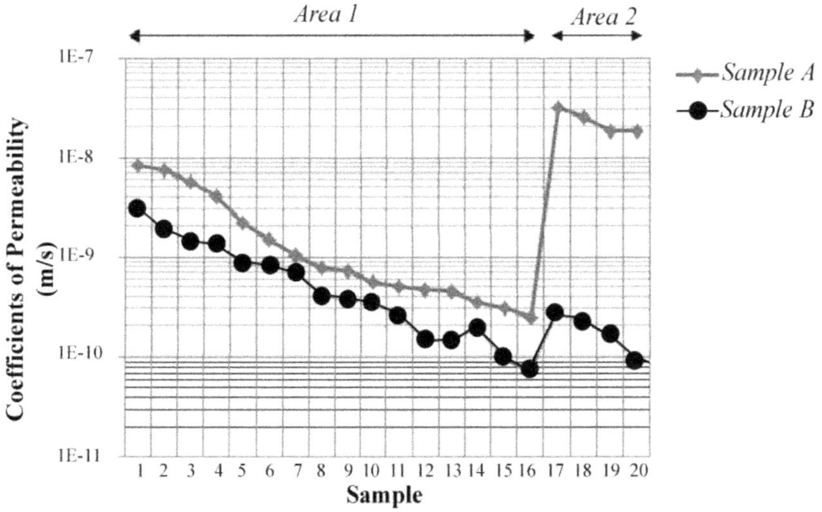

Coefficients of Permeability (m/s)

(Y-axis: 1E-7, 1E-8, 1E-9, 1E-10, 1E-11; X-axis: Sample 1–20; Area 1, Area 2)

◆ Sample A ● Sample B

Figure 3. Comparison diagrams of coefficients of permeability on samples A and B.

leachate penetration depending on GCL materials and manufacturing more than 98%.

REFERENCES

Ashmawy AK, Darwish EH, Sotelo N, Muhammad N (2002). Hydraulic performance of untreated and polymer-treated bentonite in inorganic landfill leachates. J. Clays Clay Miner. 50(5):546-552.

Egloffstein TA (2001). Natural bentonites influence of the ion exchange and partial dessication on permeability and self-healing capacity of bentonites used in GCLs. J. Geotext. Geomembr. 19:427-444.

Guyonnet D, Gaucher E, Gaboriau H, Pons CH, Clinard C, Norotte V, Didier G (2005). Geosynthetic clay liner interaction with leachate: Correlation between permeability, microstructure, and surface chemistry. J. Geotechn. Geoenviron. Eng. 131:740-749.

Guyonnet D, Touze-Foltz N, Norotte V, Pothier C, Didier G, Gailhanou H, Blanc P, Warmont F (2009). Performance-based indicators for controlling geosynthetic clay liners in landfill applications. J. Geotext. Geomembr. 27:321-331.

Islam J, Singhal N (2004). A laboratory study of landfill-leachate transport in soils. J. water Res. 38:2035-2042.

Jo HY, Benson CH, Shackelford CD, Lee JM, Edil TB, Asce M (2005). Long-term hydraulic conductivity of a geosynthetic clay liner permeated with inorganic salt solutions. J. Geotechn. Geoenviron. Eng. 131:405-417.

Jo HY, Katsumi T, Benson CH, Edil TB (2001). Hydraulic conductivity and swelling of nonprehydrated GCLs permeated with single species salt solutions. J. Geotechn. Geoenviron. Eng. 127(7):557-567.

Katsumi T, Ishimori H, Onikata M, Fukagawa R (2008). Long-term barrier performance of modified bentonite materials against sodium and calcium permeant solutions. J. Geotext. Geomembr. 26:14-30.

Kolstad D, Benson C, Edil T (2004a). Hydraulic conductivity and swell of nonprehydrated geosynthetic clay liners permeated with multispecies inorganic solutions. J. Geotechn. Geoenviron. Eng. 130:1236-1249.

Kolstad D, Benson C, Edil T, Jo H (2004b). Hydraulic conductivity of a dense prehydrated GCL permeated with aggressive inorganic solutions. J. Geosynth. Int. 11:233-240.

Lee JM, Shackelford CD, Benson CH, Jo HY, Edil TB (2005). Correlating index properties and hydraulic conductivity of geosynthetic clay liners. J. Geotechn. Geoenviron. Eng. 131(11):1319-1329.

Petrov RJ, Rowe RK (1997). Geosynthetic clay liner-chemical compatibility by hydraulic conductivity testing: factors impacting its performance. J. Can. Geotechn. 34(6):863-885.

Petrov RJ, Rowe RK, Quigley RM (1997a). Comparison of laboratory measured GCL hydraulic conductivity based on three permeameter types. J. Geotechn. Test. 20(1):49-62.

Petrov RJ, Rowe RK, Quigley RM (1997b). Selected factors influencing GCL hydraulic conductivity. ASCE J. Geotechn. Geoenviron. Eng. 123(8):683-695.

Rosin-Paumier PS, Touze FN, Pantet A (2011). Impact of a synthetic leachate on permittivity of GCLs measured by filter press and oedopermeameter tests. J. Geotext. Geomembr. 29(3):211-221.

Ruhl JL, Daniel DE (1997). Geosynthetic clay liners permeated with chemical solutions and leachates. J. Geotechn. Geoenviron. Eng. 123(4):369-381.

Shackelford CD, Benson CH, Katsumi T, Edil TB, Lin L (2000). Evaluating the hydraulic conductivity of GCLs permeated with non-standard liquids. J. Geotext. Geomembr. 18:133-161.

Shackelford CD, Sevick GW, Eykholt GR (2010). Hydraulic conductivity of geosynthetic clay liners to tailings impoundment solutions. J. Geotext. Geomembr. 28(2):149-162.

Shan HY, Lai YJ (2002). Effect of hydrating liquid on the hydraulic properties of geosynthetic clay liners. J. Geotext. Geomembr. 20:19-38.

11

Study of heavy metals pollution and physico-chemical assessment of water quality of River Owo, Agbara, Nigeria

Kuforiji Titilope Shakirat[1] and Ayandiran Tolulope Akinpelu[2]

[1]Department of Zoology, Faculty of Science, Lagos State University (LASU), P.O. BOX LASU 001, Ojo, Lagos, Nigeria.
[2]Department of Pure and Applied Biology, Faculty of Science, Ladoke Akintola University of Technology, Ogbomoso, Oyo State, Nigeria.

The various selected physicochemical and biological condition of freshwater bodies which receive varying number of outfalls of industrial and domestic effluents containing heavy metals of River Owo, Agbara Industrial Estate, a boundary town between Lagos and Ogun state, along Badagry Expressway in Nigeria, was assessed for five consecutive months. Four sampling points were chosen from the water body to reflect the effect of industrial effluent, domestic effluent and lotic habitats. Temperature, pH, salinity, total alkalinity, total hardness (TH), total settle-able solids (TSS), dissolved oxygen (DO), turbidity, biochemical oxygen demand (BOD), ammonia, phosphate, chloride level, sulphate, nitrate, cadmium, chromium, copper, iron, lead, manganese zinc, colour were analyzed monthly between November 2007 and March 2008 using standard methods and procedures. The ranges of these factors were found to be comparable to the recommended limit of the Lagos State Environment Protection Agency (LASEPA) 2001 and the Federal Environmental Protection Agency (FEPA) except for biochemical oxygen demand (BOD), ammonia, phosphate, cadmium, chromium, lead, copper and Iron which were found in higher concentrations at sampling point B (effluent discharge point) above LASEPA, FEPA and Freshwater limit. DO of the three sampling points A, C and D were between recommended limit but slightly low at point B (effluent discharge point), which is less than the recommended limit. The level of pollution is more pronounced at sampling point B due to the greater amount of BOD and the release of some heavy metals like cadmium, chromium and lead which greatly affect the water quality of the river. The study concludes that River Owo is slightly polluted while the pollution is as a result from the discharge of effluents by the companies, factory and materials from other anthropogenic sources.

Key words: Industrial and domestic effluent, biochemical physicochemical condition oxygen demand (BOD), heavy metals, monitoring, anthropogenic sources.

INTRODUCTION

Water serves as a significant utility in irrigation of agricultural lands, generation of hydro-electric power, municipal water supply, fishing, boating and body-contact recreation, communication as well as unending domestic activities of man and animals. It also serves as a receptor of industrial waste, domestic waste and waste water resulting from other uses of water (Chapman, 1996; Rosemberg and Reish, 1993). Human developmental activities have led to increased population and waste generation, which in turn have contributed to the deterioration and degradation of the environment due to increased human population and increased waste generation (Thorne and Williams, 1997). The wastes find their way into the environment as gaseous, liquid or solid materials. They apparently have impacts on the environment and the flora and fauna in the receiving media.

Environmental deterioration is then a natural outcome of rapid population growth, agricultural practices, industrialization and urbanization of society (TERI, 2000). Industrial wastes are complex mixture of different contaminants or pollutants (Ajao, 1983; Oyewo, 1998). Just as these contaminants or pollutants are different so are their effects on the receiving environment and the biota. Most gaseous emissions lead to stratospheric ozone layer depletion, acid rain and photochemical smog (Warrick, 1988; Chiras, 1998). They are the major factors responsible for premature aging of the skin, skin cancer, cataracts and immune deficiencies by mutagenicity (Treshow and Anderson, 1989; Elsom, 1992). The industrial wastes released into the aquatic environment contain harmful chemicals such as heavy metals, oil, settle-able solids, nutrients and ammonia. These pollutants have various effects on the organisms in the receiving water body (McClugge, 1991). In addition, plants and animals inhabiting the water bodies are not spared as their normal functioning is affected by pollution and also alter their populations. Depletion of their numbers and species in diversity seems to be a major effect. This will go back to man as its insatiable assumption of freshwater resources remains unending. Thus, man may be facing a great physiological threat (Clark, 1994).

Aquatic pollution by heavy metals is very prominent in industrialized and mining areas and these metals are released or leached to the water bodies (Garbarino et al., 1995; INECAR, 2000). Some heavy metals has bio-importance as trace elements but the toxic effluents of many of them on aquatic organism accumulating in their body are of great concern since they apply to the group of metals and sediments (metalloids) that are associated with contamination, potential toxicity or ecotoxity with atomic number greater than 4 g/cm or five times or more, greater than water (Garbarino et al., 1995; Hawkes, 1997). Therefore, the accurate determinations of water quality using the physic-chemical parameters and heavy metals pollution are of ultimately important for controlling their pollution, this study aims at providing additional information to existing data on water quality assessment of this water body.

MATERIALS AND METHODS

Agbara industrial estate, the study area

Agbara industrial estate is located between latitude 3°00' and 3°15'

and longitude 6°15'and 63°5'. It is approximately 31 km West of Lagos on the Badagry Expressway. It is first privately developed new town in Nigeria. It covers an area of 454 ha, which consists of industrial, commercial and private housing areas in a place that previously was only a forest. Clubs, shopping complexes, schools and medical facilities were incorporated in the master plan (Agbara Estate Limited, 1982).

Sampling sites

Four experimental (sampling) sites were chosen; upstream and downstream of effluent discharge outfall.

Sampling site A (pre-effluent discharge)

The site was located approximately 1 to 2 km upstream of the effluent entry point into River Owo. The water depth varied between 3.5 m close to the bank and 10.5 m at the centre. The speed of water-flow was sluggish.

Sampling site B (effluent discharge point)

This is the effluent discharge point into River Owo. It had a depth of between 3.8 m at the bank and 10.2 m at the centre. The flow of water was sluggish.

Sampling site C (effluent entry point)

This represented the lagoon site, the entry point of the River Owo into Olooge Lagoon. The water depth varied between 2.0 m at the bank and 4.5 m at the centre.

Sampling site D (Lagoon)

This represented the Ologe Lagoon with depth varied between 3.0 m at bank and 5.2 m at the centre.

The effluent and water temperatures were measured in-situ by dipping a thermometer bulb into the water for about 2 to 3 min and were recorded in degree Celsius; pH was determined using Griffin pH meter (model 40) and Hach test kit (Jenson and Avery, 1989). The following factors selected as water quality parameters were measured using the methods described for each factors as follows. Dissolved oxygen (DO) was determined by Azide modification of the Winkler method; Salinity was determined by use of salinometer TSI (model 33); turbidity was measured using formalin photometer set at a wavelength of 450 nm; total alkanity (TA), total hardness (TH), carbondioxide were determine by titration method (APHA, 1995); total settle-able solids (TSS) was measured by gravimetric method (APHA, 1995); biochemical oxygen demand (BOD) was measured using aluminum potassium sulphate ($AIKSO_4$) method. Nitrate, phosphate, sulphate and ammonia were determined separately by use of Hach's spectrophotometer set at the wavelength of 450 nm for sulphate, 500 nm for nitrate, 700 nm for phosphate and 620 nm for ammonia (APHA, 1995).

Water and effluent samples for heavy metal analysis was acidified using few drops of concentrated sulphuric acid to bring the pH between 2 to 3. Twenty-five milliliter of each sample was then filtered and the amount of heavy metal in each sample was determined using Atomic Absorption Spectrophotometer (AAS), (APHA, 1995). The heavy metal analysis was carried out at the University of Lagos, Chemistry Department Laboratory and the Mobil oil plc laboratory at Apapa in Lagos.

RESULTS

Nature of sampling points

The physical appearances of sampling points A, B and D were similar with a bluish green colour while sampling point C appeared different from sampling point A, B and D with a darkish green colour.

Physicochemical parameters

The physicochemical properties of River Owo at the four sampling points A, B, C and D for five consecutive months are shown in Figures 1 to 18. The values of the results were compared with standard values of LASEPA (2001) and FEPA (1991) for environmental pollution control in Nigeria. The determination of the physicochemical parameters of the sampled sites was carried out from the month of November, 2007 to March, 2008.

DISCUSSION

The physicochemical properties of River Owo at the sampling points A, C and D shows that all the parameters measure for five consecutive months were within the Federal Environmental Protection Agency (FEPA, 1991) and Lagos State Environmental Protection Agency (LASEPA, 2001) regulatory limits but slightly higher at sampling point B as a result of the effluent discharge point. The pH values of the four sampling points were ranged between 6.4 and 7.7 for the month of November (2007) to March (2008). There was no significant difference between the four sampling points and all are within FEPA and LASEPA permissible limit of 6.0 to 9.0. The result compares with findings of Ogunlaja and Ogunlaja (2007) and Uzoekwe and Oghosanine (2011). This authors found that the pH of the surface water in their environment (Ubeji creek, Warri and southern Nigeria) ranged between 6.5 and 8.5 and are within the Federal Ministry of Environment Nigeria (FMEnv.) recommended limit both nationally and internationally.

The temperature at the time of study ranged from 25.70 to 28.60°C for sampling points A, C and D. The observed higher temperature at the sampling point B (effluent discharge point) might be due to fresh effluent from the industries, homes, hospitals etc., and these values pose no threat to the homeostatic balance of the receiving water and were in agreement with the report of Jaji et al. (2007). Turbidity at the effluent discharge point ranged (48.52 to 75.20) NTU and for other sampling points (point A, C and D) ranged between 22.60 to 43.20 NTU. There were significant differences in values obtained at the effluent discharge point, in compare with other sampling points A, C and D. The turbidity values obtained for all

the locations were higher than WHO standard limit of 5 NTU (WHO, 2004). Excessive turbidity in water can cause problem for water purification process such as flocculation and filtration which may increase treatment cost. High turbid waters are associated with microbial contamination (DWAF, 1998). Again turbidity causes decrease in photosynthesis process since turbidity precludes deep penetration of light in water (Muoghalu and Omocho, 2000). Ultimately, the water receiving bodies disqualified as source of water for domestic use in the community.

The dissolved oxygen (DO) concentration of the effluent discharge point A was ranged between 0.3 to 0.7 mg/L and was observed to be lower than other sampling points A, C and D (4.4 to 6.6 mg/L) as compared with FEPA regulatory limit. The lower value at the discharge point could be attributed to the presence of degradable organic matter. Decrease in DO concentration could be attributed to breakdown of organic matter by aerobic microbes. The Oxygen required for this process is taken from the surrounding water thus diminishing its total oxygen content. Odukuma and Okpokwasili (1993) reported that it may be partially due to the displacement of dissolve oxygen by dissolved solids within the effluent.

Biological oxygen demand (BOD)

BOD test is useful in determining the relative waste loading and its higher degree therefore indicates the presence of large amount of organic pollutant and relatively higher level of microbial activities with consequent depletion of oxygen content. The value measured at sampling point B (effluent discharge point) was 333.2 to 524.2 mg/L and this was higher than 4.2 to 24.20 mg/L for the remaining sampling point A, C and D. The four sampling points A, B, C and D has low salinity might be due to the presence of calcium and magnesium ions (Boyd, 1979). There is no significant different between salinity range of the four sampling point of 0.025 to 1.280 mg/L. With regard to total hardness, no distinctly defined levels of what constitute a hard or soft water supply. The general accepted classification for hardness of water is 75 to 150 mg/L of $CaCO_3$ for soft and 150 mg/L and above for hard water (Deat, 2000). There was significant difference between the point of effluent discharge (119 to 133 mg/L) and values measured for sampling points A, C and D (27.6 to 55.6 mg/L).

Muoghalu and Omocho (2000) observed that when waste are heavily laden with pollutant and dissolved solids gain access to water bodies, they need large dose of oxygen for decomposition. The value of nutrients (total phosphate, nitrate, sulphate, ammonia) differs significantly between bodies. Unpolluted water usually contain only minute amount of nitrate (Jaji et al., 2007). Nitrate is a very important nutrient was observed to have been slightly high among the four sampling points with

Temperature °C

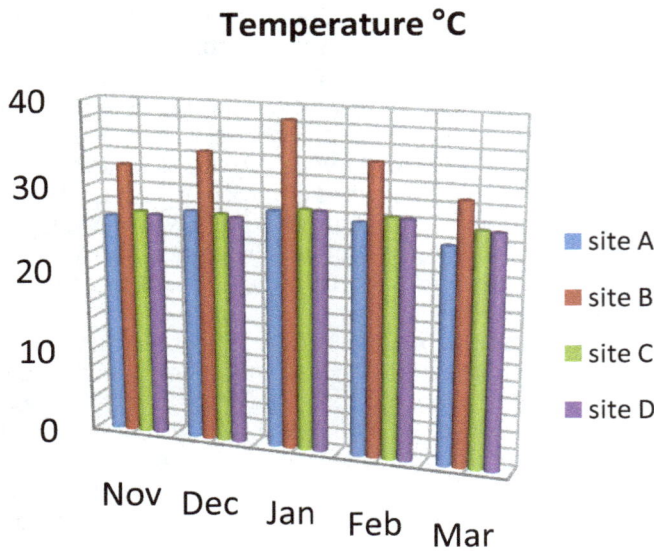

Figure 1. The effect of temperature at the four sampling points in River Owo for five consecutive months.

pH at 25°C

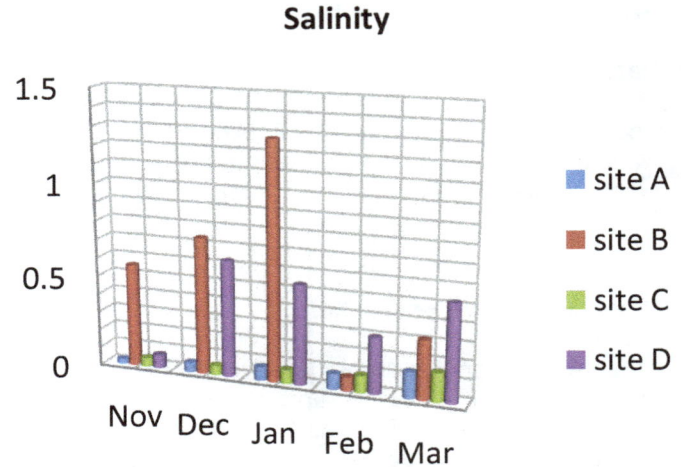

Figure 2. The effect of pH at the four sampling points in River Owo for five consecutive months.

Salinity

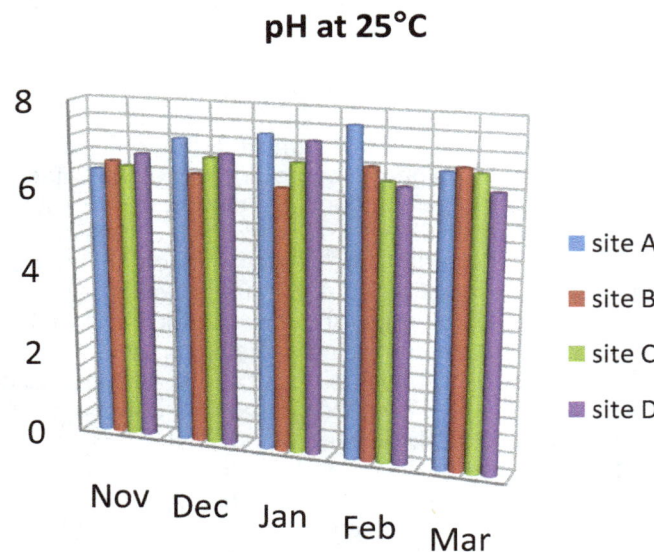

Figure 3. The effect of salinity at the four sampling points in River Owo for five consecutive months.

Alkalinity (mg/L)

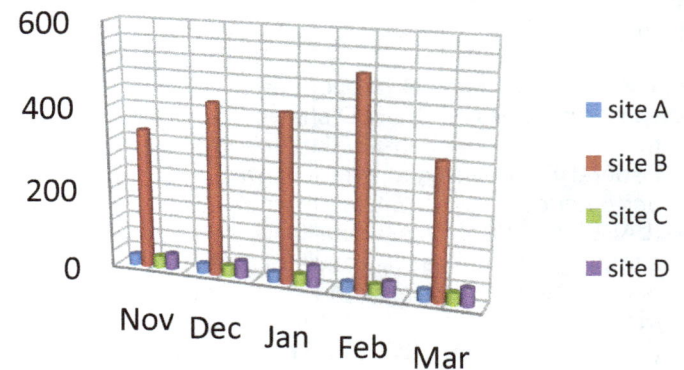

Figure 4. The effect of alkalinity at the four sampling points in River Owo for five consecutive months.

the range 2.3 to 6.84 mg/L. It was observed that the locations were slightly polluted. Elevated levels in nitrate have been reported to exhibit delayed reactions to light and sound stimuli (Robillard et al., 2003a, b) and can cause methemoglobinemia (Fatoki et al, 2008). Phosphate was also found to be low in all the sampling points. However, phosphate are essential nutrients to plants life, but when found in excess quantities (point B), stimulates excessive plant growth such as algae bloom (Igbinosa and Oko, 2009).

Heavy metal concentrations in water samples measured were slightly high for the four sampling locations but higher at sampling point B. Generally, higher level of iron, copper, lead, cadmium and chromium were observed at sampling point B (Fe > Cu > Cr > Pb > Cd) which as a result of effluent discharge point. Lead exposure has been associated with hypochromic anaemia with basophilic stifling of erythrocytes (Emory et al., 2001). Cadmium is highly toxic and accumulates in the body and eventually causes effects such as disturbances in calcium homeostasis and metabolism (Emory et al., 2001). Most chromium compounds are carcinogenic, long exposure may cause kidney, liver and nerve damage (Aremu et al., 2000). Generally, high level of metals may also be attributed to the discharge of effluents by the companies, factory, materials and other relevant occupational fields (steel making, welding, cutting, glass

Total hardness

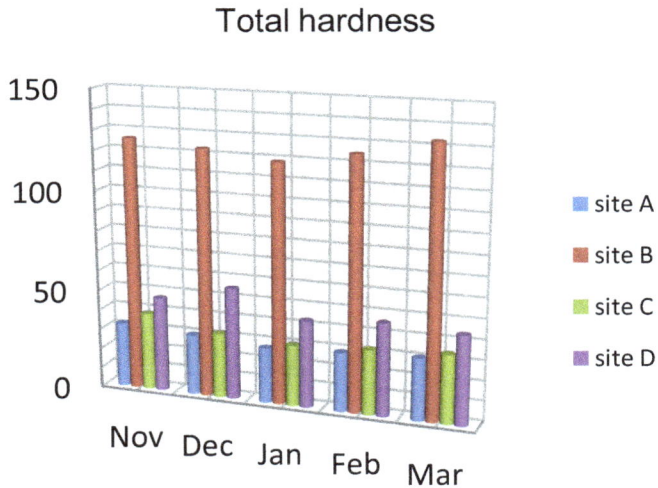

Figure 5. The effect of total hardness at the four sampling points in River Owo for five consecutive months.

Dissolved oxygen (mg/L)

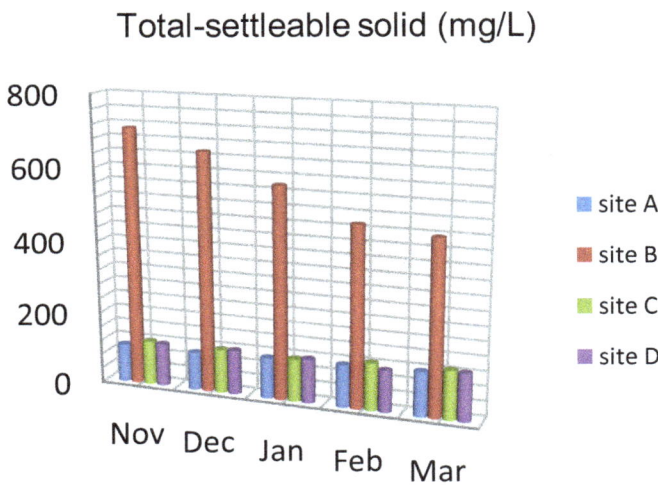

Figure 7. The effect of dissolved oxygen at the four sampling points in River Owo for five consecutive months.

Total-settleable solid (mg/L)

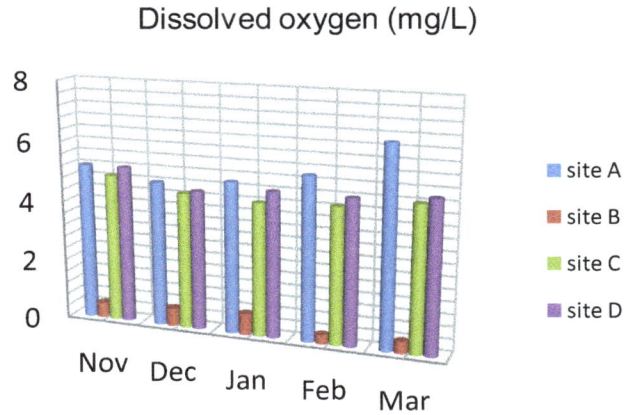

Figure 6. The effect of total-settle-able solid at the four sampling points in River Owo for five consecutive months.

Turbidity

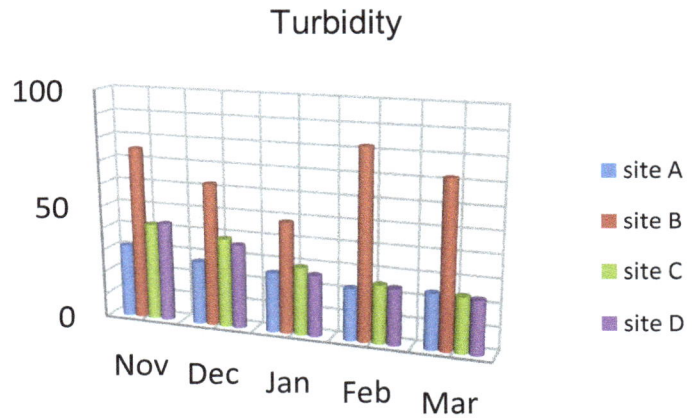

Figure 8. The effect of turbidity at the four sampling points in River Owo for five consecutive months.

and ceramic production etc.) (Vilia-Elena, 2006). The level of pollution is more pronounced at sampling point B due to the greater amount of biochemical oxygen demand (BOD) and the release of some heavy metals like cadmium, chromium and lead which greatly affect the water quality of the river.

RECOMMENDATIONS AND CONCLUSION

Metallic pollution which was noticed to be a threat to the water quality of River Owo should be arrested at the nick of time by the regulatory agencies to encourage and

Biochemical oxygen demand (BOD) (mg/L)

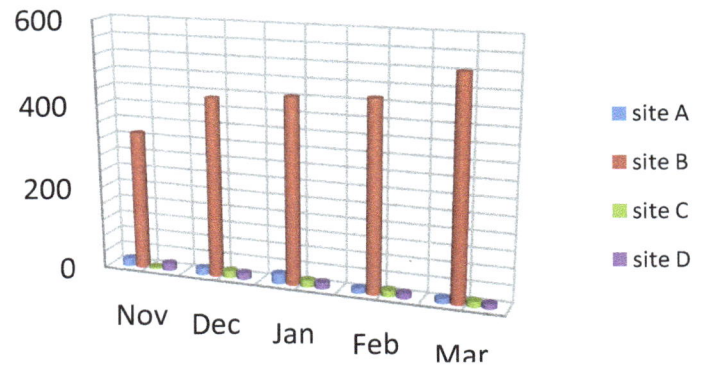

Figure 9. The effect of biochemical oxygen demand (BOD) at the four sampling points in River Owo for five consecutive months.

Ammonia (NH_3)

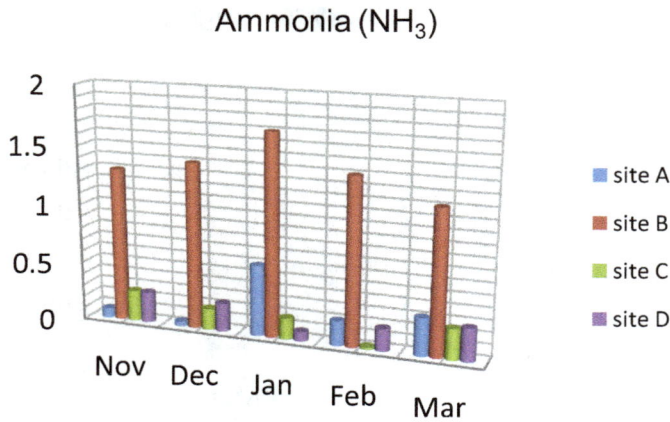

Figure 10. The effect of ammonia at the four sampling points in River Owo for five consecutive months.

Phosphate (PO_4^{3-})

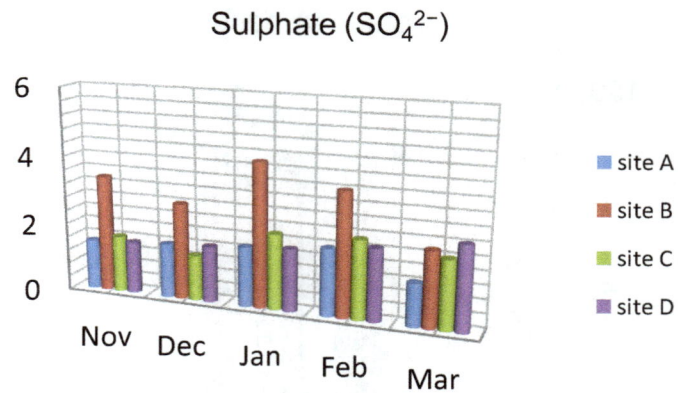

Figure 11. The effect of phosphate at the four sampling points in River Owo for five consecutive months.

Chloride Level

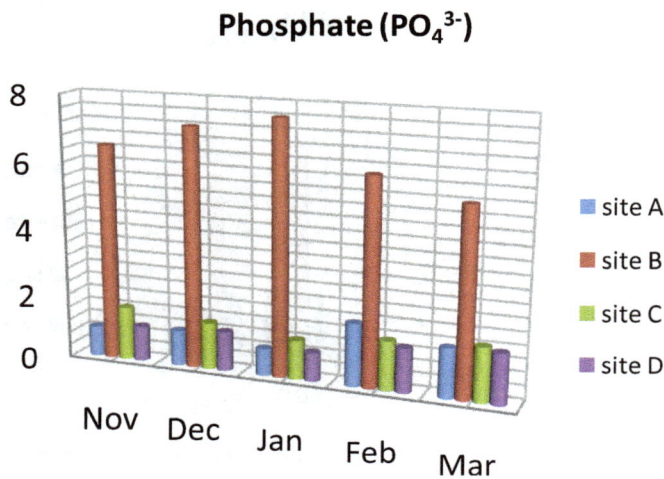

Figure 12. The effect of chloride at the four sampling points in River Owo for five consecutive months.

Sulphate (SO_4^{2-})

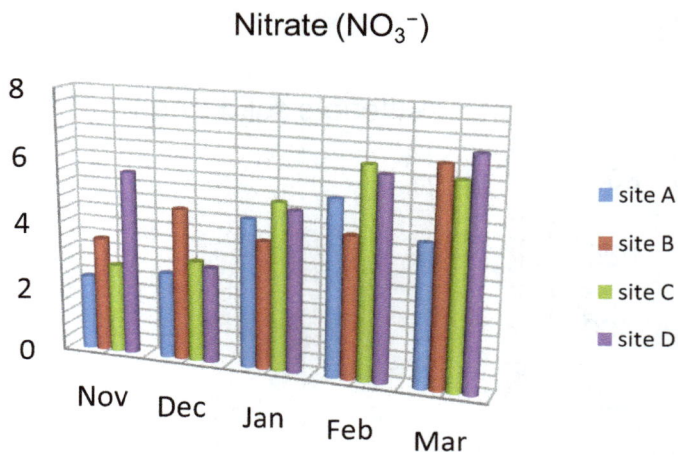

Figure 13. The effect of sulphate at the four sampling points in River Owo for five consecutive months.

Nitrate (NO_3^-)

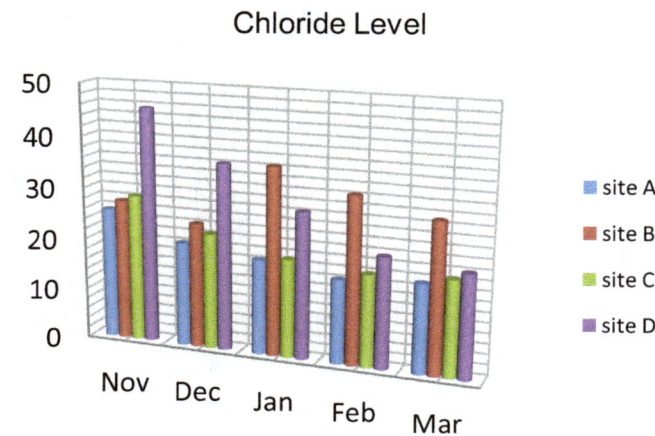

Figure 14. The effect of nitrate at the four sampling points in River Owo for five consecutive months.

Lead (mg/L)

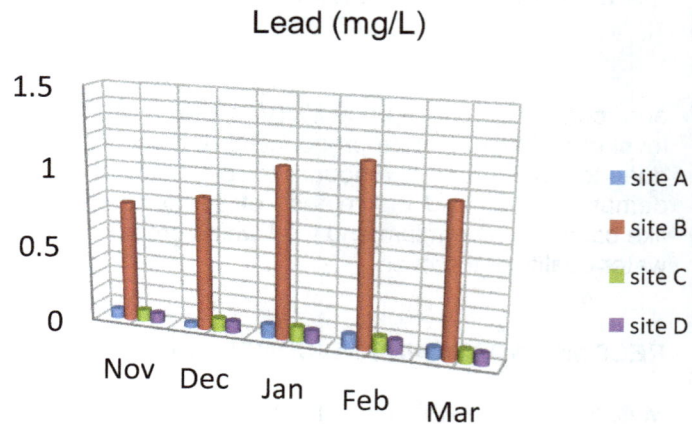

Figure 15. The effect of lead at the four sampling points in River Owo for five consecutive months.

Cadmium

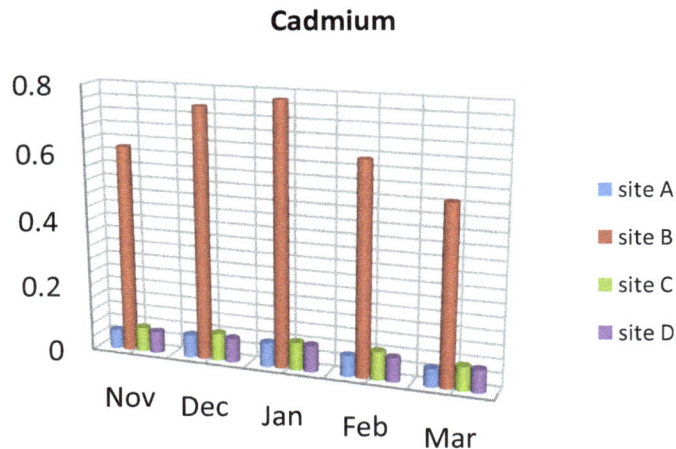

Figure 16. The effect of cadmium at the four sampling points in River Owo for five consecutive months.

Chromium

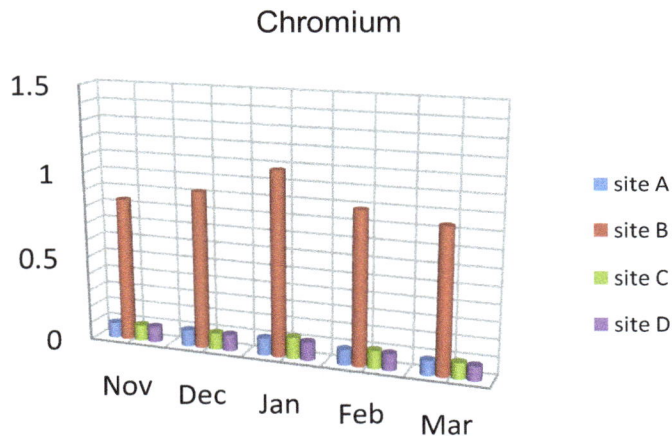

Figure 17. The effect of chromium at the four sampling points in River Owo for five consecutive months.

Colour

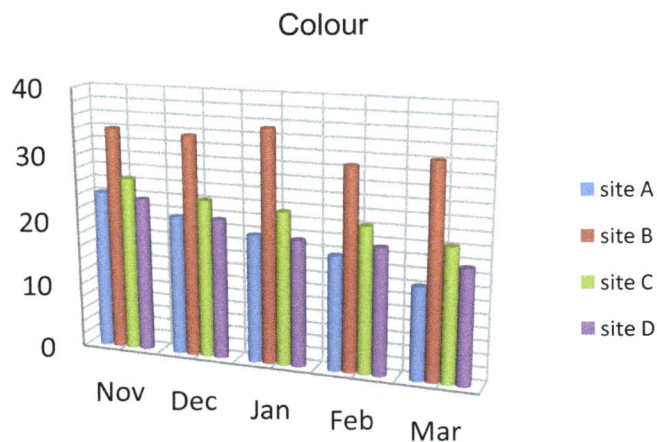

Figure 18. The effect of colour at the four sampling points in River Owo for five consecutive months.

compelled manufacturing industries/companies to treat their effluents before discharging into receiving water bodies. Also, the water quality of River Owo should be continuously monitored so that the level of pollution will be known and also to determine if the water is safe for agricultural practices. The Federal Ministry of Environment and its parastatal like Federal Environmental Protection Agency (FEPA) and other state owned and nongovernmental organizations like Lagos State Environmental Protection Agency (LASEPA), World Health Organization (WHO), should try and prevent discharge and channeling of waste into the water bodies such as River Owo in order to prevent pollution of the water.

In addition to these, manufacturing industries/companies should be encouraged and compelled by the regulatory agencies to treat their effluents before discharging into receiving water bodies. Also, the water quality of River Owo should be continuously monitored so that the level of pollution will be known and also to determine if the water is safe for agricultural practices and other benthic macro-invertebrates to live in to avoid the eradication of useful benthic macro-invertebrates.

ACKNOWLEDGEMENT

The author expresses her profound gratitude and appreciation to her able supervisor Mr. Kusemiju Victor of the Department of Zoology, Lagos State University (LASU), for his unique style of intellectual stimulation and invaluable assistance which was put at my disposal throughout the period of this project.

REFERENCES

America Public Health Association (APHA) (1995). Standard methods for the examination of water and waste water. 19th edition. America Public Health Association Inc., New York, pp. 1193.

Aremu D, Olaswuyi F, Metshitsuka S, Sridhar K (2000). Heavy metal analysis of groundwater from Warri, Nigeria. Int. J. Environ. Health Res. 12:61-72.

Boyd CE (1979). Water quality in warm water fish ponds. Craftmaster, Printers Inc. Auburn, Alabama, USA, pp. 353.

Chapman D (1996). Water Quality Assessment. A guide to the use of Biota Sediments and water in Environmental monitoring: The use of biological materials and River E IFN SPON, London, New York pp. 175-241, 243-315.

Chiras DD (1998). Enviromental Science: A system approach to sustainable development 9th ed. Wads-worth Publishing Company Behmount C.A p. 608.

Clark RB (1994). Marine Pollution. Clarendon press. Oxford, 172 pp.

Deat A (2000). White paper on integrated pollution and waste management for South Africa. A policy on pollution prevention, waste minimization, impact management and remediation. Dept. Environ. Affairs Tourism 80:274-275.

DWAF (1998). Quality of Domestic Water Supplies. Assessment Guide. 1 (2nd. Ed.) Department of Water Affairs and Forestry, Department of Health and Water Research Commission.

Elsom DM (1992). Atmospheric pollution: A global problem. Blackwell. Cambridge p. 422.

Emory E, Pattole R, Archiobold E, Bayorn M, Sung F (2001). Neurobehavioral effects of low level exposure in human. Neonates.

Am. J. Obstet. Gynecol. 181:5-11.

Fatoki SO, Muyima NYO, Lujiza N (2001). Situation analysis of water quality in the Umtata River catchment. Water SA. 27(4):467-474.

Fatoki SO, Gogwana P, Ogunfowokan AO (2003). Pollution assessment in the Keiskamma River and in the impoundment downstream. Water SA. 29(3):183-187.

Garbarino JR, Hayes H, Roth D, Antweider R, Briton TI, Taylor H (1995). Contaminants in the Mississippi River, U.S. Geological Survey Circular 1133, Virginia, U.S.A.

Hawkes JS (1997). Heavy Metal. J. Chem. Edu. 7(11):1374.

Igbinosa EO, Oko AI (2009). Impact of discharge wastewater effluents on the physiscochemical qualities of a receiving watershed in a typical rural community. Int. J. Environ. Sci. Technol. 6(2):175-182.

Institute of Environmental Conservation and Research INECAR (2000). Position paper Against Mining in Rapu-Rapu, Published by INECAR, Ateneo de Naga University, Philippines.

Jaji MO, Bamgbose O, Odukoya OO, Arowolo TA (2007). Water quality assessment of Ogun River, south west Nigeria. Environ. Monit. Assess. 133(1-3):447-482.

Lagos State Environment Protection Agency (LASEPA) (2001). Final Draft Report of Lagos State Effluent Limitation Standard and Guidance's Report of the Committee on Effluents Limitation Standards for Lagos State, p. 70.

Muoghalu LN, Omocho V (2000). Environmental Health Hazards Resulting from Awka Abattoir. Afr. J. Environ. Stud. 2:72-73.

Odukuma LO, Okpokwasili GC (1993). Seasonal Influence on Inorganic Anion Monitoring of New Calabar River, Nigeria. Environ. Manage 17(4):491-496.

Ogunlaja A, Ogunlaja OO (2007). Physicochemical analysis of water sources in Ubeji Communities and their Histological impact on organs of albino mice. J. Appl. Sci. Environ. Manag. 11(4):91-94.

Oyewo EO (1998). Industrial sources and distribution of heavy metals in Lagos Lagoon and their biological effects on Estuarine animals. (Ph.D. Thesis.) University of Lagos, 274 pp.

Robillard PY, Hulsey TC, Dekker GA, Chaouat G (2003a). Preeclampsia and human reproduction: An essay of a long term reflection. J. Reprod. Immunol. 59(2):93-100.

Robillard PY, Chaline J, Chaouat G, Hulsey TC (2003b). Eclampsia, Preeclampsia, and the evolution of the human brain. Curr. Anthropol. 44:130-135.

Teri (2000). Even as industrial effluents ravage the environment. Indian Energy Sector Key Issues, p. 6.

Thorne RS, Williams WP (1997). The response of benthic Macro-invertebrates to pollution in developing countries: A multimetric system of bioassessment. Freshwater Biol. 37:671-686.

Treshow M, Andreson FK (1989). Plant stress from air pollution. Wiley Chichester, pp. 2-96.

Vilia-Elena S (2006). Parkinson's disease and exposure to manganese during welding. Tech D. Welding Allied Process. 2:106-111.

Warrick RA (1988). Climate change and sea level rise. Clim. Monit. 15:19-44.

12

Performance and kinetic evaluation of phenol biodegradation by mixed microbial culture in a batch reactor

Sudipta Dey[1]*and Somnath Mukherjee[2]

[1]Department of Biotechnology, Heritage Institute of Technology, Anandapur, Chowbaga Road, PO: Kolkata, PIN: 700107, West Bengal, India.
[2]Professor, Environmental Engineering Division, Civil Engineering Department, Jadavpur University, Raja S. C. Mallic Road, Kolkata, PIN: 700032, West Bengal, India.

Mixed microbial culture collected from effluent treatment plant of a coke oven industry has been studied for its phenol biodegrading potential under aerobic condition in a batch reactor. The result showed that, after acclimatization, the culture could biodegrade up to 700 mg/l of phenol. The results showed that specific growth rate of microorganisms and specific substrate degradation rate increased up to 300 mg/l of initial phenol concentration and then started decreasing. The biodegradation kinetics is fitted to different substrate inhibition models by MATLAB 7.1©. Among all models, Haldane model was best fitted (Root Mean Square Error = 0.0067) for phenol degradation. The different biodegradation constants (K_s, K_i, S_m, μ_{max}, $Y_{X/S}$, k_d) estimated using these models showed good potential of the mixed microbial culture in phenol biodegradation.

Key words: Mixed culture, phenol biodegradation, kinetics, inhibition model.

INTRODUCTION

Phenolic compounds are commonly found in various industrial waste streams emanating from resin manufacturing units, coal gasification plants, petroleum refineries, coke oven industries etc (Juang and Tsai, 2006; Yan et al., 2006). Phenolics are considered in the top of the priority pollutant list given by Environmental Protection Agency (EPA, USA) (Yan et al., 2006). These compounds have high stability, high toxicity and are carcinogenic in nature, along with odor problem even at very low concentration. Phenol containing wastewater needs careful handling before they are discharged to the receiving water bodies.

The treatment of phenolic compounds from wastewater is widely practiced, although ongoing economical cost studies have shown, physical-chemical studies are costly and caused additional production of toxic chemical

sludge. Biodegradation of such recalcitrant compounds from wastewater emerged as a challenging job to the researchers worldwide for obtaining an innocuous practice. Thus the biological phenol removal would be a useable alternative because it will produce no toxic end products and also use low cost technology.

Aerobic degradation of phenol has been studied extensively. For example *Pseudomonas putida* has been widely used for biodegradation of phenols (Agarry et al., 2008a). Study on microbial degradation of phenol have shown that phenol can be aerobically degraded by wide variety of fungi and bacteria cultures such as *Candida tropicalis* (Chang et al., 1998; Ruiz-ordaz et al., 1998; Ruiz-ordaz et al., 2001), *Acinetobacter calcoaceticus* (Paller et al., 1995), *Alcaligenes eutrophus* (Hughes et al., 1984; Leonard and Lindley, 1998), *P. putida* (Hill and Robinson, 1975; Kotturi et al., 1991; Nikakhtari and Hill, 2006) and *Burkholderia cepacia* G4 (Folsom et al., 1990 ; Solomon et al., 1994).

Since the specific single bacteria seldom available in nature and also difficult to maintain in the field, it is urged that biodegradation study of phenol and phenolic

---
*Corresponding author. E-mail: sudiptadey_80@yahoo.com.

compounds would be carried out in presence of mixed population of phenol degrading bacteria. But a very few works have been done on phenol biodegradation by indigenous mixed culture bacteria, that is, not on any specific single bacteria. In 1997, Kumaran and Paruchuri have estimated some biokinetic constants of an activated sludge based phenolic wastewater treatment systems (60 - 500 mg/l of phenol) and compared those kinetic constants for pure culture in synthetic wastewater containing phenolic compounds. Later, biokinetic parameters estimation of phenolic compounds on activated sludge (thOD 0.78 -2.58 mg /mg) was carried by respirometric method by Orupold et al. (2001). They have analyzed substrate dependent oxygen uptake data by Michaelis-Menten growth kinetic model. A further investigation was also carried out with indigenous mixed microbial culture by Tziotzios et al. (2005) in packed bed and suspended growth reactors separately. They found that, olive pulp bacteria enriched culture used as inoculums show better efficiency in packed bed reactor than suspended growth reactor corresponding to phenol degradation rate, though packed reactor was found to be more shock resistant to higher phenol concentration. Agarry et al. (2008b) have studied inhibition kinetics of phenol biodegradation by binary mixed culture of *Pseudomonas aeruginosa* and *Pseudomonas fluorescence* from steady state and washout data, where they found the maximum reaction rate to be 0.322 mg/mg/h (upto 100 mg/l of phenol). Agarry et al. (2009) have also studied the substrate inhibition kinetics of phenol by pure *P. fluorescence* in continuous bioreactor. They have observed the maximum phenol degradation rate at 100 mg/l concentration to be 0.246 mg/mg/h under washout cultivation by this organism. From their experiments it is evident that phenol biodegradation by mixed culture is more efficient than pure culture. Saravaran et al. (2008) have studied substrate inhibition kinetics of phenol by indigenous mixed culture in a batch reactor and found that substrate degradation rate is maximum at 400 mg/l of phenol with a degradation rate of 15.7 mg/L/h. They have fitted the kinetics data in both Haldane and Han-Levenspiel model, where Han-Levenspiel model was found to be better fitted for their experiment. Bajaj et al. (2009) have studied phenol degradation by mixed culture for 2.5 - 7.0 mmol/L of phenol in batch system. They have used Haldane model to estimate the kinetic parameters and calculated the yield factor to be in the range of 0.10 - 0.16 showing that phenol-containing wastewater can be treated by mixed culture used as biocatalyst.

A variety of substrate utilization and inhibition models have been used to describe the dynamics of microbial growth on phenol. The substrate inhibition models along with their mathematical forms have been described below. The earliest model on microbial growth kinetics, the Monod model (1949), relates the growth rate of microorganism to the concentration of a single growth controlling substrate represented by the following equation:

$$\mu = \frac{\mu_{max} S}{Ks + S} \qquad (1)$$

Where μ is specific growth rate of mixed microbial culture (hr^{-1}) $= \frac{1}{X} \frac{dX}{dT}$, S is limiting substrate concentration (mg/l), μ_{max} is maximum specific growth rate of the culture (h^{-1}), K_S is half saturation constant (mg/l).

Different working groups (Kumar et al., 2005; Nuhoglu and Yalcin, 2005; Okpokwasili and Nweke, 2005) have proposed several mathematical models to express the culture growth and substrate utilization. Microbial growth can also be modeled by simple Monod equation (Kovar and Egli, 1998). However this equation became unpopular for growth in presence of some inhibitory substance. In such situation Haldane model are normally used to represent the growth in both lower and higher concentration of inhibitory substance. Haldane model has the form (Wang and Loh, 1999) as:

$$\mu = \frac{\mu_{max} S}{Ks + S + \dfrac{S^2}{Ki}} \qquad (2)$$

Where Ki is the substrate inhibition constant (mg/l).

Due to its significance, this model was widely adopted by most of the researchers. Aiba et al. (1968) proposed a model to express microbial growth rate as given by Equation (3):

$$\mu = \frac{\mu_{max} S \exp(-S/Ki)}{Ks + S} \qquad (3)$$

Yano and Koga (1969) proposed a model, based on a theoretical study on the dynamic behavior of single vessel continuous fermentation subject to the growth inhibition at high concentration of rate limiting substrate, e.g. the acetic acid fermentation from ethanol, the gluconic acid fermentation from glucose, etc. The model form is given in the following equation:

$$\mu = \frac{\mu_{max}}{Ks + S + \dfrac{S^2}{K_1} + \dfrac{S^3}{K_2{}^2}} \qquad (4)$$

Where K_1, K_2 are the positive constants. Similarly,

Edward (Webb, 1970) proposed the modified form of Haldane model as given by equation (5):

$$\mu = \mu_{max} \frac{S(1 + S/K)}{S + Ks + (S^2 / Ksi)} \quad (5)$$

Where Ksi is the substrate inhibition constant in mg/l.

But it was found that this model did not show any significant improvisation to Haldane model (Mulchandani and Luong, 1989). Edward (Teisser, 1970) proposed another model to predict substrate inhibition at higher substrate concentration as given by the equation (6):

$$\mu = \mu_{max}[\exp(-S/Ki) - \exp(-S/Ks)] \quad (6)$$

The model proposed by Luong (1987) as in Equation (7) appeared to be useful for representing the kinetics of substrate inhibition. Though this model is of generalized Monod type, but accounts for substrate stimulation at its both, low and high concentrations. The model has the capability to predict the values of S_m, the maximum threshold substrate concentration, above which the growth is completely inhibited (Luong 1987).

$$\mu = \frac{\mu_{max} S}{S + Ks}[\frac{1 - S}{Sm}]^n$$

$$(7)$$

Where n is an empirical constant.

Han and Levenspiel (1988) proposed a model to express substrate degradation rate (Equation 8). This model involves a delay function, which has an exponential form and incorporates the critical product or substrate concentration corresponding to the inflection point on the growth (Han and Levenspiel, 1988; Okpokwasili and Nweke, 2005).

$$q = \frac{q_{max} S [1 - S/Sm]^n}{S + Ks - [1 - S/Sm]^m} \quad (8)$$

Where q is the specific substrate degradation rate (h^{-1}), q_{max} the maximum specific substrate degradation rate (h^{-1}), Sm is the critical inhibitor concentration (mg/l) above which the reactions stops, and m and n are the empirical constants.

The corresponding form of this equation for microorganism is:

$$\mu = \frac{\mu_{max} S [1 - S/Sm]^n}{S + Ks - [1 - S/Sm]^m} \quad (9)$$

However, a mixed microbial community is needed for complete biodegradation of phenol and phenolic compounds present in the wastewater from industries. Real wastewater treatment processes have to deal with mixture of phenolic compounds and maintenance of mixed microbial culture is easier. In view of above, the present study was carried out to investigate the biodegradation study of phenol in wastewater by mixed microbial consortium. The objective of the investigation was to study the kinetic coefficients and the rate of biodegradation of phenol by an indigenous mixed microbial system at different initial phenol concentration. It was also aimed to fit different substrate inhibition models for the phenol biodegradation and to compare the goodness of fit for those models to test for the best-fit one representing the present work. Novelty in this work is the simultaneous estimation of biodegradation kinetic coefficients by substrate inhibition models along with effective yield and decay coefficients of a mixed microbial culture taken from a real life activated sludge comprising of several microorganisms of coke oven industry wastewater treatment plant.

MATERIALS AND METHOD

Microorganisms and culture acclimatization condition

The mixed microbial sludge was collected from an effluent treatment plant of a coke oven industry situated in Durgapur, West Bengal, India. The existing effluent treatment plant is operated on the principle of suspended growth biological reactor facilitated with extended aeration system. The sludge was collected from the recirculation line as well as from the top of the sludge drying bed. The indigenous mixed microbial culture was first acclimatized to phenol, so that, microbes can produce phenol degrading enzymes in the laboratory condition. Therefore, acclimatization of the mixed culture was carried out to make the microbial cell compatible to take up phenol as the sole carbon and energy source during its biodegradation. For acclimatization of sludge with phenol, first the culture was grown at very low concentration of phenol (5 mg/l) in 250 ml conical flask containing 100 ml of Mineral Salt (MS) media with 100 mg/l of glucose and 100 mg/l of beef extract, under continuous stirring (110 rpm). The composition of MS media is given as (mg/l): $(NH_4)_2SO_4$ 230, $CaCl_2$ 7.5, $FeCl_3$ 1.0, $MnSO_4.H_2O$ 100, $MgSO_4.7H_2O$ 100, K_2HPO_4 500, KH_2PO_4 250 (pH 7.0 ± 0.2). Then glucose and beef extract concentration were gradually decreased by 20 mg/l in every batch and supplemented by increased concentration of phenol. Batch process was adopted for acclimatization of sludge. After three month of acclimatization period, the sludge was changed to mineral salt medium (MS media) with phenol as sole carbon and energy source upto a concentration of 700 mg/l.

Analytical procedure

Phenol concentration was analytically estimated by High Performance Liquid Chromatography (HPLC) (Shimadzu) equipped with Ultraviolet-Visible (UV-VIS) detector and C18 column. The mobile phase used was acetonitrile and water mixture (60:40). The flow rate of the eluent was set to 1 ml/min and the detection wavelength was 275 nm. The biomass growth in the sample was monitored by measuring its absorbance at 600 nm wavelength using UV-VIS Spectrophotometer (Shimadzu). Then biomass concentration was calculated from a standard graph plotted as dry cell mass of microbial culture vs. optical density measured at 600 nm (Saravanan et al., 2008).

Experimental set up

All phenol biodegradation experiments using the mixed microbial culture were carried out in a 3 L capacity bioreactor with provisions for air supply for necessary aeration. In every batch the total volume of wastewater was taken as 1 L. A mini air compressor was used for the purpose of air supply. Compressed air was fed in the reactor at a rate of 5L/min by filtering the air by air-filter. Samples were withdrawn at predetermined time interval, after which, the biomass concentration and the residual phenol concentrations were analyzed. The phenol concentration was analyzed in the supernatant obtained after centrifugation of the sample for 5 min.

Batch biodegradation study

The biodegradation study was performed in the laboratory (Heritage Institute of Technology, Kolkata) to biodegrade inhibitory phenolic substances in wastewater coming out from phenol handling industries. All biodegradation experiments using the mixed culture were performed in batch bioreactor containing MS media with phenol as sole carbon and energy source at concentration range of 100 - 700 mg/l. 60ml inoculum was added to bioreactor for each set of experiment. This was accomplished by transferring directly (under aseptic conditions), freshly phenol-acclimatized culture to MS media with phenol at different concentrations (100 – 700 mg/l) at step up concentration interval of 100 mg/l. For each initial concentration of phenol, experiments were carried out in triplicate under identical condition and the average values are reported. All the experiments were done until the concentration of residual phenol in reactor was found to reach at equilibrium concentration at the specific time. For each batch of reaction, specific growth rate of the culture have been calculated and fitted in several substrate inhibition models: Monod model (1949), Haldane model (1968), Aiba model (1968), Yano and Koga model (1969), Edward model (1970), Luong model (1987) and Han-Levenspiel model (1988).

Study of decay kinetics of culture for phenol degradation

The typical growth curve in microbial assays usually shows a declining trend in the cell population after a complete consumption of substrate. But in suspended growth system where mixed microbial culture is used for biodegradation, the growth model can be represented in the following way:

$$\frac{dX}{d\theta} = Y_{x/s}\frac{dS}{d\theta} - k_d X$$

$$\frac{1}{X}\frac{dX}{d\theta} = \frac{Y_{x/s}}{X}\frac{dS}{d\theta} - k_d$$

$$\frac{1}{X}\frac{(X - Xo)}{\theta} = \frac{Y_{x/s}}{X}\frac{(So - S)}{\theta} - k_d \quad (10)$$

Where $\frac{1}{X}\frac{(X - Xo)}{\theta}$ is the specific cell growth rate and $\frac{(So - S)}{X\theta}$ is specific substrate utilization rate (U day^{-1}), $Y_{x/s}$ is microbial yield coefficient (mg/mg), k_d is the death coefficient of microbial mass in

day^{-1}.

The biodegradation data of each of the initial phenol concentration were plotted as $\frac{1}{X}\frac{(X - Xo)}{\theta}$ vs. $\frac{(So - S)}{X\theta}$ and slope of plot gave value of yield coefficient ($Y_{x/s}$) and negative intercept gave endogenous death coefficient (k_d) for 100 mg/l-700 mg/l of phenol concentration.

Modeling the kinetics of phenol biodegradation

At various initial phenol concentrations, the specific growth rates (μ) have been calculated and data were tested in different deterministic models as per Equations 1 - 9. The modeled equation were solved by nonlinear regression method using MATLAB$^{\copyright}$ 7.1 and directly applied on the experimental data of specific growth rates at different initial phenol concentration. However, specific substrate degradation rate terms, q and q_{max} in the original equation of the Han-Levenspiel model have been replaced by μ and μ_{max}, to represent specific growth rate and maximum specific growth rate of the culture, respectively.

RESULTS AND DISCUSSION

Effect of initial phenol concentration on its own biodegradation

Figure 1 shows the time course profile of the phenol biodegradation by the mixed culture. It can be seen that the mixed culture is able to degrade completely upto 700 mg/l phenol in almost 36 h. It is evident from Figure 1 that the time taken by the mixed culture to degrade phenol is depended on its initial concentration. It was found that biodegradation rate increases with the increase in phenol concentration upto 300 mg/l, but then starts decreasing. A maximum rate $\frac{dS}{d\theta}$ of 24.5 mg/L/h is obtained at initial phenol concentration of 300 mg/l. At concentration 400 mg/l of phenol, degradation rate has been evaluated as 20 mg/L/h. Rate is less than 20 mg/L/h for 100 mg/l and 700 mg/l of phenol concentration.

Effect of phenol concentration on the growth of the culture

Phenol concentration has been shown to have an inhibitory effect at higher concentration (Kumar et al, 2005; Gabriela et al. 2006; Stoilova et al., 2006). Microbial growth was observed at phenol concentration upto 700 mg/l. The growth profile of the culture at different initial phenol concentration is shown in Figure 2. It was observed that phenol concentration below 300 mg/l shows almost no inhibitory effect, as lag phase of growth was very short. For phenol concentrations higher than300 mg/l, lag phase took longer times, which is also observed in Figure 1 as increasing trend of time taken to degrade phenol of higher initial concentration. At concentrations above 300 mg/l of phenol, a distinct substrate inhibition was found. This is quite similar to the result obtained by

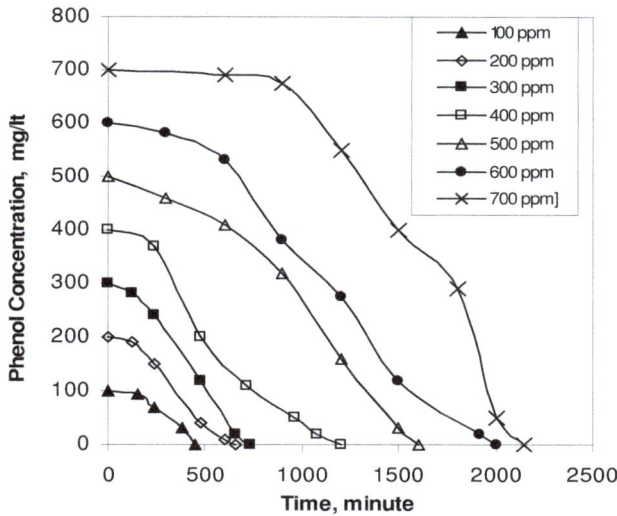

Figure 1. Phenol biodegradation profile with time.

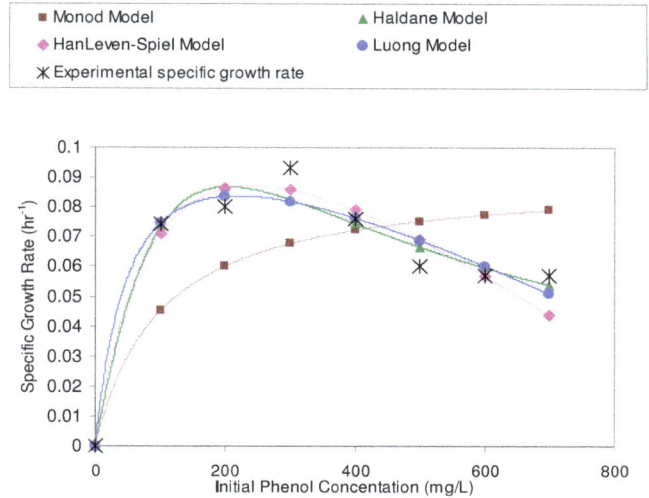

Figure 2. Microbial growth profile with time.

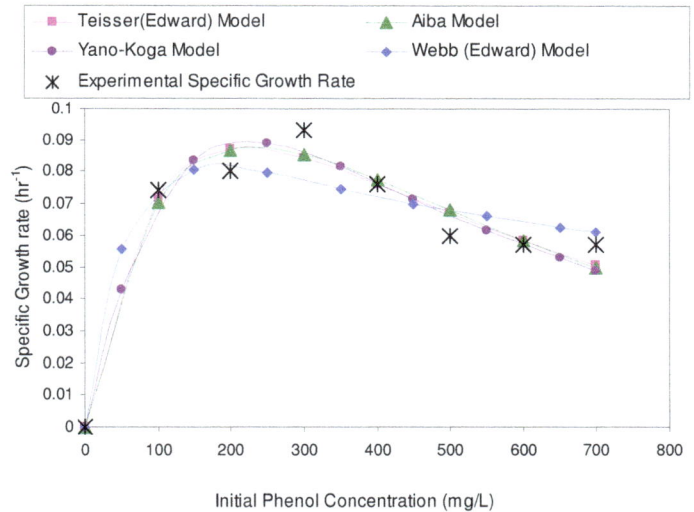

Figure 4. Predicted specific growth rate of the culture due to Monod, Haldane, Han-Levenspiel and Luong model.

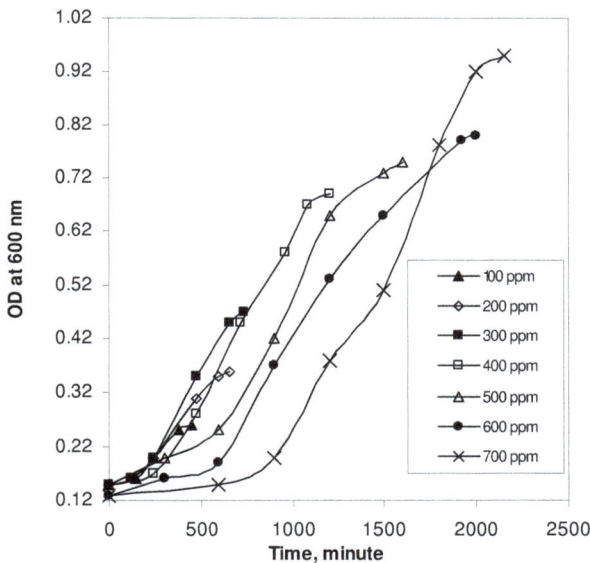

Figure 5. Experimental and predicted specific growth rate of the culture due to Teisser (Edward), Webb (Edward), Yano-Koga and Aiba model.

Bajaj et al. (2009) where they have found the mixed microbial culture is showing substrate inhibition after 4 mM (376.44 mg/l) of phenol. Saravanan et al. (2008) have also reported to have substrate inhibition above 300 mg/l of phenol as seen by the decrease in the specific growth rate. Figure 3 shows that, in the present study, until 300 mg/l of phenol concentration, the specific growth rate of the culture increased (highest μ =0.093 h^{-1}). For concentration higher than 300 mg/l, specific growth rate decreases and became almost constant at 600 mg/l (μ = 0.057 h^{-1}) and 700 mg/l (μ = 0.057 h^{-1}) of phenol.

Exploration of best-fit kinetic model for phenol biodegradation

Family of plots in Figure 1 reveals the pattern of phenol removal throughout the respective biodegradation period is similar to each other irrespective of initial phenol concentration. Figure 4 and 5 showed the comparative plots of experimental specific growth rates and the model predicted ones as given by equation 1 - 9 and solved by MATLAB© 7.1. Among several models used to fit the present experimental data for specific growth rates versus different initial phenol concentration, not only Haldane

Table 1. Summary of growth kinetics parameter values obtained from different models during biodegradation of phenol by mixed microbial culture used in present work.

Model	μ_{max} (hr^{-1}).	Ks (mg/l)	Ki	Ksi	K	S_m	n	m	RMSE
Monod	0.09	100.6	-	-	-	-	-	-	0.0208
Haldane	0.3057	257.5	162.6	-	-	-	-	-	0.0067
Hanlevenspiel	0.2901	252.1	-	-	-	720	1	1	0.0218
Luong	0.1291	59.39	-	-	-	1148	0.9	-	0.0072
Edward	0.1675	95.05	-	200	1000	-	-	-	0.0090
Aiba	0.2579	200.3	502	-	-	-	-	-	0.0078
Teisser	0.1386	95.04	699.5	-	-	-	-	-	0.0075
Yano-Koga	0.2981	286.2	-	-	$K_1 = 261.7$ $K_2 = 499.8$	-	-	-	0.0079

S = 0.08399522
r = 0.99230903

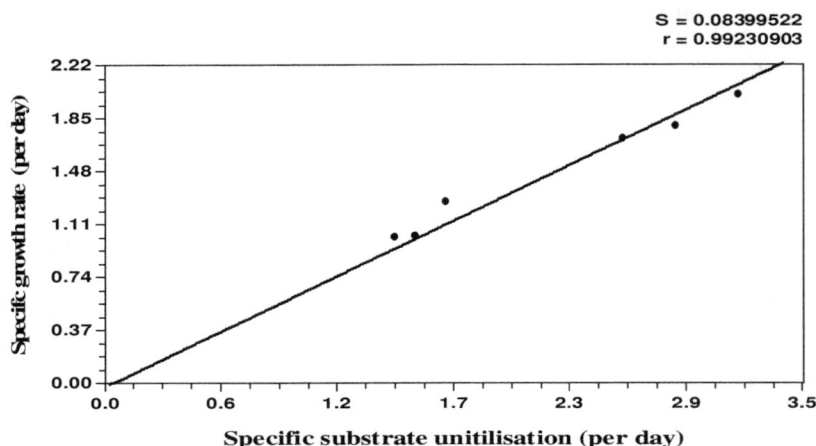

Figure 6. Plot for determining yield and death coefficient of culture for phenol biodegradation.

model but also Luong model showed fit well reasonably as determined by the root mean square error (RMSE) calculated between experimental and the model predicted specific growth rate values. This could be attributed based on the models themselves, which are considered more refined from the standpoint of development of these models.

The biokinetic constants of growth of the culture obtained from these models along with root mean square error between experimental and predicted rate values are shown in Table 1. The Table also reports the value of K_s and μ_{max} as per Monod model by nonlinear regression method. Haldane and Han-Levenspiel model predicts marginal differences in both K_s and μ_{max} values, but differs a lot in RMSE calculated between experimental and model predicted specific growth rates. This may be due to the fact that Haldane model takes care of the value of inhibition constant Ki, that is an important parameter in understanding the kinetics of the microorganism in the system. Luong and Han- Levenspiel models also predicted the critical substrate concentration (S_m) value, at which specific growth rate fall to zero (1148

and 720 mg/l, respectively). But Han- Levenspiel predicted S_m value (720 mg/l), agrees well with the experimental result. Experimental data shows that the specific growth rate at 700 mg/l of initial phenol concentration is 0.057 h^{-1} and after that, it becomes very low. Though RMSE value for Han- Levenspiel model is high enough than Luong model, but that may be because of the good fitting of Luong model at lower initial phenol concentration. The difference in the models predicted kinetic constant values for the present experiment perhaps due to the fact that the two models were originally developed for systems containing a different microorganism and substrate. Table 2 shows the comparison of biokinetic constants as obtained by different investigators as well as in the present study.

Evaluation of microbial yield and death coefficient during phenol degradation

A plot of specific growth rate vs. specific substrate consumption rate (Figure 6) according to Equation (10)

Table 2. Summary of growth kinetics obtained in various studies for the treatment of phenolic wastes.

S/No.	Authors	Bacterial strain	System	Concentration range (mg/l)	Monod model		Haldane model		
					μ_{max} (hr^{-1})	Km (mg/l)	μ_{max} (hr^{-1})	Ks (mg/l)	Ki (mg/l)
1	Powlowsky and Howell	Mixed culture I	Batch	0-900	-	-	0.260	25.4	173.0
		Mixed culture II (Filamentous organism)	Batch	0-1000	-	-	0.223	5.86	934.5
2	Livingstone and Chase	NCIB8250 *Actino bacter* sp +NCIB10535 (*Pseudomonus* sp.) +NCIB1015(*Pseudomonus* sp)	Batch	0-500	-	-	0.418	2.9	370
3	Huchinson and Robinson	*P. putida* F1 ATCC 17484	Batch	<200	-	-	0.388	1.06	903
4	Hill and Robinson	*P. putida* (ATCC 17484, Stainer 110) *A. calcoacetius* (phenol only) *P. fluoroscens* 2218 (phenol only)	Batch/ Continuous	0-700	-	-	0.534	0.015	470
5	Kumaran and Paruchuri	Pooled culture (Phenols) *P. fluoroscens*, *P. putida, P. cepacia, A. calcoacetius, C. tropicalis*	Batch	60-500	0.465	3 0.96	0.542	36.2	145.0
6	Okaygun et al.	*Pseudomonus* sp. and *Klebsiella* sp.	Batch	100-170	-	-	0.325	8.2	170.0
7	Arutchelvan, V	*Bacillus brevis*	Batch	750 - 1750	-	-	0.026 - 0.078	2.2 - 29.3	868 - 2434.7
8	Buitron G	Mixed culture	Batch	40	-	-	0.258	3.9	121.7
9	Marrot B	Mixed culture	Continuous	2500	-	-	0.438	29.5	72.4
10	Saravanan et al.	Mixed culture	Batch	100 - 800	0.37	144.68	0.3085	44.92	525.00
11	Bajaj et al.	Mixed culture	Batch	23.5 - 659	-		0.3095	74.65	648.13
12	Present study	Mixed culture	Batch	100 - 700	0.0995	35.76	0.150	51.8	404.04

was used to evaluate the yield and death coefficient for the phenol biodegrading micro-organisms. The yield coefficient in the present study (0.66 mg/mg) is found to be close to those reported in literature as $Y_{x/s}$ = 0.6 ± 0.12 mg/mg (Livingston and Chase, 1989). Bajaj et al. (2009) have reported the yield factor as 0.1 - 0.17 (A580 units or OD at 580 nm/mmol of phenol) for mixed microbial culture. The calculated decay coefficient for the present study was 0.00103 h^{-1} similar to values published by Kumaran and Paruchuri (1997), who obtained decay coefficient to be 0.005 hr^{-1} for phenol degradation by mixed culture. Kumar et al. (2005) have calculated a decay coefficient value of 0.0056 h^{-1} of *P. putida* MTCC 1194 for phenol degradation in batch culture. Arutchelvan et al. (2006) have reported decay coefficient to be 0.003 - 0.12 h^{-1} and yield coefficient of 0.293 - 0.571 for phenol removal by *Bacilus bravis*. This yield coefficient is slightly less and decay coefficient is slightly more than the value obtained in the present experiment. The kinetic data as estimated by different scientists deriving from either pure or mixed culture including the present study is listed in Table 2. The wide variation of kinetic coefficients values perhaps is due to the fact that most of the investigators have carried out the research either under pure culture system in different initial test condition. The present study deals with mixed microbial culture containing several species, whose death rate is not same, for which the overall death coefficient marginally deviated from the available literature data.

Conclusion

Kinetics of phenol degradation was studied under aerobic condition in batch reactor using an indigenous mixed microbial culture, isolated from an effluent treatment section of a coke oven plant. The culture could grow and biodegrade phenol upto 700 mg/l. However, phenol exhibited inhibition to both specific growth rate and substrate degradation rate above 300 mg/l of initial phenol concentration. Specific growth rates of the culture under different initial phenol concentration from 100mg/l to 700 mg/l have been calculated. By fitting specific growth rates on suitable substrate inhibition models, biokinetics constant that are necessary to understand the kinetics of biodegradation process were evaluated by MATLAB 7.1© software. Root Mean Square Error values between the experimental specific growth rates and the model predicted ones have been calculated for different substrate inhibition models. It is observed that the best model that fit the present study is Haldane model having lowest RMSE value of 0.0067 and predicting reasonable kinetic coefficient values. Therefore, the mixed culture used in the present work is a potential culture that can be used for phenol biodegradation under aerobic condition in real life wastewater treatment.

REFERENCES

Aiba S, Shoda M, Nagatami M (1968). Kinetics of product inhibition in alcohol fermentation, Biotechnol. Bioeng., 10: 845-864.

Agarry SE, Durojaiya AO, Solomon BO (2008a). Microbial Degradation of phenol: A Review. Int. J. Environ. Pollut., 32(1): 12-28.

Agarry SE, Solomon BO, Layokun SK (2008b). Substrate inhibition kinetics of phenol degradation by binary mixed culture of Pseudomonas aeruginosa and Pseudomonas fluorescence from steady state and wash- out data, Afr. J. Biotechnol., 7(21): 3927-3933.

Agarry SE, Audu TOK, Solomon BO (2009). Substrate inhibition kinetics of phenol degradation by Pseudomonas fluorescence from steady state and washout data, Int. J. Environ. Sci. Tech., 6(3): 443-450.

Arutchelvan V, Kanakasabai V, Elangovan R, Nagarajan S, Muralikrishnan V (2006). Kinetics of high strength phenol degradation using Bacillus brevis, J. Haz. Mat. B129: 216-222.

Bajaj M, Gallert C, Winter J (2009). Phenol degradation kinetics of an aerobic mixed culture, Biochem Eng J., 46: 205-209.

Buitron G, Gonzalez A, Lopez-Marin LM (1998). Biodegradation of phenolic compounds by an acclimated activated sludge and isolated bacteria, Water Sci.Technol., 37:371–378.

Chang YH, Li CT, Chang MC, Shieh WK (1998). Batch phenol degradation by Candida tropicalis and its fusant. Biotechnol. Bioeng., 60: 391-395.

Edward VH (1970). The influence of high substrate concentrations on microbial kinetics. Biotechnol. Bioeng., 12: 679-712.

Folsom BR, Chapman PJ, Pritchard PH (1990). Phenol and trichloroethylene degradation by Pseudomonas cepacia G4: Kinetics and interactions between substrates. Appl. Environ. Microbiol., 57: 1279-1285.

Gabriela Vazquez-Rodriguez B, Youssef Cherif, Waissman Julio (2006). Two-step modeling of the biodegradation of phenol by an acclimated activated sludge. Chem. Eng. J., 117: 245- 252.

Han K, Levenspiel O (1988). Extended Monod kinetics for substrate, product, and cell inhibition, Biotechnol. Bioeng., 32:430–437.

Hill GA, Robinson CW (1975). Substrate inhibition kinetics: Phenol degradation by Pseudomonas putida, Biotechnol. Bioeng., 17: 599-615.

Hutchison DH, Robinson CW (1990). A microbial regeneration process for granular activated carbon II regeneration studies. Water Res., 24: 1217–1223.

Hughes EJ, Bayly RC, Skurray RA (1984). Evidence for isofunctional enzymes in the degradation of phenol, m – and p – toluene, and p – cresol via catechol meta cleavage pathways in Alcaligenes eutrophus. J. Bacteriol., 158: 79-83.

Juang RS, Tsai SY (2006). Growth kinetics of Pseudomonas putida in the biodegradation of single and mixed phenol and sodium salicylate. Biochem. Eng. J., 31: 133-140.

Kotturi G, Robinson CW, Inniss WE (1991). Phenol degradation by a psychrotrophic strain of Pseudomonas putida. Appl. Environ. Microbiol., 34: 539-543.

Kumar A, Kumar S, Kumar S (2005). Biodegradation kinetics of phenol and catechol using Pseudomonas putida MTCC 1194. Biochem. Eng. J., 22, 151–159.

Kovar KK, Egli T (1998). Growth kinetics of suspended microbial cells: from single substrate- controlled growth to mixed-substrate kinetics, Microb. Mol. Biol. Rev., 62: 646–666.

Kumaran P, Paruchuri K (1997). Kinetics of Phenol Transformation. Water Res., 31(1): 11-22.

Leonard D, Lindley ND (1998). Carbon and energy flux constraints in continuous cultures of Alcaligenes eutrophus grown on phenol. Microb., 144(1): 241-248.

Livingston AG, Chase HA (1989). Modeling phenol degradation in a fluidized bed bioreactor, AIChE J., 35: 1980-1992.

Luong JHT (1987). Generalization of Monod kinetics for analysis of growth data with substrate inhibition. Biotechnol. Bioeng., 29: 242-248.

Marrot B, Barrios-Martinez A, Moulin P, Roche N (2006). Biodegradation of high phenol concentration by activated sludge in an immersed membrane bioreactor, Biochem. Eng. J., 30 174-183.

Mulchandani A, Luong JHT (1989). Microbial inhibition kinetics revisited, Enzyme Microb. Technol., 11: 66-73.

Monod J (1949). The growth of bacterial cultures. Annu. Rev. Microb. 3: 371-394.

Nikakhtari H, Hill GA (2006). Continuous bioremediation of phenol polluted air in an external loop airlift bioreactor with a packed bed. J. Chem. Technol. Biotechnol. 81(6): 1029-1038.

Nuhoglu N, Yalcin B (2005). Modeling of phenol removal in a batch reactor, Process Biochem., 40: 1233-1239.

Okaygun MS, Green LA, Akgerman A (1992). Effects of consecutive pulsing of an inhibitory substrate on biodegradation kinetics. Environ. Sci. Technol. 26: 1746-1752

Orupold K, Masirin A, Tenno T (2001). Estimation of Biodegradation Parameters of Phenolic Compounds on Activated Sludge on Respirometry. Chemosphere, 44:1273-1280.

Okpokwasili GC, Nweke CO (2005). Microbial growth and substrate utilization kinetics, Afr. J. Biotechnol., 5 (4): 305-317.

Paller G, Hommel RK, Kleber HP (1995). Phenol degradation by Acinetobacter calcoaceticus, NCIB 8250. J. Basic Microbiol, 35: 325-335.

Pawlowsky U, Howell JA (1973). Mixed culture biooxidation of phenol: Determination of kinetic parameters. Biotechnol. Bioeng., 15: 889-896.

Ruiz-ordaz N, Ruiz-Lagunez JC, Castanou-Gonzalez JH Hernandez-Manzano E, Cristiani-Urbina E, Galindez-Mayer J (1998). Growth kinetic model that describes the inhibitory and lytic effects of phenol on *Candida tropicalis* yeast. Biotechnol. Prog., 14: 966-969.

Ruiz-ordaz N, Ruiz-Lagunez JC, Castanou-Gonzalez JH, Hernandez-Manzano E, Cristiani-Urbina E, Galindez-Mayer J (2001). Phenol biodegradation using a repeated batch culture of *Candida tropicalis* in a multistage bubble column. Revista Lat. Microbiogy, 43: 19-25.

Saravanan P, Pakshirajan K, Saha P (2008). Growth kinetics of an indigenous mixed microbial consortium during phenol degradation in a batch reactor, Bioresour. Technol., 99: 205-209.

Solomon BO, Posten C, Harder MPF, Hecht V, Deckwer WD (1994). Energetics of *Pseudomonas cepacia* growth in a chemostat with phenol limitation, J. Chem. Technol. Biotechnol. 60: 275-282.

Stoilova A, Krastanov V, Stanchev D, Daniel M, Gerginova Z, Alexieva (2006). Biodegradation of high amounts of phenol, catechol, 2,4-dichlorophenol and 2,6-dimethoxyphenol by *Aspergillus awamori* cells. Enzyme Microb. Technol., 39(5): 1036-1041.

Tziotzios G, Teliou M, Kaltsouni V, Lyberatos G, Vayenas DV (2005). Biological phenol removal using suspended growth and packed bed reactors, Biochem. Eng. J., 26: 65-71.

Wang SJ, Loh KC (1999). Modeling the role of metabolic intermediates in kinetics of phenol biodegradation. Enzyme Microb. Technol. 25: 177-184.

Yan J, Jianping W, Bai J, Daoquan W, Zongding H (2006). Phenol biodegradation by the yeast Candida tropicalis in the presence of m-cresol. Biochem. Eng. J., 29: 223–227.

Yano S. Koga (1969). Dynamic behavior of the chemostat subject to substrate inhibition, Biotechnol. Bioeng., 11:139–153.

List of symbols:

S	mg/l	Substrate concentration
So	mg/l	Initial substrate concentration
X	mg/l	Biomass concentration
Xo	mgl	Initial biomass concentration
μ	hr^{-1}	Specific growth rate
μ_{max}	hr^{-1}	Maximum specific growth rate
K_1, K_2	mg/l	Positive constants
Ks	mg/l	Half saturation constant
Ki	mg/l	Substrate inhibition constant
n, m	-	Empirical constant
Yx/s	mg/mg	Microbial yield coefficient
K_d	hr^{-1}	Microbial decay coefficient
Θ	Day	Batch time

Comparison of spectrophotometric methods using cuvette tests and national standard methods for analysis of wastewater samples

Sonya Dimitrova, Nadejda Taneva, Kapka Bojilova, Vesela Zaharieva, Svetlana Lazarova, Mariana Koleva, Rumen Arsov and Tony Venelinov

University of Architecture, Civil Engineering and Geodesy, 1 Christo Smirnenski blvd., 1046 Sofia, Bulgaria.

The suitability of spectrophotometric methods using cuvette tests (CT) for determination of ammonia (NH_4^+-N), nitrates (NO_3^--N), total phosphates (PO_4^{3-}-P) and chemical oxygen demand (COD) in real wastewater were evaluated by comparison with corresponding Bulgarian national standard methods (BS). The CT methods are based on measuring of ready-to-use cuvettes (bar-coded reagent vials) in wavelength range of 340 to 900 nm. Nine wastewater samples from the inlet of the wastewater treatment plant of Sofia city were collected in the period of three months and the above mentioned quality indicators were measured in parallel both with CT and BS methods. Mean values of ten replicate determinations for each of the samples were compared statistically using Dixon's t-test and linear regression model. Excellent linear correlation ($R^2 > 0.99$) was found. As another mean of comparison, all the methods (CT and BS) for the determination of all the quality indicators were validated and uncertainties were estimated. Based on this, all data were compared and proved statistically equivalent.

Key words: Method comparison, method validation, uncertainty estimation, cuvette tests, linear regression, spectrophotometer, wastewater.

INTRODUCTION

The management of the urban water systems is relied generally on chemical analysis of a set of wastewater quality indicators. Chemical oxygen demand is an integrated parameter, which gives information for the level of organic contamination of wastewater. COD tests are traditionally used for assessing the effectiveness of the biological wastewater treatment and the organic load of the treated wastewater. Nitrogen and phosphorus are also among the important indicators which are controlled at the wastewater treatment plants.

There are many analytical methods for measuring the above mentioned indicators. COD, NH_4^+-N, NO_3^--N and PO_4^{3-}-P at laboratory scale set-up of samples collected manually and/or by automatic samplers, are usually carried out according to the national and international

standard methods. These methods are well studied and documented, but they are too sophisticated and time consuming for operational control and require considerable practical experience and skill to get reproducible results. Furthermore, most of them require the use of toxic substances, which can be harmful to the analyst and their subsequent utilization can be dangerous to environment. Therefore, it is an emerging need for new rapid, low cost and nono-toxic substance consuming, reliable and precise analytical techniques and instruments applicable for real-time control.

Over last decades many new automated analytical methods and instruments have been developed and evaluated in respect of their applicability for wastewater control - UV/Visible spectroscopy (Ferree and Shannon,

2001; Langergrabe et al., 2003), ion chromatography (Karmarkar, 1999) and others.

Spectrophotometric methods for water analysis, based on ready-to-use cuvette tests, can be an alternative to the time-consuming reference methods. Practically all compounds and indicators of water environment could be measured directly or after suitable preliminary treatment using contemporary spectrophotometers and portable photometers. The cuvette tests quality is demonstrated by the fact that for the first time a COD cuvette test has been accepted as a reference method (ISO/IEC 15705, 2002).

The aim of this study was to evaluate the comparability of the NH_4^+-N, NO_3^--N, PO_4^{3-}-P and COD data obtained by cuvette tests and the corresponding Bulgarian standard methods (BDS 17.1.4.02-77, 1977; BDS 17.1.4.10-79, 1979; BDS ISO 7890-3:1997, 1998; BDS EN ISO 6878:2004, 2005) for analysis of real wastewater samples.

MATERIALS AND METHODS

The study was carried out on nine wastewater samples collected at the inlet of the wastewater treatment plant of Sofia city, which serves about 1 200 000 habitants. Samples were collected weekly during a three-month period. Each sample was analyzed on the day upon receipt. Ten replicate determinations were performed on each of the 9 samples, both by CT and BS. The cuvette tests were performed according to the user's manual issued by instrument manufacturer. The spectrophotometric measurements of the PO_4^{3-}-P and NO_3^--N by the Bulgarian standard methods were realized in quartz cuvettes on the same spectrophotometer after appropriate calibration at 680 and 410 nm, respectively.

Equipment

DR 2800 - portable spectrophotometer (Hach Lange GmbH) with 340 to 900 nm wave length range (tungsten halogen lamp) and referent ray to compensate lamp wear and power fluctuation was used. The devise has integrated system for barcodes reading of the prepared tests, with ten measurements for rotation and elimination of wrong reading caused by prepared cuvettes wasting.

LT 200 - thermo-reactor (Hach Lange GmbH). TenSette plus - Electronic pipette 0.2 to 5 mL (Hach-Lange GmbH).

Determination methods

Chemical oxygen demand (COD)

The method for the determination of COD in wastewater samples according to the Bulgarian National Standard (BSS 17.1.4.02-77) is based on the titrimetry, whereby to 20 mL of the samples Ag_2SO_4, $K_2Cr_2O_7$ and H_2SO_4 are added. After boiling under reflux for 2 h, the samples are cooled down and indicator is added, before titration with $FeNH_4(SO_4)_2$.

The method for the determination of COD in wastewater samples using cuvette tests is based on addition of 2 mL of sample to the cuvette, which is heated in thermo-reactor for 2 h at 148 ± 50°C. After cooling, the cuvette is inserted into the spectrophotometer and measured. Depending on the concentration, cuvette test for COD LCK 314 (15 to 150 mgO_2/L) and LCK 114 (150 to 1000 mgO_2/L)

are used.

Stock solution of 1000 ± 1 mg/L COD (Hach-Lange GmbH) and CRM (RTC COD 500-500) with certified value for COD of 500.00 ± 7.65 mgO_2/L -(LOT No. 016203) were used for the validation studies.

Nitrates

The method for the determination of nitrates in wastewater samples according to the Bulgarian National Standard Bulgarian National Standard (BDS ISO 7890-3:1998) is based on spectrophotometric determination at 410 nm of the color intensity of the formed substance between the nitrates and the sulfosalicylic acid in presence of alkali base and Na_2EDTA and NaN_3.

The method for the determination of Nitrates in wastewater samples using cuvette tests is based on the reaction between nitrate ions with 2,6-dimethylphenol in presence of sulfuric acid and phosphorus acid. Cuvette tests for nitrates used were LCK 339 (0.23 to 13.50 mg/L). Stock solution of 10 ± 0.1 mg/L NO_3-N (Hach-Lange GmbH) and CRM (CertiPrep) with certified value for NO_3-N of 1005 ± 3 mg/L (LOT No. 2-78NO3N-2) were used for the validation studies.

Total phosphates

The method for the determination of Phosphates in wastewater samples according to the Bulgarian National Standard (BDS EN ISO 6878:2004) is based on spectrophotometric determination at 880 nm of the color intensity of the formed substance between orthophosphates, ammonium molibdate and antimony in presence of ascorbic acid and sulfuric acid.

The method for the determination of total phosphates in wastewater samples using cuvette tests is based on the reaction between phosphate ions and molybdate ions and subsequent reduction by ascorbic acid. Cuvette tests for total phosphates used were LCK 348 (0.5 to 5.0 mg/L). Stock solution of 1000 ± 1 mg/L P tot (Hach-Lange GmbH) and CRM (RTC TPO 1000 to 500 ML) with certified value for P total of 1000.0 ± 15.5 mg/L (LOT No. 017605) were used for the validation studies.

Ammonia

The method for the determination of ammonia in wastewater samples according to Bulgarian National Standard (BDS 17.1.4.10-79) is based on distillation in presence of phosphate buffer and boric acid and titration using sulfuric acid and methylrod and methileneblou as mixed indicator.

The method for the determination of ammonia in wastewater samples using cuvette tests is based on the reaction between the ammonium ions and hypochlorite ions and salicylic ions in presence of sodium nitroprucide. Cuvette tests for ammonia used were LCK 303 (2 to 47 mg/L]). Stock solution of 1000 ± 1 mg/L NH_4-N (Hach-Lange GmbH) and CRM (SPEX Ammonium Standard) with certified value for NH_4-N of 1002 ± 3 mg/L (LOT No. 2-95NH4N-2) were used for the validation studies. Distilled water was used throughout.

Statistical data treatment

The least squares method and Dixon's t-test were used to establish the relationships between measurements and data obtained by the two methods, as well as for evaluation of correlation and significance of any founded discrepancies.

Table 1. Mean concentrations for NH_4^+-N, NO_3^--N, PO_4^{3-}-P and COD with respective RSDs (n = 10).

Sample	NH_4^+-N (mg/L)		NO_3^--N (mg/L)		PO_4^{3-}-P (mg/L)		COD (mgO_2/L)	
	BS	CT	BS	CT	BS	CT	BS	CT
1	6.75±0.33	7.17±0.07	0.49±0.05	0.54±0.03	2.04±0.18	1.97±0.06	368±15	372±8
2	6.78±0.25	7.19±0.08	0.81±0.02	0.90±0.02	2.13±0.01	1.95±0.08	104±18	101±8
3	3.83±0.14	3.98±0.12	2.49±0.06	2.55±0.03	1.07±0.07	0.93±0.03	73±7	69±14
4	5.24±0.27	6.07±0.28	0.79±0.02	0.88±0.02	2.00±0.14	1.95±0.05	40±5	46±3
5	13.74±0.34	15.39±0.24	0.18±0.04	0.25±0.04	5.04±0.21	4.92±0.10	62±10	68±2
6	13.89±0.32	14.40±0.76	0.16±0.01	0.11±0.07	6.97±0.19	6.81±0.18	277±8	273±7
7	13.62±0.16	14.21±0.21	0.22±0.01	0.22±0.04	4.95±0.27	4.39±0.07	266±7	284±7
8	14.78±0.34	15.48±0.27	0.25±0.02	0.30±0.03	5.18±0.33	4.86±0.16	303±9	323±10
9	14.07±0.30	15.03±0.55	0.28±0.04	0.31±0.02	5.10±0.22	4.91±0.14	276±13	274±4

RESULTS AND DISCUSSION

Table 1 shows mean values and their standard deviations (n=10) for ammonia (NH_4^+- N), nitrates (NO_3^- -N), total phosphates(PO_4^{3-}-P) and chemical oxygen demand (COD) for cuvette test (CT) and standard methods (BS) for replicate measurements of the 9 sampling weeks without any outliers removal.

As can be seen, generally the RSDs of BS and CT methods overlap, but 60% of COD, 80% of NO_3^- -N and all NH_4^+-N mean values measured by CT methods are higher than those measured by standard methods. This is not the case with phosphorus measurements.

The method performance was inspected for any potential systematic errors as data obtained for NH_4^+-N, NO_3^--N, PO_4^{3-}-P and COD by cuvette tests were plotted against the corresponding mean values of the standard methods (Figures 1 to 4). The equations of the best-fit lines trough the data were determined using the method of least squares. It was found that the relationships between the measurements by both methods are approximated very well by linear regression equation:

Y (CT) = a + b x (BS)

The coefficients of determination (R^2) are close to 1 (>0.99): for NH_4^+-N - 0.9936; for NO_3^--N - 0.9970; for PO_4^{3-}-P - 0.9948 and for COD – 0.9923. The obtained regression lines for each one of measured indicators were statistically evaluated by their comparison with a hypothetic ideal line the slope of which is unit and the intercept is zero. The zero-hypothesis H_0: $a = 0$; $b = 1$ was revised according to the t- criteria:

$t_{statistic} < t_{critcal}$ (P = 95 %);
$t_{statistic}$ (b) = (b-1)/S_b;
$t_{statistic}$ (a) = (a-0)/S_a,

where P is significance level; S_a and S_b –standard deviations of the estimators a and b, respectively.
Regression lines parameters (slope, y intercept and the

standard error of estimate in y direction) provide specific estimatation of errors, but only in case of approved linear relationships (Simeonov, 1997; Westgard and Hunt, 1973). A study of the regression straight line gives a possibility to evaluate at least three kinds of errors: random, proportional and constant. If the zero-hypothesis is not realized for the line slope it is an indication for a proportional systematic error occurrence. If the zero-hypothesis is not realized for the line intercept, this suggests a constant systematic error.

The results from statistical assessment of the regression lines are given in Table 2. Two methods are admitted as statistically equivalent if zero-hypothesis is realized, therefore it can be concluded that there are no systematic errors for the data obtained by the both methods.

As a second means of comparison, Dixon's t-test was performed on both the CT and BS data (as dependent data-sets), and compared with the tabular value. Here, the zero-hypothesis H_0: $\overline{x}_d$ = 0 is checked against H_{alt}:

$\overline{x}_d \neq 0$ using Equation 1:

$$t_{exp} = \frac{\overline{d}}{\frac{s_d}{\sqrt{n}}} \tag{1}$$

where $d_i = x_{1i} - x_{2i}$, s_d – standard deviation of the mean, n - number of samples.

The zero-hypothesis is realized if the experimentally obtained value is lower than the tabular value for the same degrees of freedom. Calculations show experimental values of -4.79 for NH_4^+–N; -3.04 for NO_3^--N; 3.82 for PO_4^{3-}-P and -1.50 for COD with tabular value of 2.31 (df=8). The calculations show that only for COD the H_0 hypothesis proved true.

According to the EU Application note 1 (European

NH₄-N comparison

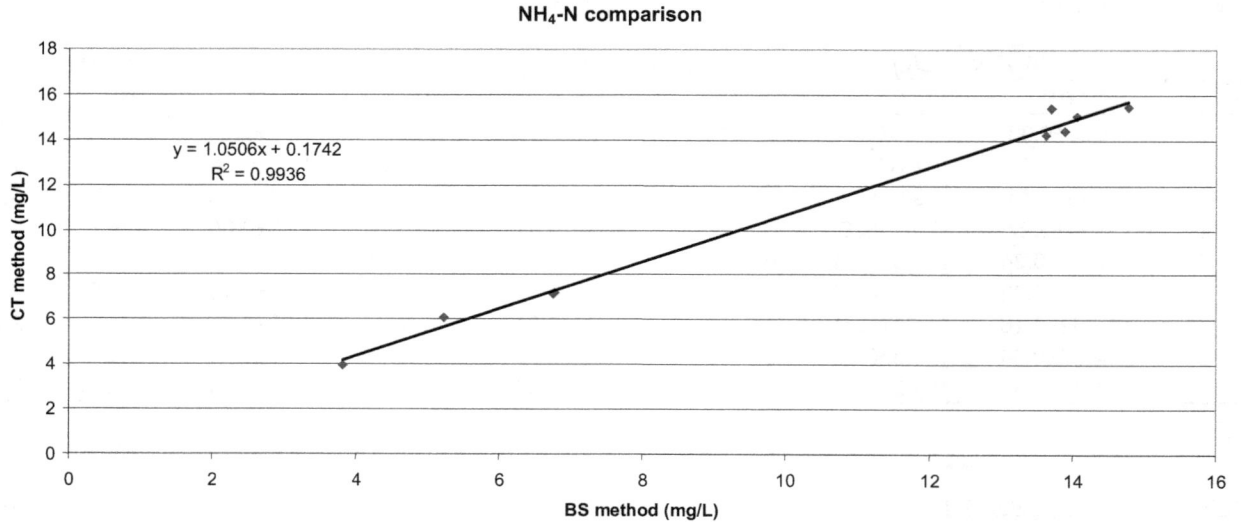

Figure 1. Comparison of mean values for NH₄⁺-N using a cuvette test method and the Bulgarian standard method.

NO₃-N comparson

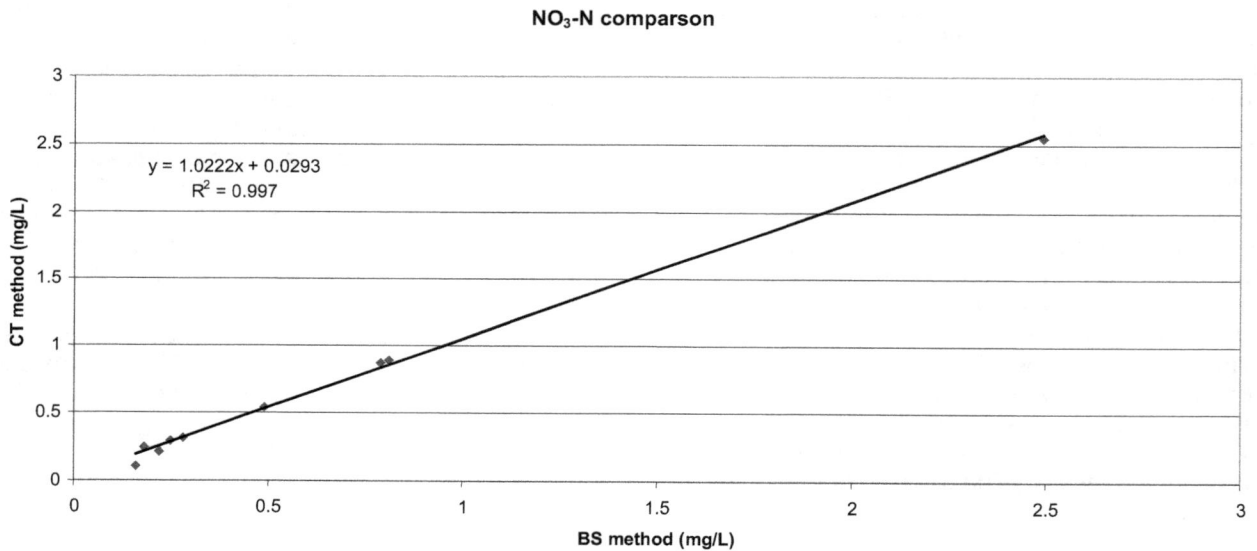

Figure 2. Comparison of mean values for NO₃⁻-N using a cuvette test method and the Bulgarian standard method.

Commission, 2005), two mean values are best compared using their uncertainties. Uncertainty is best estimated during method validation. The experimental design was set up in a way so that the repeatability, reproducibility and trueness estimates are used for measurement uncertainty estimation (ISO/IEC 21748:2010, 2010).

Trueness was proven by measurement of three independent samples of a certified reference material on two different days. From these data the uncertainty of trueness and method bias were calculated. Repeatability and intermediate precision were determined by replicate analysis and assessment of between-day effects. This was achieved by preparation of three independent

samples of a certified reference material on three extra days. Combination from these data and the data obtained for trueness were used for calculation of the uncertainties of repeatability and due to intermediate precision.

Measurement uncertainty components of repeatability and due to intermediate precision can easily be calculated using the ANOVA function in the Microsoft Excel (Equations 2 and 3):

$$u_f = \frac{s}{\sqrt{n}} = \frac{\sqrt{MS_{within}}}{\bar{x}} \tag{2}$$

PO$_4$-P comparison

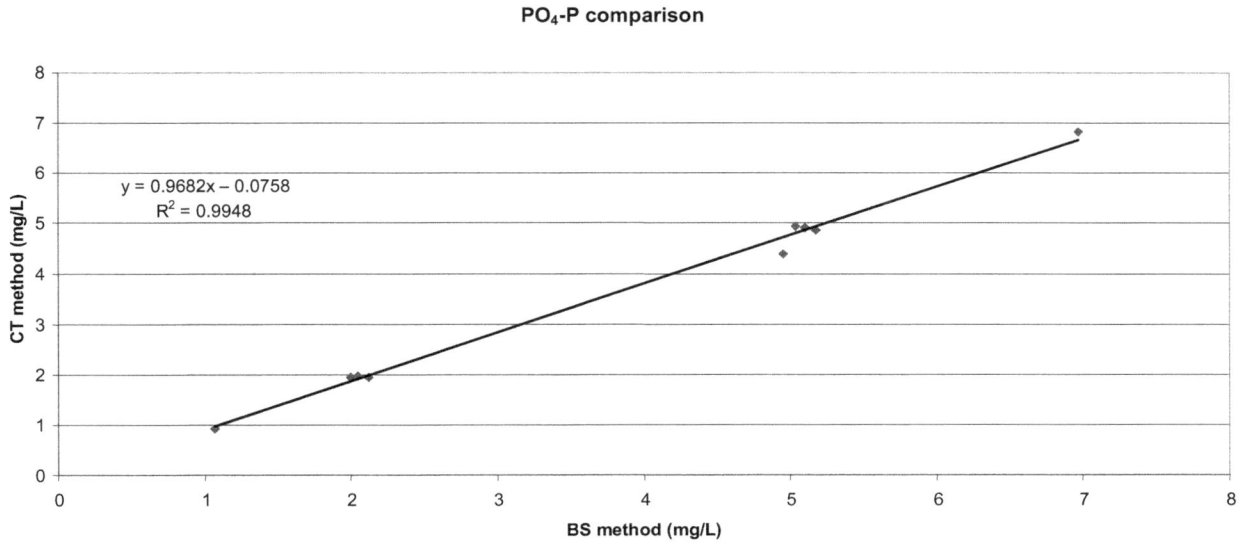

Figure 3. Comparison of mean values for PO$_4^{3-}$-P using a cuvette test method and the Bulgarian standard method.

COD comparison

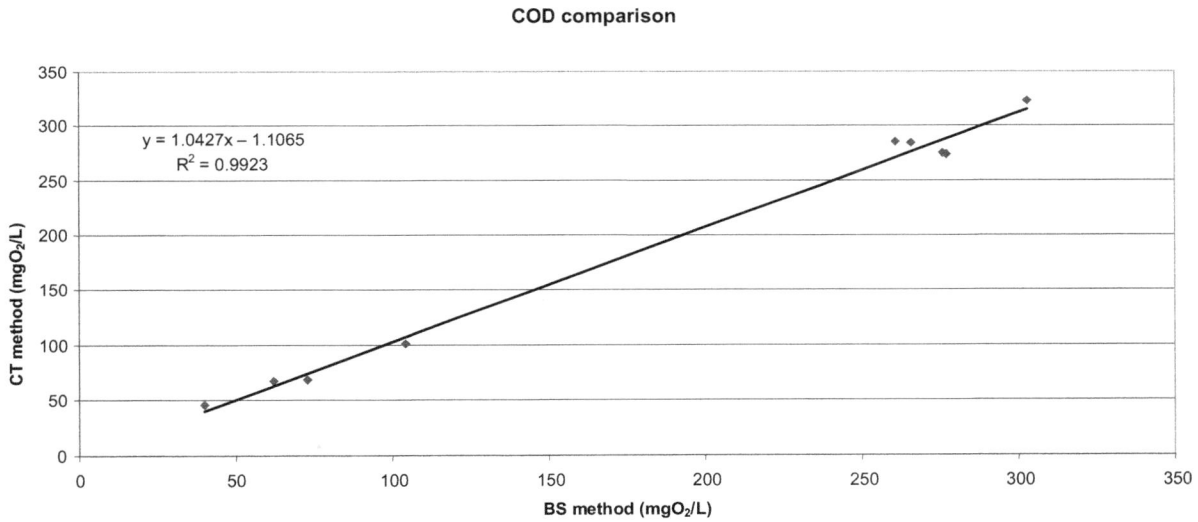

Figure 4. Comparison of mean values for COD using a cuvette test method and the Bulgarian standard method.

where u_r is the uncertainty of repeatability, s_r is the SD of all the repeatability measurements, and n is the number of replicates and $\overline{x}$ is the mean of all measurements performed.

where u_{ip} is the uncertainty due to intermediate precision, s_d is the day-to-day variation, d is the number of measurement days, and n is the number of replicates and $\overline{x}$ is the mean of all measurements performed.

The uncertainty of trueness (u_t) is calculated using the Equation 4:

$$u_{ip} = \frac{s_d}{\sqrt{d}} = \frac{\sqrt{MS_{between} - MS_{within}}}{\overline{x}} \cdot \frac{n_{bergroup}}{} \qquad (3)$$

$$u_t = \frac{s}{\sqrt{n}} = \sqrt{\frac{s^2}{n} + \frac{\sum u^2}{n_{mc}^2}} \qquad (4)$$

Table 2. Estimation of the regression coefficients $T_c = t$ (95%; f_1, f_2)

Parameter	NH_4^+-N	NO_3^--N	PO_4^{3-} P	COD
n	9	9	9	9
Intercept (a)	0.177	0.032	-0.079	0.592
Slope (b)	1.050	1.021	0.969	1.024
Sa	0.172	0.016	0.091	5.511
Sb	0.173	0.017	0.021	0.024
t statistic (a)	0.511	1.595	-0.693	0.086
t statistic (b)	1.611	0.985	-1.182	0.781
$t_{critcal}$	2.36	2.36	2.36	2.36
Proportional systematic error	No	No	No	No
Constant systematic error	No	No	No	No

Table 3. Mean concentrations for NH_4^+-N, NO_3^--N, PO_4^{3-}-P and COD with respective uncertainties (k=2).

Samples	NH_4^+-N (mg/L)		NO_3^--N (mg/L)		PO_4^{3-}-P (mg/L)		COD (mgO$_2$/L)	
	BS	CT	BS	CT	BS	CT	BS	CT
1	6.75±0.44	7.17±0.49	0.49±0.05	0.54±0.03	2.04±0.29	1.97±0.32	368±15	372±11
2	6.78±0.44	7.19±0.49	0.81±0.90	0.90±0.05	2.12±0.30	1.95±0.31	104±3	101±14
3	3.83±0.25	3.98±0.27	2.49±0.26	2.55±0.15	1.07±0.15	0.93±0.15	73±2	69±9
4	5.24±0.34	6.07±0.41	0.79±0.08	0.88±0.05	2.00±0.28	1.95±0.31	40±1	46±6
5	13.74±0.89	15.39±1.05	0.19±0.02	0.25±0.02	5.04±0.71	4.92±0.79	62±2	68±9
6	13.89±0.90	14.40±0.98	0.16±0.02	0.11±0.01	6.97±0.98	6.81±1.09	277±9	273±8
7	13.62±0.89	14.21±0.97	0.22±0.02	0.22±0.01	4.95±0.69	4.39±0.70	261±9	285±9
8	14.78±0.96	15.48±1.05	0.25±0.03	0.30±0.02	5.18±0.73	4.86±0.78	303±10	323±10
9	14.07±0.91	15.03±1.02	0.28±0.03	0.31±0.02	5.10±0.71	4.91±0.79	276±9	274±8

where s_t is the SD, n_t is the number of replicates, u_{mat} is the uncertainty of the certified value of the CRM used and n_{mat} is the number of the CRMs used.

The combined uncertainty (u_c) is then calculated using Equation 5:

$$u_c = \sqrt{u_r^2 + u_{t_b}^2 + u_t^2} \qquad (5)$$

After calculation of the measurement uncertainties for all the methods, the mean values for all the parameters were compared using Equation 6:

$$\overline{x_1} \pm t_2 \sqrt{u_{meas}^2 - u_{crm}^2} \qquad (6)$$

where $\overline{x}$ - average content, μ - certified value, u_{meas} - combined expanded uncertainty of the measurement and u_{CRM} - combined expanded uncertainty of CRM used.

The results from the 9 weeks of samplings with their uncertainties are presented in Table 3. In all chases the results obtained by the two methods yield statistically equivalent data. Only for NO_3^--N two of the nine samples

do not fulfill Equation 6, since they are marginally out.

In order to find which of the methods (CT and BS) is more precise and reliable, method accuracy and trueness were estimated by performing students' t-test comparing the methods' performance on a certified reference material (CRM). Data from repeatability assessment was used. Here, the zero-hypothesis H_0: $\mu_{BS} = \mu_{CRM}$ and $\mu_{CT} = \mu_{CRM}$ is checked against H_{alt}: $\mu_{BS} \neq \mu_{CRM}$ and $\mu_{CT} \neq \mu_{CRM}$ using Equation 7:

$$t_{exp} = \frac{|\overline{x} - \mu|}{\dfrac{s}{\sqrt{n}}} \qquad (7)$$

where $\overline{x}$ is the mean of all measurements performed, μ is the certified value of the used CRM, s is the standard deviation of the mean and n=15.

The zero hypothesis is realized if the calculated $t_{exp} < t_{tabl}$ and the data is presented in Table 4. Calculations show that both CT and BS methods yield statistically equivalent data and can be used as an alternatives for the measurement of PO_4^{3-}-P and COD. Based on the statistical data treatment, it is advised to use the CT method for the determination of NH_4^+-N and NO_3^--N.

Table 4. Calculated t_{emp} using repeatability data and Equation 7.

Parameter	NH_4^+-N		NO_3^- -N		PO_4^{3-}-P		COD	
	BS method	CT method	BS method	CT method	BS method	CT method	BS method	CT method
Mean Value ± standard deviation	986.8±27.0	1000.4±31.0	1077.3±47.3	1011±29	0.54±0.03	999.7±68.7	498.3±6.7	503.2±5.9
Certified value of a CRM	1002	1002	1005	1005	0.52	1005	500	500
t_{exp}	2.18	0.21	6.40	0.80	1.50	0.30	0.96	2.13
t_{tabl}	2.14	2.14	2.14	2.14	2.14	2.14	2.14	2.14

Conclusion

Different wastewater quality indicators - ammonia (NH_4^+-N), nitrates (NO_3^--N), total phosphates(PO_4^{3-}-P) and chemical oxygen demand (COD) were measured using spectrophotometric method using cuvette tests (CT) and Bulgarian standard methods (BS), and compared. The methods showed generally similar results for real wastewater samples obtained at the inlet of this wastewater treatment plant. Statistical evaluation indicated that the CT and BS methods are comparable. The spectrophotometric methods with cuvette tests are convenient and easy to use with some advantages (that is, lower sample volumes, lower chemical reagents needs and respectively less waste production.

ACKNOWLEDGEMENTS

The authors would like to thank Hach Lange Ltd company for consultation and technical assistance and Bulgarian Science Fund, contract DMU03/82, for the financial support.

REFERENCES

Ferree MA, Shannon RD (2001). Evaluation of a second derivative UV/visible spectroscopy technique for nitrate and total nitrogen analysis of wastewater samples. Water. Res. 35(1):327-332.

BDS 17.1.4.02-77 (1977). Environment preservation. Hydrosphere. Water quality – method for the determination of oxidability, Bulgarian Institute for Standardization, Sofia.

BDS 17.1.4.10-79 (1979). Environment preservation. Hydrosphere. Water quality – methods for the determination of ammonia, Bulgarian Institute for Standardization, Sofia.

BDS EN ISO 6878:2004 (2005). Water quality – Determination of phosphorus – Spectrometric method with ammonium molibdate, Bulgarian Institute for Standardization, Sofia.

BDS ISO 7890-3:1997 (1998). Water quality – Determination of nitrate – part 3: Spectrometric method using sulfosalicylic acid, Bulgarian Institute for Standardization, Sofia.

European Commission (2005). ERM Application note 1.

ISO/IEC 15705:2002 (2002). Water quality - Determination of the chemical oxygen demand index (ST-COD) -- Small-scale sealed-tube method, International Organization for Standardization/International Electrotechnical Commission (IEC), Geneva.

ISO/IEC 21748:2010 (2010). Guidance for the use of repeatability, reproducibility and trueness estimates in measurement uncertainty estimation, International Organization for Standardization/International Electrotechnical Commission (IEC), Geneva.

Karmarkar SV (1999). Analysis of wastewater for anionic and cationic nutrients by ion chromatography in a single run with sequential flow injection analysis. J. Chrom. A. 850(1-2):303-309.

Langergrabe G, Fleischmann N, Hofstaedter F (2003). A multivariant calibration procedure for UV/VIS spectrometric quantification of organic matter and nitrate in wastewater. Water Sci. Technol. 47(2):63-71.

Simeonov V (1997). Principles of chemical analysis data. University Press "St. Kliment Ohridsky", Sofia.

Westgard J, Hunt M (1973). Use in interpretation of common statistical tests in method-comparison studies. Clin. Chem. 19(1):49-57.

14

Removal of phosphorus from Nigeria's Agbaja iron ore through the degradation ability of *Micrococcus* species

O. W. Obot[1]* and C. N. Anyakwo[2]

[1]Department of Mechanical Engineering, Faculty of Engineering, University of Uyo, Nigeria.
[2]Department of Metallurgical and Materials Engineering, Federal University of Technology, Owerri, Nigeria.

Study on the potential of *Micrococcus* species to remove phosphorus (P) from Nigeria's Agbaja iron ore was carried out by submerged culture technique. Findings reveal that *Micrococcus* species which was originally isolated from the ore samples solubilized phosphorus with 69.66% phosphorus removal rate. The microbe also completely accumulated iron (Fe) and cadmium (Cd) ions found in the medium while the uptake of copper (Cu), zinc (Zn) and manganese (Mn) were equally remarkable. However, microbial mortification occurred over time as a consequence of over-accumulation of trace metals and other antimicrobials which reduced further solubilization. The study shows that phosphorus can reasonably be removed by the agent *Micrococcus* species but its capacity was acutely hampered by a rapid decline in microbial population after the 4th week of experimentation. Further work is suggested with respect to possibilities of microbial metabolic wastes timely removal and disposal which may prolong the phosphorus removal capability of the microbe.

Key words: Ore, culture, microbes, biodegradation, phosphorus, serial, dilution.

INTRODUCTION

The Nigeria's Agbaja iron ore reserve, which according to Uwadiale (1991) is over 1.2 billion tons, is part of a much larger formation called the 'Lokoja Ironstone', covering a surface area of 400 km^2 and contains at minimum 2,300 million tons (Astier et al., 1989). The Agbaja iron ore reserve with an estimated 47% Fe content is, however, also associated with high phosphorus (P) content and has been categorized as nonbeneficial (Amadi et al., 1982; Uwadiale and Nwoke, 1983). Phosphorus is a deleterious inclusion in steel as it causes brittleness and fracture at low stress values. Allowable phosphorus concentration in high quality steel is in the range of 0.03 to 0.02 wt.% or less (Kudrin, 1985). The twin problems of the high-phosphorus content and beneficiation difficulties, which were subjects of sustained investigation by many researchers in the early 1980s were not addressed and it led to the abandonment of the reserve.

The purpose of this study is to remove phosphorus from Nigeria's Agbaja iron ore through degradation ability of *Micrococcus* species. The works of investigators on mineral processing through the use of microorganisms abound and are well documented (Rawlings, 2002; Delvasto et al., 2005; Anyakwo and Obot, 2008, 2010). The choice of *Micrococcus* species of bacteria as the removing agent was prompted by the fact of their common environmental association with the ore, and it is expected that for the same reason, it should be able to remove it. The future prospects of this approach are enormous both to the environment and metallurgy for as long as wastes from this approach pose no threats and phosphorus can be removed in a comparatively less expensive way.

MATERIALS AND METHODS

Raw iron ore was obtained from Agbaja in Kogi State of Nigeria. It was crushed with hammer and anvil and sieved with Shital test kits to generate 0.50/0.25 mm particle size distribution. Precisely, 20 g of this sample was subjected to compositional analysis. The concentrations of Fe, MgO, Cu_2O, ZnO and MnO_2 were determined

*Corresponding author. E-mail: obotowo2004@yahool.com.

Table 1. Composition (%) of Nigeria's Agbaja iron ore

Component	FeT	SiO$_2$	P$_2$O$_5$	MgO	Cu$_2$O	ZnO	S	MnO$_2$	Al$_2$O$_3$
Level	51.50	0.57	1.25	0.08	0.005	0.091	3.25	0.001	34.77

using AAS (UniCam 939). The level of sulfur (S) was determined by Eschaka method, while Al$_2$O$_3$ by titrimetry. Also measured were SiO$_2$ and P$_2$O$_5$ by colorimetry method using ammonium molybdate and ammonium vanadate, respectively.

To isolate microbes with strong solubilizing potential, 10 g of sample was placed in 90 ml distilled water in 250 ml conical flask and then serially diluted to 10^{-6}, in order to decongest possible available microbes and allow for a moderate growth to occur. At the end of 14 days bacteria found on the ore surface were cultured by spread plate technique on Bacto - Nutrient Agar (NA). The NA plates were incubated appropriately at room temperature (28±2°C). Specific colonies were then sub-cultured on freshly prepared basal media and the pure cultures were characterized according to recommended procedures (Sneath and Holt, 1986; Alsina and Blanch, 1994).

The ability of the bacterial isolates to utilize the ore as their sole source of energy for growth was determined by the method of Okpokwasili and Okorie (1988) and Itah and Essien (2005) using the mineral salt medium (MSM) of Zajic and Supplison (1972). Briefly, 10 ml of MSM were dispensed into each test tube and then 2% (0.2 g) of the milled ore sample was added. The ore-supplemented medium was then sterilized by autoclaving at 121°C for 15 min under 15 psi atmospheres. Thereafter, 1 ml of 18 h old tryptone soya broth culture of the bacterial isolates were aseptically seeded into the ore-supplemented MSM and then incubated un-disturbed at 28±2°C for 3 weeks. Un-inoculated tubes were included for each test isolate to serve as controls.

The population dynamics of the bacteria test isolates was used as the index of ability to utilize the ore medium for growth. The growth rate of the isolates was graded as high (+++), moderate (++), minimal (+) and no growth (-). Among the isolates with strong capability to utilize ore based substrate for growth, *Micrococcus* species was one of the most prevalent and was subsequently selected for the phosphorus removal studies.

In order to determine the phosphorus solubilizing capability of *Micrococcus* species, the submerged culture technique was adopted. In this procedure, 100 ml of nutrient broth (NB) was dispensed into 250 ml conical flasks and thereafter supplemented with milled ore. The ore supplemented NB was sterilized as before and allowed to cool after which 1 ml of the solubilizing agent was seeded into each flask. Un-inoculated flasks and flasks inoculated with the test organisms but without ore samples served as the control. They were left to stand for 10 weeks, during which representative samples were removed at weekly basis for the determination of phosphorus content, pH and growth of the solubilizing agent.

The amount of phosphorus in ore-NB was determined by volumetric analytical technique (Jain, 1982) in which, ammonium phospho-molybdate precipitate was obtained and phosphorus concentration in it analyzed by titrating the precipitate with 0.1 N HCl using 4 to 5 drops of phenolphthalein as indicator. The pH of the fermentation broth was determined using a pH meter (EIL 7020, Kent Industrial measurement Ltd). The growth of the solubilizing agent, *Micrococcus* species, in the ore supplemented broth culture was determined by pour plate using freshly prepared NA. The NA plates were incubated for 24 h at room temperature after which the number of cells in colony forming unit per mili-litre was determined with the aid of colony counter.

Also determined before and at the end of phosphorus removal experiment, was the trace metals levels in the ore-supplemented medium (ore+NB) and in ore+NB inoculated with *Micrococcus* species. The concentrations of iron (Fe), copper (Cu), cadmium (Cd), zinc (Zn), nickel (Ni), manganese (Mn) and lead (Pb) in the substrates were determined with AAS after digestion with a solution of concentrated HNO$_3$ (0.3 ml) and HCl (6.0 ml) (Binning and Baird, 2001).

RESULTS AND DISCUSSION

The results of ore compositional analysis data are shown in Table 1. Loss on ignition was taken at 939°C. The results revealed that Agbaja ore was rich in Fe content and as well confirmed the high-phosphorus and high-alumina status of the ore which had earlier been reported by Uwadiale, (1989).

The curve of the weight percent phosphorus content during phosphorus removal from the ore by *Micrococcus* species for 10 weeks is shown in Figure 1. It shows that the phosphorus removal started rapidly and slowed down between 1 and 2 weeks.

Thereafter, a sinusoidal removal tendency was observed possibly due to the effect of the life cycle of the bacterium on the removal process, with the upper and lower curves coinciding for the periods of maximum and minimum cells population, respectively. Not much removal was observed from 8th week till the end of the experiment.

The growth curve of *Micrococcus* species during phosphorus removal for 10 weeks is shown in Figure 2. The cells population grew exponentially to a maximum at about 4th week and soon after declined till the 6th week. A period of poor growth was observed from the 6th to 8th week when the population suddenly declined and finally remained stagnant to the end.

The substrates pH obtained during phosphorus removal from the ore for 10 weeks is shown in Figure 3. As shown in Figure 3, the pH which began in a relatively weak acid region gradually progressed into a completely basic region approaching pH 10 by the end of the experiment in 10th week.

The variation of phosphorus content, substrate pH and the log of density of *Micrococcus* species during phosphorus removal from the ore for 10 weeks is shown in Figure 4. The Figure 4 shows that the initial phosphorus removal was less rapid at the end of 1st week. Phosphorus seemed to have reverted to the ore in 2nd week thereafter, the phosphorus removal continued smoothly becoming insignificant from 9th week and terminating at 0.267 wt.% by the end of the experiment. Figure 4 also shows that the cells population grew rapidly and climaxed in 4th week. The maximum phosphorus

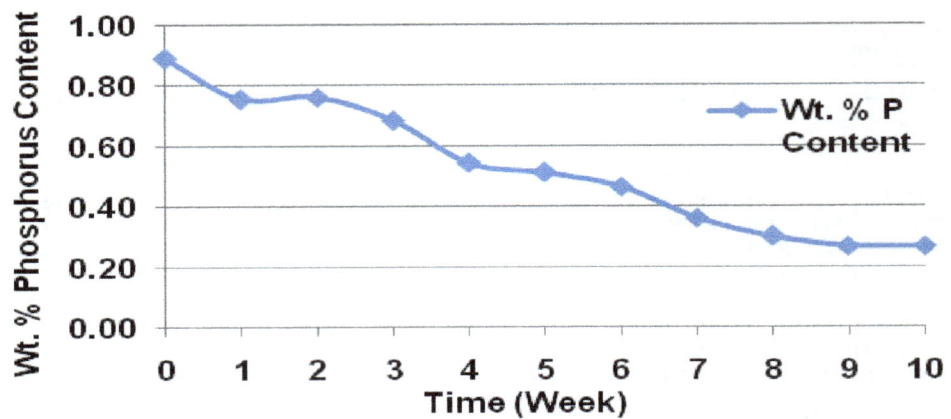

Figure 1. Curve of weight percent phosphorus content versus time during phosphorus removal from Nigeria's Agbaja iron ore 0.50/0.25 mm by *Micrococcus* species for 10 weeks.

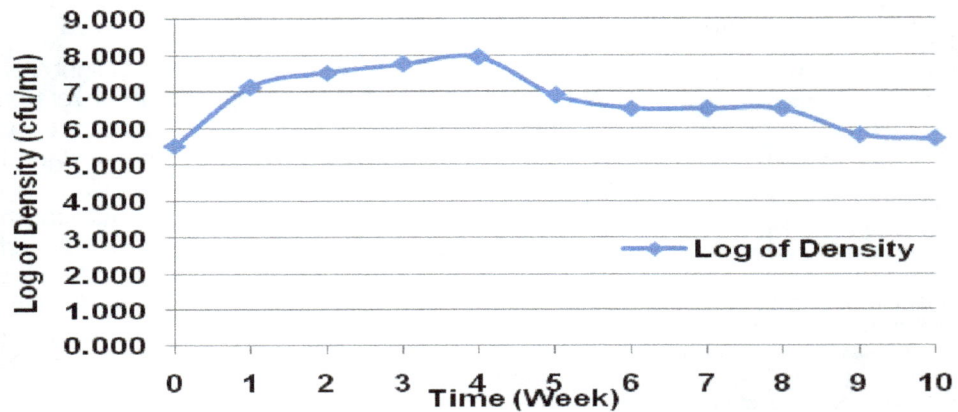

Figure 2. Growth of *Micrococcus* species during phosphorus removal from Nigeria's Agbaja iron ore 0.50/0.25 mm for 10 weeks.

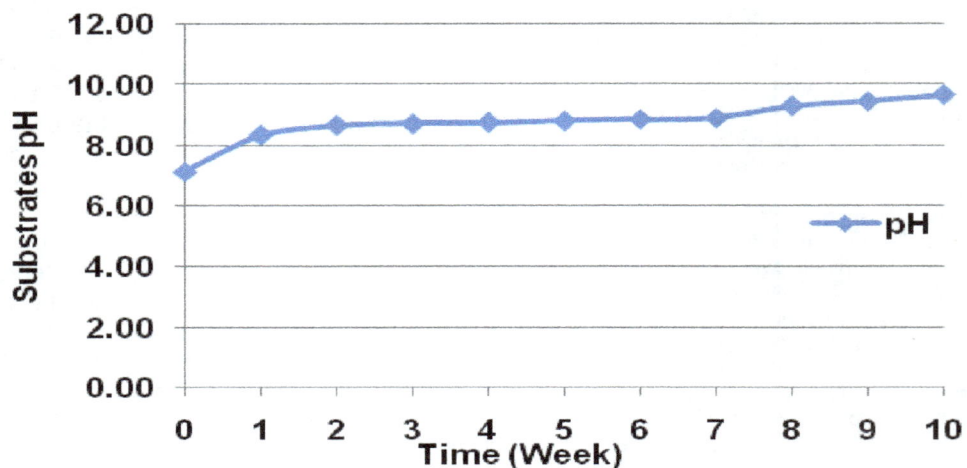

Figure 3. Curve of substrates pH versus time during phosphorus removal from Nigeria's Agbaja iron ore 0.50/0.25 mm by *Micrococcus* species for 10 weeks.

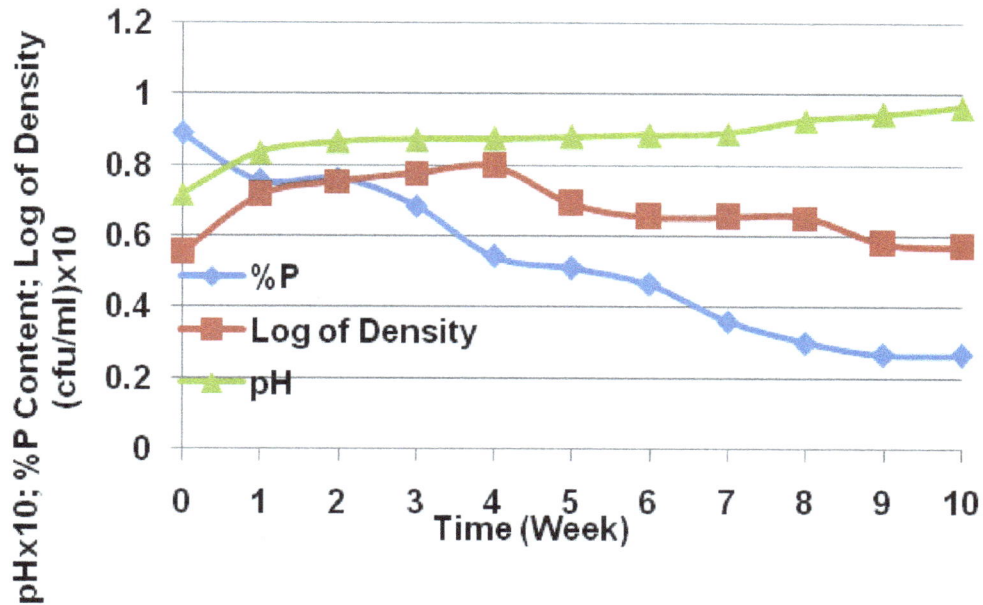

Figure 4. Variation of phosphorus content, substrates pH and log of density of *Micrococcus* species during phosphorus removal from Nigeria's Agbaja iron ore 0.50/0.25 mm for 10 weeks.

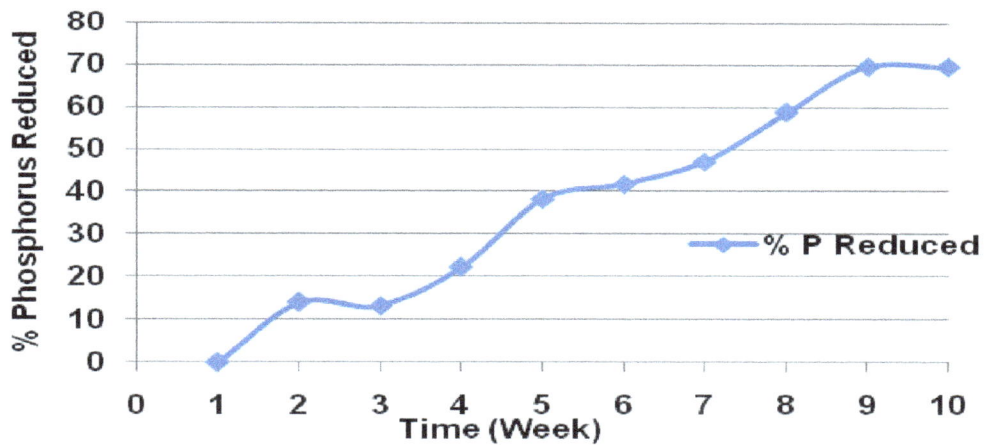

Figure 5. Curve of percent phosphorus removed versus time for Nigeria's Agbaja iron ore 0.50/0.25 mm by *Micrococcus* species for 10 weeks.

removal was sustained during the period of cumulative cells growth which lasted till 9th week. The pH of the NB medium after the initial week shows that the removal proceeded in a basic medium permanently till the end of the experiment in 10th week.

Figure 5 shows the percentage phosphorus removed by *Micrococcus* species in the course of 10th weeks and 69.66% was the maximum. It is observed that apart from the smooth removal gaps encountered between 2nd and 3rd weeks, and also between 9th and 10th weeks when the cells population might have suffered some set back, the bacterium progressively removed phosphorus from the ore sample during the period of experimentation.

The fluctuation in trace metals concentration in the ore-supplemented NB for 10 weeks during phosphorus removal by *Micrococcus* species is shown in Figures 6 and 7. Comparing the analytical results of control (NB-pure) and the NB+*Micrococcus* species without ore (control), it is observed that *Micrococcus* species merely accumulated 0.80% Zn and released more ions of Fe, Cu, Cd which led to a respective increase in their concentrations from 3.0782 to 10.1399 ppm, 0.0907 to

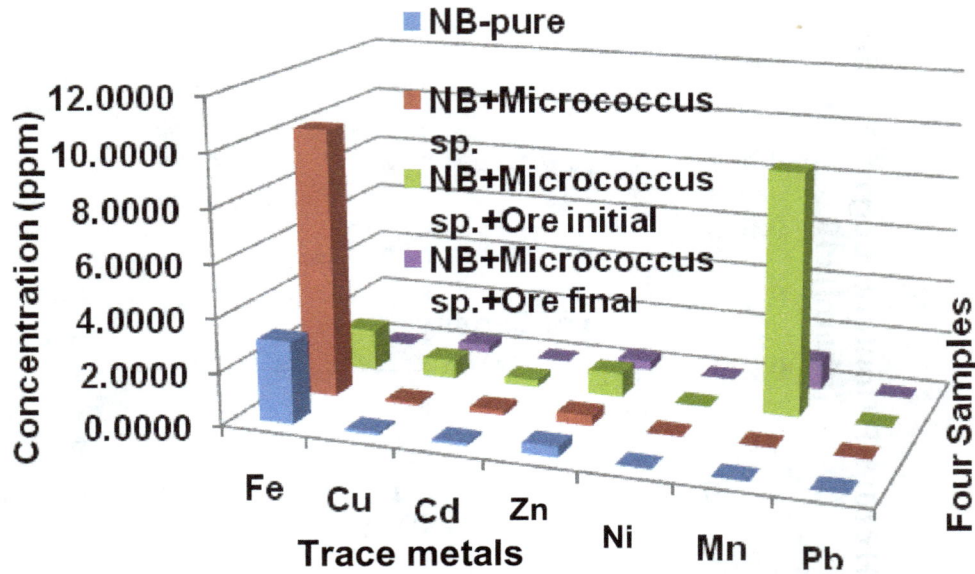

Figure 6. Analytical results of NB medium cultures during phosphorus removal from Nigeria's Agbaja iron ore 0.50/0.25 mm by *Micrococcus* species for 10 weeks

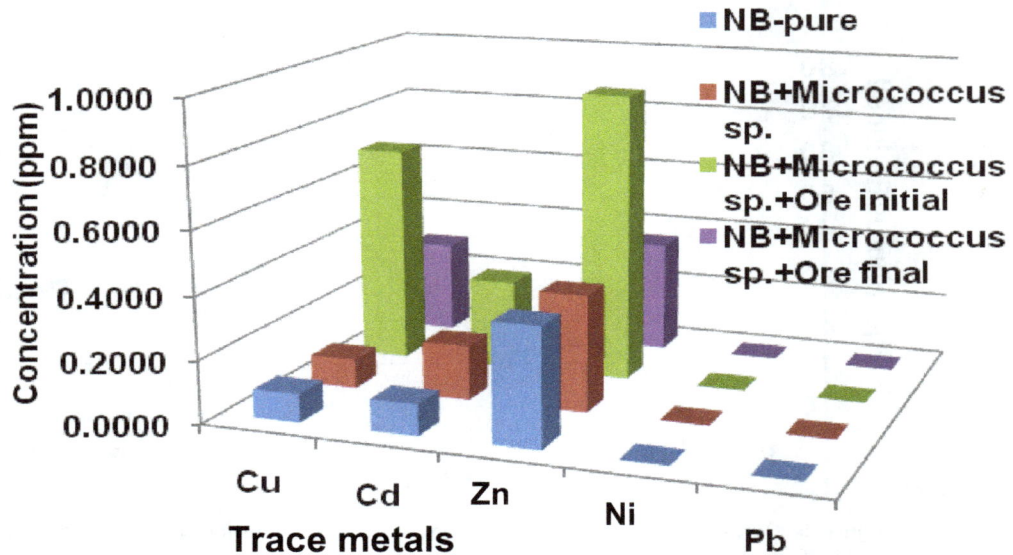

Figure 7. Analytical results of NB medium cultures during phosphorus removal from Nigeria's Agbaja iron ore 0.50/0.25 mm by *Micrococcus* species for 10 weeks.

0.0929 ppm and 0.0995 to 0.1736 ppm. Ni, Mn and Pb ions were absent in the controls. In comparing the NB + *Micrococcus* species + ore's initial and final results, it was also apparent that the same microorganism which demonstrated very poor sensitivity earlier on to Fe, Cu and Cd ions after adjusting to the medium's environment, became highly sensitive to and accumulated 100% Fe from its initial 1.5614 ppm, 58% Cu from 0.7014 ppm, 100% Cd from 0.2869 ppm, 61.49% Zn from 0.9247 ppm and 88.15% Mn from 9.0823 ppm. Ni and Pb ions were

absent in the medium. A fact which the above comparisons has established is that *Micrococcus* species actually accumulated most of the trace metals in the fermented broth medium and in the test medium and in some cases accumulated all available ions of a metal. This development therefore, may be the reason the cells population was declining which consequently may have affected the removal process due to cells lysis (Mohapatra, 2008). Figure 7 is a scale-up modification of Figure 6 without the values for Fe and Mn.

CONCLUSION AND RECOMMENDATION

The present study on the phosphorus removal from Nigeria's Agbaja iron ore samples using a biological agent, *Micrococcus* species has revealed positive effects. 69.66% of phosphorus was solubilized by the microorganism. Phosphorus utilization by the bacterium resulted in growth and concomitant production of basic metabolic products and adsorption of detectable concentrations of Fe, Cu, Pb, Zn, Cd and Mn from the samples. The metabolic activity of the bacterium in the fermentation broth was remarkable but later was reduced plausibly as a result of over-accumulation of toxic metabolites and or exhaustion of available nutrients. These resulted in cells lysis and death. What this means is that the microbial route for the removal of phosphorus is successful and encouraging provided the metabolic wastes associated with the process can be well managed. It is possible that removal can be continuous in view of microbial exponential growth rate.

ACKNOWLEDGEMENT

We thank all those who in one way or the other supported this study especially, Dr. Joseph P. Essien for his invaluable suggestions on the microbiology approach to this work, also the staff of the Departments of Microbiology and Chemical Engineering of University of Uyo. We equally thank the staff of Ministry of Science and Technology, Akwa Ibom State. Deserving special appreciation is Dr. A. O. Ano of the Nigerian Root Crops Research Institute, Umuahia, Abia State, for lending equipment worth millions of naira free of charge in order to complete this research. And finally, Engr. Peter Asangausung, the Senior Technologist in-charge of the Chemical Engineering Laboratory where the bulk of this work was done, we thank him for his resourcefulness, best wishes, support and demonstrated concern.

REFERENCES

Alsina M, Blanch AR (1994). A set of keys for biochemical identification of environmental Vibrios species. J. Appl. Bacteriol., 74: 79-85.

Amadi NJ, Odunaike AA, Mathur JP (1982). Preliminary Bench Scale Beneficiation Studies with Three Lumps of Iron Ore Sample from Agbaja. (Tech. Rep.). Central Metallurgical Research and Development Centre, Jos, Nigeria.

Anyakwo CN, Obot OW (2008). Phosphorus Removal from Nigeria's Agbaja Iron Ore by Aspergillus Niger. IREJEST. 5(1): 54-58.

Anyakwo CN, Obot OW (2010). Phosphorus Removal Capability of *Aspergillus terreus* and *Bacillus subtilis* from Nigeria's Agbaja Iron Ore. JMMCE. 9(12): 1131-1138.

Astier JE, Donzeau M, Uwadiale GGOO (1989). The Loklja Oolitic Ironstone Deposit: Possible Use in the Ajaokuta Iron and Steel Plant. J. Min. Geol., 25: 1-2.

Binning K, Baird D (2001). Survey of heavy metals in the sediments of the Swatkop River Estuary, Port Elizabeth South Africa. Water SA 24(4): 461-466.

Delvasto P, Ballester A, Muñoz JA, González F, Blázquez ML, García-Balboa C (2005). Exploring the Possibilities of Biological Beneficiation of Iron-ores: The Phosphorus Problem. Proceedings of the 15th Steelmaking Conference, 5th Ironmaking Conference & 1st Environment and Recycling Symposium IAS (CD-ROM). Argentinean Steelmaking Institute (IAS). San Nicolás, Buenos Aires, Argentina, pp.71-82.

Itah AY, Essien JP (2005). Growth profile and hydrocarbonoclastic potential of microorganisms isolated from tarballs in the Bight of Bonny, Nigeria, World J. Microbiol. Biotechnol., 21: 1317-1322.

Jain SK (1982). An Introduction to Metallurgical Analysis: Chemical and Instrumental, India, New Delhi, Vikas Publishing House.

Kudrin V (1985). Steel Making. Moscow: MIR Publishers.

Mohapatra PK (2008). Textbook of Environmental Microbiology. I.K. International Publishing House Pvt Ltd, New Delhi.

Okpokwasili GC, Okorie BB (1988). Biodeterioration potentials of microorganisms isolated from car engine lubricating oil. Tribol. Int. 21: 215-220.

Rawlings DE (2002). Heavy Metals Mining using Microbes. Annu. Rev. Microbiol., 56: 65-91. Downloaded 2 December, 2010 from www.ncbi.nlm.nih.gov/pubmed/12142493.

Sneath HAP, Holt GJ (1986). Bergey's Manual of Systematic Bacteriology, Vol. 2, Williams and Wilkins, Baltimore, London, 1599pp.

Uwadiale GGOO (1989). Upgrading Nigerian Iron Ores. J. Miner. Metallurgical Process., AIME, 117-123.

Uwadiale GGOO (1991). Electrolytic Coagulation and Selective Flocculation of Agbaja Iron Ore, J. Min. Geol., 27(1): 77-85.

Uwadiale GGOO, Nwoke MAU (1983). Beneficiation of Agbaja Iron Ore by Reduction Roasting- Magnetic Separation: Semi Pilot Plant Scale-up and Establishment of Residence Point of Phosphorus, National Steel Council, Metallurgical Research and Tests Division, Jos, Nigeria.

Zajic JE, Supplison B (1972). Emulsification and Degradation of Bunker C. fuel oil by microorganisms. Biotechnol. Bioeng., 14: 331-343.

15

Industrial sludge based adsorbents/ industrial by-products in the removal of reactive dyes – A review

A. Geethakarthi* and B. R. Phanikumar

Department of Civil Engineering, VIT University, Vellore 632014, India.

Adsorption has been an effective separation process for non-biodegradable pollutants. Study of recovery of dyes reveals adsorption as an efficacious process. Many textile industries use commercial activated carbon for the treatment of dye waste. The current research is focused on the need to alternate commercial activated carbon with a cost effective, potential adsorbent. The major limitations in using commercial activated carbon in large scale are its regeneration and high cost of operation. Many researchers have studied the feasibility of using low cost adsorbents derived from natural materials, industrial waste materials, agricultural products and biosorbents as precursors. Numerous works have been reported on these adsorbents being used in the removal of heavy metals and dyes. This paper reviews the development of different industrial sludge/by products as adsorbents under various activation methods. The decolourization of reactive dye solutions by the developed adsorbents under batch mode is also discussed. Based on the reviews, development of activated carbon from preliminary tannery sludge is suggested by various activation methods.

Key words: Adsorption, reactive dye, activated carbon, tannery sludge.

INTRODUCTION

Dyes are natural or synthetic colorants used in various industries such as textiles, tanneries, paints, pulp and paper. Even if a small amount of dye is present in water (for example, even less than 1 ppm for some dyes), it is highly visible and therefore undesirable (Sun and Yung, 2003). During the dyeing process, 10 to 15% of the dye is lost in the effluent (Ravikumar et al., 2005). Effluents discharged from dyeing industries are highly colored and are toxic to aquatic life in the receiving water bodies. Some dyes are carcinogenic which require separation and advanced treatment before being discharged into conventional systems. Based on the chromophore group, 20 to 30 different groups of dyes can be discerned. Dyes

may be classified according to chemical constituents, application class and end use. Azo, monoazo, diazo, nitro, azine, thioazine, anthroquinone, quinoline, sulphur, xanthene, phthalocyanine and nitoros are classification of dyes based on their chemical constituents. Based on their applications, dyes are classified as acid dyes, basic dyes, direct dyes, mordant dyes, vat dyes, reactive dyes, disperse dyes and sulfur dyes. There are many structural varieties of dyes that fall into either the cationic, nonionic or anionic type.

REACTIVE DYES

Reactive dyes are the most used azo dyes combined with different types of reactive groups. They bind to the textile fibres to form covalent bonds and have the favorable characteristics of bright color, water-fast, simple application techniques and have low energy consumption. Reactive dyes pose the greatest problem in terms of color, which is exacerbated by the dominance of cotton in today's fashion industry. The human eye can detect concentrations of 0.005 mg/L of reactive dye in water, and

*Corresponding author. E-mail: geethakarthi@yahoo.com.

Abbreviations: KOH, Potassium hydroxide; **K₂CO₃,** Potassium carbonate; **NaOH,** Sodium hydroxide; **Na₂CO₃,** Sodium carbonate; **ZnCl₂,** Zinc chloride; **MgCl₂,** Magnesium chloride; **H₂O₂,** Hydrogen peroxide; **H₃PO₄,** Orthophosporic acid; **NaCl,** Sodium chloride; **N₂,** Nitrogen; **SBET,** BET Surface area.

therefore, presence of dye exceeding this limit would not be permitted on aesthetic grounds (Pierce, 1994). After the reactive dyeing process is complete, up to 800 mg/L of hydrolyzed dye remains in the bath (Steankenrichter and Kermer, 1992). Fixation rates for reactive dyes tend to be in the range of 60 to 70% although the values tend to be higher in dyes containing two reactive groups (Carr, 1995).

Therefore, up to 40% of the color is discharged in the effluent from reactive dyeing operation, resulting in a highly colored effluent. An additional problem is that, the reactive dyes in both ordinary and hydrolyzed forms are not easily biodegradable, and thus, even after extensive treatment, color may still remain in the effluent. Reactive dyes cannot be easily removed by conventional waste-water treatment systems since they are stable to light, heat and oxidizing agents and are biologically non-degradable, so they have therefore been identified as problematic compounds in textile effluents. Hence, their removal is also of great importance (Sun and Yung, 2003; Ravikumar et al., 2005; Ozdemir et al., 2004). Techniques for dye removal include coagulation (Tan et al., 2000), ozonation (Chu and Ma, 2000), membrane process (Tan and Sudak, 1992), filtration with coagulation (Graham et al., 1992), ozonation with coagulation (Lin and Lin, 1993) and adsorption (Al-Degs et al., 2000; Nicolet and Rott, 1999). Among these, adsorption process yields the best results, as it can be used for removing different types of coloring materials.

LOW COST ADSORBENTS AND ACTIVATION METHODS

Among the various methods available for removing dyes from effluents, adsorption by commercial activated carbon is the most effective. The efficiency of activated carbon in adsorption is due to its structural characteristics, porous texture and chemical nature. The use of activated carbons derived from expensive starting materials is not satisfactory for most pollution control applications. This has led research towards economic adsorbents. A convenient alternative treatment for the discoloration of wastewaters from the textile industry is the usage of non-conventional adsorbents with lower cost and higher colour removal efficiency (Crini, 2006; Forgacs et al., 2004; Pearce et al., 2003; Shukla et al., 2002). Industrial wastes represent unused resources and cause serious disposal problems. These waste materials are to be turned into a useful resource for the color removal of reactive dyes. Solid wastes such as metal hydroxide sludge, fly ash and red mud can be developed into adsorbents through pretreatment for use in dye removal (Mohan et al., 2002). Developments of various low cost adsorbents for color removal are often not compared with activated charcoal which shows high removal efficiencies (Jain et al., 2003). This is due to the fact that the precursors used for the preparation of the commercial activated carbon are origin of natural products, but that for the industrial sludge based carbons are derived from the unused waste resources. The surface area of the activated carbon is higher for commercial carbons than the industrial sludge based carbons; hence, the adsorption capacity is higher for the former compared to the later. This paper reviews industrial sludge based adsorbents/industrial by - products in the removal of reactive dyes. The different activation methods involved in the development of various industrial sludge into useful sorbents are also discussed.

REVIEW OF INDUSTRIAL SLUDGE BASED ADSORBENTS AND INDUSTRIAL BY PRODUCTS IN THE REMOVAL OF REACTIVE DYES

Development of industrial sludge based adsorbents/ industrial by products

Low cost materials, in their natural and modified forms have been extensively studied as alternative adsorbents for dye removal (Ahmad et al., 2007; Gurses et al., 2006; Aksu and Tezer, 2000; Acemioglu, 2004). Numerous research works are carried out throughout the world using waste materials, in order to avert an increasing toxic threat to the environment and to streamline the present waste disposal techniques. Different activation methods are involved in the development of industrial sludge into useful sorbents. Activation can be carried out by chemical or physical means. The physical activation method consists of carbonization and activation. Carbonization or thermal decomposition is the first step through which the raw materials undergo major physical and chemical changes producing char. The thermal decomposition of the carbonaceous material in the absence of air or oxygen through oxidation by steam, carbon dioxide, inert gases like nitrogen or a combination of these gases in the temperature range from 800 to 1100 °C results in the formation of solids, liquids and gaseous products (Goutam et al., 2006). The variation in the pyrolysis time and temperature was studied to evaluate the changes in the structural and chemical properties of a metal sludge carbon derived from a galavanization industry (Karifala et al., 2008).

The waste oil precursor provided active carbonaceous phase responsible for development of micropores and formation of active surface chemistry. The final pyrolysis temperatures were 650 and 950 °C with holding time 30, 60 or 120 min. The long holding time and high-temperature of pyrolysis stabilize both, organic and inorganic phases. These phases undergo aromatization, carbonization, thermal decomposition, incorporation of nitrogen and undefined solid-state reactions. In chemical activation, the starting materials are impregnated with an activating agent such as $ZnCl_2$, $NaCl$, Na_2CO_3, K_2CO_3, KOH, $NaOH$, Al_2O_3, H_3PO_4, H_2SO_4, NH_4Cl. Orthophosporic acids (Molina- Sabio et al., 1995), potassium hydroxide (Olivares et al., 2006), zinc chloride (El - Nabarawy et al., 1997) and sulphuric acid (Cureda et al., 2006) are some of the major chemical impregnants. The chemical reagents may promote the formation of a rigid matrix, less prone to volatile loss and volume contraction upon heating to high temperature. Lower temperatures (500 to 800°C) with low energy cost, less activation time, higher adsorbent yields and high development of porosity are the advantages of chemical activation. Type of activating agent, impregnation ratio, activation temperature and activation time is related with different physical and chemical characteristic of the products. KOH and $ZnCl_2$ are the widely used chemical agents to obtain high surface area in the preparation of activated carbon. The influence of different chemical - strong base (KOH) and Lewis acid ($ZnCl_2$) - on the structure of the activated carbon products resulting in higher surface areas and higher carbon yields are the advantages in using KOH and $ZnCl_2$ as the precursors.

The chemicals left in the carbonized samples after being washed out lead to the formation of porosity, especially microporosity, and the distribution of the chemicals in precursor prior to carbonization process govern the pore size distribution of the final carbon products (Dastgheib and Rockstraw, 2001; Girgis and El-Hendawy, 2002). There are a few reports relating to the production of activated carbon from sewage sludge by KOH activation. All sludge-based activated carbons produced by chemical activation were pyrolyzed under nitrogen atmosphere. The technique of pyrolyzing KOH-impregnated sludge in steam in a single step has the potential to reduce the cost of the process and further improve the adsorptive properties of the activated carbon. Xiaoning et al. (2008) chose potassium hydroxide as an activating agent.

The merit of this modification is that, steam acting as an oxidant in high temperature can assist KOH to produce and widen the micropores and mesopores in the activation process. Depending on the activation method, partial oxidation takes place and the carbon surface becomes rich in a variety of functional groups with a broad range of concentrations determined by the method of activation, chemicals used and tem-perature of preparation. The effect of acidic treatments of activated carbons on dye adsorption was investigated by Shaobin and Zhu (2007). The physico-chemical properties of activated carbons were characterized by N_2 adsorption, mass titration, temperature-programmed desorption (TPD), and X-ray photoelectron spectrometry (XPS). It was evident that, BET surface area was slightly enhanced due to acid treatment, although the pore volume was relatively unchanged when compared to the original activated carbon. Acid treatment can remove some mineral matter while changes of surface groups will block the micropores, resulting in little change in pore structure. The increase in the acidity of carbon upon acid treatment, may be attributed to the removal of inorganic compounds

leaving sites on the carbon surface which can chemisorb oxygen in air at room temperature.

Activated carbons, when stored in ambient air, are oxidized, which results in further increase in the number of surface functional groups such as phenols, lactones, and carboxylic acids and the formation of new groups. It is difficult to determine the type and quantity of each surface group, since they influence the carbon adsorptive behavior to a great extent (Salame and Bandoz, 1999). The nature of the surface groups can be modified through physical, chemical, and electrochemical treatments. The most common are liquid phase treatments using HNO_3 and H_2O_2, gas phase oxidation with O_2 or N_2O and heat treatment under inert gas to selectively remove some of the functional groups (Figueiredo et al., 1999). Different studies have shown that, acid oxygen-containing surface groups decrease the adsorption of organic compounds in aqueous solution, while their absence favors adsorption, independently of the polarity of the compounds (Franz et al., 2000). This was justified by the fact that, water is mainly adsorbed on the surface establishing hydrogen bonds with the oxygen surface groups, which can produce clusters that may block the passage of the adsorptive molecules into the micropores. Qing et al. (1995) conducted experimental studies on pyrolysis of digested sewage sludge. To understand the mechanism of the development of adsorptive capacity, the surface area and pore structure were characterized for various chars pyrolyzed under different conditions. The effects of process parameters, temperature and hold time were found to be significant.

Generally, the surface area of the resultant char increases with temperature and hold time. However, the surface area is reduced for chars obtained at 550 and 650°C, possibly owing to the pore enlargement phenomenon as a result of loss of volatiles in the intermediate thermoplastic phase. The resultant chars from pyrolysis are found to possess about 23 to 30% of carbon and the rest is all ash. However, owing to the nature of the carbon, the total surface area of the chars is mostly attributed to the carbon present. Reduction in mesopore volume, due to sintering at high temperatures and prolonged times was also observed. In this study, the optimum temperature and hold time for maximum surface area development of the resultant char were found to be 850°C and 2 h, respectively. The performance of the developed low cost adsorbents depends on different factors like texture (surface area, porosity), surface chemistry (surface functional groups) and ash content. It also depends on adsorptive characteristics like molecular weight, polarity, molecular size and functional groups. Many techniques have been used for the characterization of activated carbons. This includes infraspectroscopy (Li and Lin, 1999), X - ray diffraction (Galiatsatou et al., 2001), scanning electron microscopy (Galiatsatou et al., 2001), transmission electron microscopy (Rouzaud, 1990), optical microscopy (Rodriguez et al., 1986), iodine number (Li and Lin, 1999), ion exchange capacity (Gierak, 1996) and apparent surface area estimation by nitrogen adsorption (Diaz - Teran et al., 2001). Hojamberdiev et al. (2008) produced activated carbon from paper sludge by mechano chemical grinding and physical activation in the presence of nitrogen atmosphere. The physical activated sample showed relatively higher BET surface area of 70 m^2/g at 600°C. According to Khalili et al. (2000), paper sludge can successfully be converted into activated carbon with SBET > 1000 m^2/g by chemical activation using $ZnCl_2$. Similarly, a number of studies have so far been focused on the bagasse pith for the removal of dyes from wastewater (Tsai et al., 2001). Thus, low cost adsorbents can be produced from industrial solid wastes and by products. The activation methods involved in the development of various industrial based sludge/industrial by-products as adsorbents are mentioned in Table 1. The methods adopted for the analysis of surface and textural properties are also summarized along with the surface area and the pore volume of the activated carbons.

Industrial sludge based adsorbents/ industrial by-products in the removal of reactive dyes

Literature reports several studies on the effective exploration of industrial sludges and industrial by-products as adsorbents and the preparation of activated carbons from solid wastes and its application on dye removal (Kadirvelu et al., 2003; Onal et al., 2007). Industrial metal hydroxide sludge was used as a low-cost adsorbent for removing a reactive textile dye. Reactive Blue 19 in aqueous solution (Silvia et al., 2008). The maximum adsorption capacities varied between 275 mg/g (at 25°C and pH 4) and 21.9 mg/g (at 25°C and pH 10). Additionally, a simulated real effluent containing the selected dye, salts and dyeing auxiliary chemicals, was also used in equilibrium and kinetic experiments. The batch adsorption experimental conditions for the simulated real effluent adsorption were maintained as the same that of the Reactive Blue 19 dye solution. The presence of salts, like Na_2CO_3, NaCl and auxiliary dyeing chemicals present in the simulated effluent showed negative effect on adsorption. The negative adsorption was due to the coverage of the adsorbent surface, the possible blockage of pores and the possibility of existing competition (among dye and wetting anionic species and the sequestrant agents, for example). Hence, simulated textile effluent used as adsorbate showed a decrease in the adsorption capacity from 91.0 mg/g for Reactive Blue19 to 31.0 mg/g for simulated effluent.

Equilibrium data were described by both Langmuir and Freundlich models. Adsorption kinetic and adsorption equilibrium studies of three reactive dyes namely, Remazol Brilliant Blue (RB), Remazol Red 133 (RR) and Rifacion Yellow HED (RY) from aqueous solutions at various initial dye concentration (100 to 500 mg/L), pH (2 to 8), particle size (45 to 112.5 µm) and temperature (293 to 323 K) on fly ash were studied in a batch mode operation (Dizge et al., 2008). The Scanning Electron Microscopy (SEM) image showed that, fly ash particles are composed of irregular and porous particles. The effects of initial dye concentration and pH on the reactive dye removal were determined with the experimental data mathematically described using intraparticle and external mass transfer diffusion models. The experimental data showed conformity with an adsorption process, with the removal rate dependent on both intraparticle and external mass transfer diffusion. The adsorption capacities of commercial activated carbon for reactive dyes vary from 7.69 to 1179 mg/g (Crini, 2006). Among the reactive dyes, the adsorption of less charged dyes by an anionic adsorbent is unfavourable because of the weaker attraction toward the site of opposite charge and the dye molecules may escape from the solid phase to the bulk phase with the increasing temperature. The kinetic energy and the mobility of the large dye ions to be adsorbed on the adsorbent may increase with an elevated temperature.

The adsorption of Reactive Blue 222 from synthetic aqueous solution onto granular activated carbon (GAC) and coal-based bottom ash (CBBA) (Ali et al., 2007) were studied. Both adsorbents were very effective in removing the dye at all the pH values at low initial dye concentration. But, the dye adsorption by GAC and CBBA was affected by pH at high initial dye concentration. Dye removal efficiency was higher for low dye concentrations because of availability of unoccupied binding sites on the adsorbents. Percent color removal decreased with increasing dye concentrations because of nearly complete coverage of the binding sites at high dye concentrations. At lower pH, the surface of the adsorbents became positively charged and facilitated sorption of the color cation probably by exchange sorption (Mohan and Karthikeyan, 2004). The monolayer adsorption capacities of GAC and CBBA were 6.53 and 4.02 mg/g of the dye solution. This was due to the microporous nature of the CAC. The industrial sludge developed carbons/industrial by-products used in the removal of reactive dyes are shown in Table 2.

CONCLUSIONS

Adsorbents derived from industrial wastes demonstrated outstanding capability in removing dyes compared to commercially activated carbons. The selection of activation depends on the nature, reactive group, pH, toxic compounds of the dye. Undoubtedly, industrial sludge based adsorbents offer a great promise for commercial purposes. Despite considerable research, there is little information about the combination of the activation methods involved in the development of

Table 1. Various activation conditions of industrial sludges/ Industrial by-products.

Industrial sludge/ by-products	Activation condition	BET Surface area (m^2/g)	Total pore volume (cm^3/g)	Instrumental analysis	References
Paper mill sludge	$ZnCl_2$ followed by pyrolysis at 800°C for 2h.	1249	1.153		Khalili et al., 2000
Granular molded waste paper	Pyrolysis at 1100°C for 60 mins.	1241	0.226		Masahiro et al., 2004
Paper sludge		1204	1.08	SEM	Khalili et al., 2002
Molasses	Sulphuric acid	1200	1.098		Legrouie et al., 2005
Waste newspaper	$ZnCl_2$ and H_2O_2 followed by pyrolysis under N_2 atmosphere	1740	1.15	SEM, XRD, FT-IR	Kiyoshi et al., 2003
Coal fly ash	Monoplanetry high energy ball mill	938	0.344	SEM, EDS, XRD	Teresa and Karin, 2006
Bituminous coal ash	Steam activation	863.50	0.469	SEM	Emad et al., 2008
Olive oil mill residue	Steam/N_2 mixture at 800°C and 850°C	800	0.57		Pala et al., 2006
Bitiminuous coal ash	10% by weight $ZnCl_2$	231.5	0.122	SEM	Emad et al., 2008
Waste oil sludge	Pyrolysis under N_2 atmosphere at 650 to 950°C	202	0.839	XRD, DG – DTA	Teresa and Karin, 2006
Baggase fly ash		168.8	0.101	SEM, XRD, FT-IR	Mall et al., 2005
Galvanised metal based sludge	Pyrolysis under Nitrogen atmosphere at 650 and 950°C	127	0.295	FT-IR, XRD, XRF[a], ICP[b], SEM, TGA	Karifala et al., 2008
Basic oxygen furnace sludge		78.54		XRD, SEM, TGA	Tarun et al., 2009
Paper sludge	Powdered sample under N_2 atmosphere	70			Hojambadiev et al., 2008
Paper sludge	Pelletized sample	37			Hojambadiev et al., 2008
Paper sludge	Powdered sample in air	23			Hojambadiev et al., 2008
Fly ash		3.62		SEM, XRD, FT-IR	Kara et al., 2007
Coal based bottom ash		1.77			Ali et al., 2007

Table 1 Cont.

Fly ash		0.342	SEM, XRD, FT-IR	Demirbas and Nas, 2009
Fly ash		0.342	SEM	Dizge et al., 2008
Sugar beet bagasse	zinc chloride followed by activation temperature of 400 to 900°C under N_2 atmosphere		FT– IR, SEM, DTA[a], TGA[b]	Onal et al., 2007
Oxygen furnace slag	(i) Treated by milling (ii) Treated with hydrochloric acid		FT-IR, SEM	Yongjie et al., 2009
Petrochemical sludge	Combined activation with $ZnCl_2$ and pyrolysis in N_2 atmosphere at 500°C			Chiang et al., 2009
Bagasse pith from sugar industry	Chemical activation with phosphoric acid			Nevine, 2008
Bagasse pith from sugar industry	Chemical activation with phosphoric acid			Nevine, 2008
Paint industry sludge	Hydrogen peroxide			Fikret and Sinem, 2006
Olive oil mill residue	Physical activation by steam/N_2 mixture at 800 and 850°C			Pala et al., 2006
Fertilizer sludge				Jain et al., 2003

a - Differential thermal analysis; b- Thermo gravimetric analysis.

sorbents. Based on the above review works, a study on the sorbent utilizing tannery sludge collected from a common effluent treatment plant may be developed. The sludge is subjected to a combination of physical and chemical activation to be developed into an effective adsorbent.

The activation method is optimized based on the maximum monolayer adsorption capacity, surface and textural properties of the adsorbent in the removal of the reactive dyes. Not much literature is available on the relationship between the surface chemical properties of activated carbons and their capacity to adsorb dyes. Hence, instrumental analyses are performed in revealing the constituents and surface complex bonding of the adsorbents with the dye molecules. Batch/discontinuous kinetic studies are also to be studied under different operational parameters. Suitable statistical and experimental design needs to be employed to evaluate the effects of major process variables, dyestuff and adsorbent concentrations and sludge retention time on decolorization efficiency. Furthermore, a comparative analysis of performance and cost of the activated sorbents is to be carried out to understand process efficiency and mechanism.

ACKNOWLEDGEMENT

The proposed work is financially supported by the Government of India, Ministry of Science and Technology, Department of Science and Technology, New Delhi.

REFERENCES

Acemioglu B (2004). Adsorption of Congo red from aqueous

Table 2. Adsorption of reactive dyes by various industrial sludge adsorbents/ by products.

Industrial sludge/ by-products	Dye	Equilibrium time	Adsorption capacity (mg/g)	Instrumental analysis	References
Metal hydroxide sludge	Reactive Red 19		275 ± 45	XRF[a], pH at zero pH$_{ZPC}$[b]	Silvia et al., 2008
Petrochemical sludge	Reactive orange II		270		Chiang et al., 2009
Petrochemical sludge	Chrysophenine		191		Chiang et al., 2009
Fly ash	Reactive Blue 19		179.64	FT-IR, SEM, XRD	Dizge et al., 2007
Fly ash	Reactive Blue 19	48 h	129	FT-IR, SEM, XRD	Kara et al., 2007
Oxygen furnace slag	Reactive Blue 5	180 min	109.5	FT-IR, SEM	Yongjie et al., 2009
Fly ash	Reactive blue 21	16 h	106.71	SEM	Demirbas and Nas, 2009
	Reactive Black 19		103.4		
Oxygen furnace slag	Reactive Red 120		84.5	FT-IR, SEM	Yongjie et al., 2009
	Reactive Blue 5	180 min.	74.4		
Fly ash	Reactive Red 198		87.41		Dizge et al., 2007
Olive oil mill residue	Reactive blue		70.42		Pala et al., 2006
Metal hydroxide sludge	Reactive Red 2	60 min	48 – 62	FT-IR, XRD, pH$_{ZPC}$[b]	Netpradit et al., 2004
	Reactive Red 120				
	Reactive red 141				
Fly ash	Reactive yellow 84		61.24		Dizge et al., 2007
Olive oil mill residue	Reactive red	20 min	59.88		Pala et al., 2006
	Reactive Yellow	20 min	50		
Fly ash	Reactive Yellow 84		61.24		Dizge et al., 2007
Olive oil mill residue	Reactive red	20 min	59.88		Pala et al., 2006
	Reactive Yellow	20 min	50		
Oxygen furnace slag	Reactive Black 19		49.4	FT-IR, SEM	Yongjie et al., 2009
Oxygen furnace slag	Reactive Red 120		44.1	FT-IR, SEM	Yongjie et al., 2009
Fly ash	Reactive Yellow 84		28		Kara et al., 2007
Baggase fly ash	Basic violet 1	4 h	26.3	SEM, TG – DTA[c]	Mall et al., 2005
	Acid Orange 10	4 h	18.8		
Fly ash	Reactive Red 198		17		Kara et al., 2007
Coal based bottom ash	Reactive blue 222	90 min	6.35		Ali et al., 2007
Bagasse pith from sugar industry	Reactive Orange		3.48		Nevine, 2008
Coal based charfines	Reactive red 2	80 min	0.735		Venkatamohan and karthikeyan 2004

a - X - Ray Fluorescence; b - Point charge; c - Thermo gravimetric differential thermal analysis.

solution onto calcium-rich fly ash. J. Colloid Interface Sci., 274: 371–379.

Ahmad AA, Hameed, BH, Aziz N (2007). Adsorption of direct dye on palm ash: kinetic and equilibrium modeling. J. Hazard. Mater., 141: 70 - 76.

Aksu Z, Tezer S, (2000).Equilibrium and kinetic modeling of biosorption of Remazol Black B by Rhizopus arrhizus in a batch system: effect of temperature. Proc. Biochem., 36: 431 - 439.

Al - Degs Y, Kharaisheh MAM, Allen SJ, Ahmad MN (2000). Effect of carbon chemistry on the removal of reactive dyes from textile effluent. Wat. Res., 34: 927- 935.

Ali RD, Yalcin G, Nurset K, Elcin G (2007). Comparison of activated carbon and bottom ash for removal of reactive dye from aqueous solution. Bioresour. Tech., 98: 834- 839.

Carr K (1995). Reactive dyes, especially bi-reactive molecules: structure and synthesis. In: Peters AT Freeman HS. (Eds.), Modern Colourants: Synthesis and Structure. Blackie Academic and Professional, London, pp. 87 - 122.

Chiang H, Chen T, Pan S, Chiang H (2009). Adsorption characteristics of Orange II and Chrysophenine on sludge adsorbent and activated carbon fibers. J Hazard. Mater., 161: 1384 -1390.

Chu W, Ma C (2000). Quantitative prediction of direct and indirect dye ozonation kinetics. Wat. Res., 34: 3153 - 3160.

Crini G (2006). Non-conventional low-cost adsorbents for dye removal: a review. Bioresour. Technol., 97: 1061–1085.

Cureda - Correa EM, Diaz – Diez MA, Macias - Gracia A Ganan G J (2006). Preparation of activated carbons previously treated with sulfuric acid - A study of their adsorption capacity in solution. Appl. Surface Sci., 252: 6042 - 6045.

Dastgheib SA, Rockstraw DA (2001). Pecan shell activated carbon: synthesis, characterization, and application for the removal of copper from aqueous solutions. Carbon, 39: 1849 -1855.

Demirbas E, Nas MZ (2009). Batch kinetic and equilibrium studies of adsorption of Reactive Blue 21 by fly ash and sepiolite. Desalination, 243: 8 - 21.

Diaz - Teran J, Nevskaia DM, Lopez - Peinado AJ, Jerez A (2001). Porosity and adsorption properties of an activated charcoal. Colloidal and surfaces A: Physicochem. Engg. Aspects, 187: 167- 175.

Dizge N, Aydiner C, Demirbas E, Kobya M, Kara S (2008). Adsorption of reactive dyes from aqueous solutions by fly ash: Kinetic and equilibrium studies. J. Hazard. Mater., 150: 737 - 746.

El - Nabarawy TH, Mostafa MR, Youssef AM (1997). Activated carbons tailored to remove different pollutants from gas stream and from solution. Adsorption Sci. Technol., 15: 61 - 68.

Emad NEIQ, Stephen JA, Gavin MW (2008). Influence of preparation conditions on the characteristics of activated carbons produced in laboratory and pilot scale systems. Chem. Eng. J., 142: 1 -13.

Figueiredo JL, Pereira MFR, Freitas MMA, Orfao JJM (1999). Modification of the surface chemistry of activated carbons. Carbon, 37: 1379 - 1389.

Fikret K, Sinem C (2006). Biosorption of zinc (II) ions onto powdered waste sludge (PWS): Kinetics and isotherms. Enzyme Microb. Technol., 38: 705 - 710.

Forgacs E, Cserhati T, Oros G (2004). Removal of synthetic dyes from wastewaters: a review. Environ. Int., 30: 953 - 971.

Franz M, Arafat HA, Pinto NG (2000). Effect of chemical surface heterogeneity on the adsorption mechanism of dissolved aromatics on activated carbon. Carbon, 38: 1807 - 1819.

Galiatsatou P, Metaxas M, Kasseloui-Rigopoulou V (2001). Mesopores activated carbon from agricultural by products. Mikrochim. Acta, 136: 147-152.

Gierak A (1996). Application of activated carbon for the sorption of some heavy metals from aqueous solution and their determination by atomic spectroscopy. Sci. Tech., 14: 47 - 57.

Girgis BS, El-Hendawy AA (2002). Porosity development in activated carbons obtained from date pits under chemical activation with phosphoric acid. Microporous and Mesoporous Mater,, 52: 105 -117.

Goutam C, Douglas GM, Narendra NB, Jafar SS, Mohammadzadeh AKD (2006). Preparation and characterization of chars and activated carbons from Saskatchewan lignite. Fuel Processing Technol., 87: 997 -1006.

Graham NJD, Brandao CCS, Luckham PF (1992). Evaluating the removal of color from water using direct filtration and dual coagulants. J. Am. Wat. Works Assoc., 84: 105.

Gurses A, Dogar C, Yalcin M, Acikyidiz M, Bayrak R, Karaca S (2006). The adsorption kinetics of the cationic dye methylene blue onto clay. J. Hazard. Mater., 131: 217 - 228.

Hojamberdiev M, Kameshima Y, Nakajima A, Okada K, Kadirova Z (2008). Preparation and sorption properties of materials from paper sludge. J Hazard. Mater., 151: 710 - 719.

Jain AK, Gupta VK, Bhatnagar A, Suhas (2003). Utilization of industrial waste products as adsorbents for the removal of dyes. J. Hazar. Mater., B101: 31 - 42.

Kadirvelu K, Kavipriya M, Karthika C, Radhika M, Vennilamani N, Pattabhi S (2003). Utilization of various agricultural wastes for activated carbon preparation and application for the removal of dyes and metal ions from aqueous solutions. Bioresour. Technol., 87: 129 -132.

Kara S, Aydiner C, Demirbas E, Kobya M and Dizge NX (2007). Modeling the effects of adsorbent dose and particle size on the adsorption of reactive textile dyes by fly ash. Desalination, 212: 282 – 293.

Karifala K, Jieshan O, Zongbin Z, Yu C, Teresa JB (2008). Development of surface porosity and catalytic activity in metal sludge/waste oil derived adsorbents: Effect of heat treatment. Chem. Eng. J., 138: 155 - 165.

Khalili NR, Campbella M, Sandi G, Golas J (2000). Production of micro- and mesoporous activated carbon from paper mill sludge. I. Effect of zinc chloride activation. Carbon, 38: 1905 - 1915.

Khalili NR, Vyas JD, Weangkaew W, Westfall SJ, Parulekar R (2002). Sherwood. Synthesis and characterization of activated carbon and bioactive adsorbent produced from paper mill sludge. Separation Purification Technol., 26: 295 - 304.

Kiyoshi O, Nobuo Y, Yoshikazu K, Atsuo Y (2003). Porous properties of activated carbons from waste newspaper prepared by chemical and physical activation. J. Colloid Interface Sci., 262: 179 -193.

Legrouri K, Khouya E, Ezzine M, Hannache H, Denoyel R, Pallier R, Naslain R (2005). Production of activated carbon from a new precursor molasses by activation with sulphuric acid. J. Hazard. Mater., 118: 259 - 263.

Li BQ, Lin ZY (Eds.) (1999). Preparation of advanced activated carbon with ultra-clean coals, Prospects for Tenth International Conference on Coal Science in the 21st Century, Taiyuen, China, Shanxi, September 12 -17, Science and Technology press, China.

Lin SH, Lin CM (1993). Treatment of textile waste effluents by ozonation and chemical coagulation. Wat. Res., 27: 1743 - 1748.

Mall ID, Srivastava VC, Agarwal NK, Mishra IM (2005). Adsorptive removal of malachite green dye from aqueous solution by bagasse fly ash and activated carbon - kinetic study and equilibrium isotherm analyses. Coll. Surf. A. 264: 17 - 28.

Mall ID, Srivastava VC, Agarwal NK, Mishra IM (2005).Removal of congo red from aqueous solution by bagasse fly ash and activated carbon: Kinetic study and equilibrium isotherm analyses. Chemosphere. 61: 492 – 501.

Masahiro S, Takahiko I, Kensuke K, Yoshifumi C, Toshihiro M, Takayuki O (2004). Pore structure and adsorption properties of activated carbon prepared from granular molded waste paper. J Mater Cycles Waste Manage., 6: 111 - 118.

Mohan D, Singh KP, Singh G, Kumar K (2002). Removal of dyes forms wastewater using fly ash, a low cost adsorbent. Ind. Eng. Chem. Res., 41: 3688 - 3695.

Mohan SV, Karthikeyan J (2004). Adsorptive removal of reactive azo dye from an aqueous phase onto charfines and activated carbon. Clean Tech. Environ. Policy. 6: 196 - 200.

Molina-sabio M, Rodriguez - Reinoso F, Caturla F, Selles MJ (1995). Porosity in granular carbons activated with phosphoric acid. Carbon, 33: 1105 -1113.

Netpradit S, Thiravetyan P, Towprayoon S (2004). Adsorption of three azo reactive dyes by metal hydroxide sludge: effect of temperature, pH and electrolytes. J Colloid Interface Sci., 270: 255 - 261

Nevine KA (2008) .Removal of reactive dye from aqueous solutions by adsorption onto activated carbons prepared from sugarcane bagasse pith. Desalination, 223: 152–161.

Nicolet L, Rott V (1999). Recirculation of powdered activated carbon for the adsorption of dyes in municipal wastewater treatment plants. Wat. Sci. Technol., 40: 191- 198.

Olivares – Marin M, Fernandez – Gonzalez C, Macias – Gracia A, Gomez – Serrano V (2006). Preparation of activated carbons from cherry stones by activation with KOH. Applied surface Sci., 252: 5980 - 5983.

Onal Y, Akmil - Bas C, Sarici - Ozdemir C, Erdogan S (2007). Textural development of sugar beet bagasse activated with ZnCl2. J. Hazard. Mater., 142: 138 -143.

Ozdemir O, Armagan B, Turan M, Celik MS (2004). Comparison of the adsorption characteristics of azo – reactive dyes on mezoporous minerals. Dyes Pigments., 62: 49 - 60.

Pala A, Galiatsatou P, Tokat E, Erkaya H, Israilides C, Arapoglou D (2006). The use of activated carbon from Olive oil mill residue for the removal of colour from textile wastewater. Eur. Water,, 13: 29 -34.

Pearce CI, Lloyd JR, Guthri JT (2003). The removal of color from textile wastewater using whole bacterial cells: a review. Dyes Pigments, 58: 179 - 196.

Pierce J (1994). Colour in textile effluents - the origins of the problem. J. Society Dyers Colourists., 110: 131 - 133.

Qing Lu G, Low JCF, Liu CY, Lua AC (1995). Surface area development of sewage sludge during pyrolysis. Fuel, 74: 344 - 348.

Ravikumar K, Deebika B, Balu K (2005). Decolorization of aqueous dye solutions by a novel adsorbent: Application of statistical designs and surface plots for the optimization and regression analysis. J. Hazard. Mater., 122: 75 - 83.

Rodriguez - Reinoso F (1986). Preparation and characterization of activated carbons. Proceeding of the NATO advanced study institute on carbon and coal gasification. NATO ASI series E. Martinus Nijhoff publishers, Netherlands, 105: 603.

Rouzaud J (1990). Contribution of transmission electron microscopy to the study of the coal carbonization processes. Fuel Processing Technol., 24: 55 - 69.

Salame II, Bandosz, TJ (1999). Experimental study of water adsorption on adsorption on activated carbons. Langmuir, 15: 587 - 593.

Shaobin W, Zhu, ZH (2007). Effects of acidic treatment of activated carbons on dye adsorption. Dyes Pigments, 75: 306 - 314.

Shukla A, Zhang YH, Dubey P, Margrave JL, Shyam S (2002). The role of sawdust in the removal of unwanted materials from water, J. Hazard. Mater., 95: 137–152.

Silvia CR, Santos, Victor JP, Vilar, Rui, AR, Boaventura (2008). Waste metal hydroxide sludge as adsorbent for a reactive dye. J. Hazard. Mater., 153: 999 - 1008.

Steankenrichter I, Kermer WD (1992). Decolourising textile effluents. J. Society Dyers Colourists., 108: 182 - 186.

Sun Q, Yung L (2003). The adsorption of basic dyes from aqueous solution on modified peat- resin particle. Wat. Res., 37: 1535 - 1544.

Tan BT, Teng TT, Omar AKM (2000). Removal of dyes and industrial dye wastes by magnesium chloride. Wat. Res., 34: 597 - 601.

Tan L, Sudak RG (1992). Removing color from groundwater source. J. Am. Wat. Works Assoc., 84: 79 - 87.

Tarun KN, Ashim KB, Sudip KD (2009). Clarified sludge (basic oxygen furnace sludge) - an adsorbent for removal of Pb (II) from aqueous solutions - kinetics, thermodynamics and desorption studies. J. Hazard. Mater., 252 - 292.

Teresa JB, Karin B (2006).Effect of pyrolysis temperature and time on catalytic performance of sewage sludge/industrial sludge-based composite adsorbents. Applied Catalysis B: Environ., 67: 77 – 85.

Tsai WT, Chang CY, Lin MC, Chien SF, Sun HF, Hsieh MF (2001). Adsorption of acid dye onto activated carbons prepared from agricultural waste bagasse by ZnCl2 activation. Chemosphere, 45: 51 - 58.

Venkatamohan S, Karthikeyan J (2004). Adsorptive removal of reactive azo dye from an aqueous phase onto charfines and activated carbon, Clean tech. Environ. Policy. 6: 196 - 200.

Xiaoning W, Nanwen Z, Bingkui Y (2008). Preparation of sludge-based activated carbon and its application in dye wastewater treatment. J. Hazard Mater., 153: 22 - 27.

Yongjie X, Haobo H, Shujing Z (2009). Adsorption removal of reactive dyes from aqueous solution by modified basic oxygen furnace slag: Isotherm and kinetic study. Chem. Eng. J., 147: 272 - 279.

16

Assessment of health risks associated with wastewater irrigation in Yola Adamawa State, Nigeria

Hussaini I. D.[1]*, Aliyu B.[2], Bassi A. A[1]. Abubakar S. I.[3] and Aminu M.[1]

[1]Adamawa State College of Agriculture, P. M. B. 2088 Ganye, Adamawa State, Nigeria.
[2]Modibbo Adama University of Technology, Yola Adamawa State, Nigeria.
[3]Federal Polytechnic Bali, Taraba State, Nigeria.

The study was conducted to assess the health risks from the use of wastewater for irrigation to ensure a sustainable and safer agricultural production. Samples of wastewater and lettuce wash-water were analyzed to determine the faecal coliform count and the presence of helminth eggs. Besides, the opinion of the public was seeked through the use of questionnaire. The data collected were analysed using some statistical analysis (ANOVA), correlation analysis and simple statistics. The result shows that there were significant differences ($p \geq 0.05$) among the faecal coliform counts in the wastewater of the study area. However, in the case of the lettuce wash-water, there were no significant differences ($P \geq 0.05$) among the values of the faecal coliform in the area except that of Bajabure which differed significantly from the rest. The result also shows that there was no correlation ($r = -0.15$) between the faecal coliform in the wastewater and that in the lettuce wash-water. Helthminth eggs were found to be present in some of the wastewater and the lettuce wash-water of the study area.

Key words: Wastewater, health risk, coliform, helminth, irrigation, lettuce.

INTRODUCTION

The use of wastewater in growing crops is normally associated with some problems. One of such problems is the health hazards posed to the farmers/farm workers, crop handlers, consumers and residents around the wastewater irrigated fields. It is to be noted that wastewater normally carry a potentially dangerous loads of pathogenic organisms and other contaminants such as heavy metals that have the potentials to pose some risks to public health.

Health risks from the use of wastewater can include; the spread of infectious disease by bacteria (typhoid fever, dysentery, and tetanus), worm infections (round worm, whipworm, and tapeworm) and other diseases (Saxena and Frost, 1992). Raw wastewater frequently contains high number of faecal coliform bacteria. Blumenthal et al. (1996) reported figures as high as 200 to 300 eggs/L in Japan, 38 to 670 eggs/L in Brazil and 18 to 840 eggs/L in Morocco.

These figures indicate that the health risks associated with wastewater irrigation varies with different communities depending on the population, industrialization and life style of the people.

In communities where wastewater treatment is inadequate or non-existent (as in Adamawa state), people could be infected by drinking contaminated water, juices made with contaminated water or other beverages made with contaminated water or ice; eating food improperly handled by infected people or carriers; eating vegetables and fruits contaminated by irrigation with polluted water or fertilized with untreated sewage; eating meat or drinking milk from animals that grazed on contaminated pasture or drank contaminated water; eating fish grown, caught or harvested in contaminated water; and eating food exposed to flies or vermin that feed on or come into contact with sewage (Taylor et al., 1996).

Farmers, their families and crop consumers might be at risk, because the vegetable grown in wastewater irrigated fields are mostly eaten uncooked. Therefore, the pathogens present in wastewater can contaminate these

---
*Corresponding author: E-mail: hussainidanjuma@gmail.com.

vegetables and pose a health risk to consumers (Blumenthal et al., 1996). In a study in Pakistan (Feenstra et al., 2000) reported that, untreated wastewater used for irrigation contained a concentration of helminth eggs, faecal coliforms and bacteria that exceeded the WHO guideline, (WHO, 1996). These posed a high potential health risk to both farmers and crop consumers. Farmers are at higher risk because in addition to the consumption of some of the uncooked products, they had intensive contact with wastewater, as they do most of the fieldwork manually and barefooted. A significantly higher prevalence of Entamoeba hystolitica infections and diarrhoea in children was also observed in a wastewater-irrigated area in Mexico (Feenstra et al., 2000).

In evaluating health impacts of wastewater, it should be known that, it is the actual risk that make people fall ill that must be quantified and not merely the presence of pathogens in water (potential risk). Whilst potential risk may be high, the actual risk according to Feenstra et al. (2000), in addition to personal hygiene, societal and environmental factors, could depend on the time of survival of pathogen in waste water; infective dose; host immunity; and the microorganism's ability to cause disease (Pathogenicity). The health risks are in the following order; helminthes infection > bacteria > protozoa and lastly virus (Swartzbrod, 1995; Blumenthal et al., 2000; van der Hoek et al., 2000; Dalsgard, 2001; van der Hoek et al., 2002).

In addition to the risks posed to public health by pathogens, ponds and canals used for handling wastewater may also serve as habitats for disease vectors such as mosquitoes and snails. All of these health issues have to be addressed in order to make agricultural wastewater use safer and sustainable.

In Adamawa State, irrigated agriculture is not new to the various communities that inhabit the riverbanks. They practice irrigation during the dry season to provide food crops and fresh vegetables (Ray and Bashir, 1999). However, due to increase in the discharge of wastewater and the competing demands of freshwater, the use of wastewater irrigation has become more pronounced and is now very common at the outskirts of some major towns in the state. Crops grown are mostly vegetables, such as; lettuce, cabbage amaranthus, spinach, and okra.

During the peak consumption periods of these products, there used to be outcry of rampant cases of water-related diseases and this has always been attributed to lettuce and other products from the wastewater-irrigated fields. This study is therefore conducted to access the health risks from wastewater irrigation.

MATERIALS AND METHODS

Study area

Yola (Jimeta) the Adamawa state capital is located between longitude 12° 26' E and Latitude 9° 16' N (http://www.en.wikipedia .org/wiki/Jimeta) along the banks of the River Benue (Adebayo, 1999). The state is in the Sahel region of Nigeria generally Semiarid with low rainfall, low humidity and high temperature. The climate is also characterized by high evapotranspiration especially during the dry season (Adebayo, 1999). Yola the state capital being an urban centre has an estimated population of about 200,000 people. There is high water demand for domestic as well as industrial and agricultural purposes. This explains the large amount of wastewater being produced from the town. The sources of the wastewater for the Shinco, Doubeli, Geriyo and Mayanka is from Yola (Jimeta) through some interconnected drains to the floodplain near River Benue where the wastewater is used for irrigation, while that of Bajabure is partly from the town and partly from the River Benue. The samples were collected at the point of application. The number of farmers could not be ascertained due to lack of records and the farmers practiced basin method of irrigation.

Data collection

Two methods were used for this purpose

(i) Laboratory analysis.
(ii) Determination of (excess) case of associated diseases in the study area.

Laboratory analysis

Five sampling areas were identified; they are Shinco, Doubeli, Geriyo, Mayanka and Bajabure. These areas were designated as A, B, C, D, and E.

Measurement of faecal coliform counts

From wastewater samples

From each sampling area, twenty (20) samples of one litre each were randomly collected from different points. The collected samples from each sampling points were then thoroughly mixed to get a representative sample of each area. The samples were collected during the irrigation season once a week for four weeks (26 January, 2006 to 16th February, 2006). From each representative sample, three samples of one litre each were collected in some sterile screw capped bottles rinsed with the wastewater. The samples were taken to Microbiology Laboratory of Federal University of Technology (now Modibbo Adamawa University of Technology) Yola, Nigeria in less than twenty minutes.

The samples were analysed using the membrane filtration method. Samples were filtered through a membrane filter with a pore size of 0.45 µm. the membrane filter was then placed on a pad saturated with growth medium incubated for 24 h. During the incubation, each faecal coilform bacterium developed into a yellow colony that was visibly seen. After the incubation, the yellow colonies were counted and the count per 100 ml was calculated (Table 1).

From lettuce wash-water sample analysis

Twenty (20) grams of lettuce were randomly collected from five points in each of the sampling areas. The collected samples for each of the sampling area were mixed together as representative samples of the respective sampling areas. These representative samples were taken in some sterile plastic bags immediately (less than 20 min) to the Microbiology Laboratory of Modibbo Adama University of Technology Yola. In the Laboratory each representative sample of the lettuces were washed in four litre of

Table 1. Faecal coliform bacteria in wastewater samples.

Location	Number of faecal coliform / 100 ml				
	Week 1	Week 2	Week 3	Week 4	Mean/week
Shinco	370	270	280	287	300[b]
Doubeli	337	327	317	284	316[a]
Geriyo	234	240	190	234	225[e]
Mayanka	277	254	227	247	251[d]
Bajabure	301	275	258	307	285[c]
LSD(0.05)	46.0	41.0	60.0	34.3	45.3

Mean with the same letters are not significantly different at $P \geq 0.05$.

Table 2. Faecal coliform bacteria in lettuce washwater.

Location	Number of faecal coliform / 100 ml				
	Week 1	Week 2	Week 3	Week 4	Mean/week
Shinco	90	110	124	97	105[a]
Doubeli	90	104	120	104	104[a]
Geriyo	97	97	104	104	100[a]
Mayanka	97	107	110	97	103[a]
Bajabure	27	37	37	37	35[b]
LSD(0.05)	19.3	16.9	16.7	16.7	17.3

Means with the same letters in the same column are not significantly different at $P \geq 0.05$.

distilled water. Three potions of the washed-water from each representative sample were collected and analysed as described for the wastewater. The result is presented in Table 2.

Determination of helminth eggs

For this purpose the presence-absence tests was adopted to determine the level of contamination.

Determination of (excess) cases of associated diseases in the study area

Questionnaire administration

Two hundred questionnaires aimed at seeking the public opinions on the health risk related to wastewater irrigation were administered with the help of some trained personnel. The questionnaires were directed towards the following target groups, Farmers/farm workers; crop handlers; health personnel; and general public (consumers).

Fifty questionnaires were randomly distributed to each of the above target groups.

Method of data analysis

Analysis of variance ($p \geq 0.05$), correlation analysis and simple percentages were used to analyse the data collected from which inferences were drawn.

RESULTS AND DISCUSSION

Faecal coliform counts

The faecal coliform counts in the water samples of the study area (Table 1) indicated that there were significant differences ($p \geq 0.05$) among the values of coliform counts. Wastewater from Doubeli had the highest value of faecal coliform counts (316 FC /100 ml) while Geriyo had the lowest value (225 FC/100 ml), this is similar to the results obtained by Ensink et al. (2002).

The faecal coliform counts in the lettuce wash-water (Table 1) ranged between 35 and 105 FC/100 ml. The result indicates that lettuce samples collected from Bajabure recorded the lowest value (35 FC/100 ml) and this differ significantly from the ones collected in other sampled areas. Shinco lettuce had the highest coliform counts (105 F.C/100 ml) but the value does not differ significantly from those in Doubeli, Geriyo and Mayanka.

Microbial analysis carried out also indicated presence of helminth eggs in the wastewater and the corresponding lettuce wash-water except the one from Bajabure.

Though it was gathered from the public in the study area that wastewater irrigated vegetables could be among the major causes of water borne diseases (Table 5); the correlation analysis (Table 3) shows that there is a

Table 3. Correlation analysis between wastewater samples and Lettuce wash-water samples.

Source	Water sample	Lettuce sample
Wastewater sample	1	-0.149
Lettuce wash-water sample	- 0.149	1

Table 4. Presence of Helminth eggs.

Source	Shinco	Doubeli	Geriyo	Mayanka	Bajabure
Wastewater	Present	Present	Present	Present	Absent
Lettuce wash-water	Present	Present	Present	Present	Absent

Table 5. Data from the questionnaires on Health risk of Wastewater Irrigation.

Respondents	Level of agreement				
	Strongly agree	Agree	Disagree	Strongly disagree	Row total
Farmers	16(28.75)	10(6.25)	20(11.00)	4(4.0)	50
Crop handlers	14(28.75)	8(6.25)	22(11.00)	6(4.0)	50
Health personnel	45(28.75)	3(6.25)	2(11.00)	0(4.0)	50
Consumers	40(28.75)	4(6.25)	0(11.00)	6(4.0)	50
Total	115	25	44	16	200

Source: Field Survey (2006).

weak or near no correlation (r = - 0.15) between the faecal coliform counts in the wastewater samples and the amount of coliforms in the corresponding lettuce. This may mean that faecal coliform in the wash-water could be as a result of contamination during handling or due to the method irrigation practiced. Hence consumption of the wastewater irrigated vegetables could not be one of the causes of the water bone diseases.

Evaluation of heath impact of wastewater as irrigation water

The values of the faecal coliforms in wastewater in all the study area are within the WHO guideline for restricted irrigation of <=1000 F.C./100 ml (WHO, 1996; Pescod, 1992). The values are, however above the potential risk values of 100 to 105 F.C. / 100 ml for the unrestricted irrigation (WHO, 1996, Blumenthal et al., 2001). This indicates that the wastewater in the study areas has low potential risk values that could pose risk to public health. However, the faecal coliform of the lettuce wash-water (taken as the actual risk) in most of the areas are below the World Health Organization potential risk values (WHO, 1996). This shows that the wastewater can be safely used for irrigation under good management, such as; the use of furrow method of irrigation, the use of waste stabilization pond or the use of wetlands with aquatic plants, which should be supplemented by other health protection measures (Blumenthal et al., 2001).

From the survey, 70% of the respondents are of the opinion that wastewater irrigation posed risk to public health, this result is similar to that obtained in Sangodoyin (1992). Most of those that agree with the statement "wastewater irrigation pose risk to public health" are the health personnel and the consumers. This can be attributed to their knowledge of human health and their experiences in water related ill-health. Those that disagreed with the statement are mostly the farmers and crop handlers. These are into the business that involves irrigating with wastewater.

From the study (Tables 1 to 5 and Figure 1), farmers, consumers, health personnels have diverse opinion on the health risk of using wastewater for irrigation.

Conclusion

The prevalence of water borne diseases during the irrigation season is a major concern. But this could not be linked to the consumption of wastewater irrigated crops, unless assessments of other risk factors are made.

The study revealed that, wastewater in the study area has low actual risk values and could not be the major cause of the water borne and water related diseases during the irrigation season. Hence the wastewater could be used for wastewater irrigation and the crops so produced can be consumed with proper handling and good preparation. The study also revealed that the faecal coliform found in the wash-water (actual risk values) could be as a result of contamination through handling and the use of basin irrigation method by the farmers in

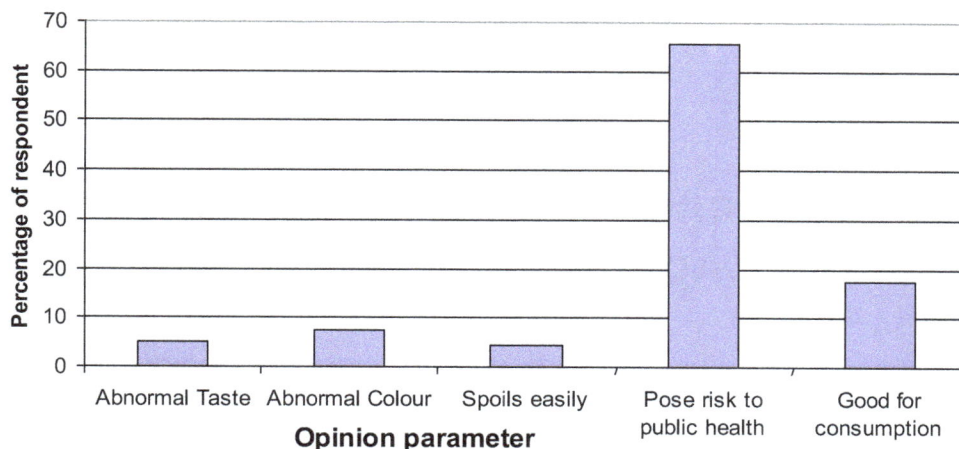

Figure 1. Personal opinion on crops irrigated with wastewater.

the study area. This encourages intimate contact between the lettuce and the wastewater.

RECOMMENDATIONS

(i) Furrow methods of irrigation are recommended instead of the present basin (the sunken bed) method to reduce the level of product contamination.

(ii) Harvesting of crops irrigated with wastewater should be done four days to one week after irrigation. This will ensure that harvesting is done when the soil is not wet.

(iii) Farmers and crop handlers should ensure frequent use of protective covers (for example, hand gloves) and other health protective measures.

(iv) Sound environmental and personal hygiene should be observed.

(v) Ensure that vegetables from the wastewater irrigated areas are washed thoroughly (with brine) before consumption.

(vi) Similar study could be undertaken in almost every community for sustainable wastewater irrigation.

REFERENCES

Adebayo AA, Tukur AL (1999). Adamawa States in Maps. Paraclete, Yola.

Blumenthal UJ, Mara DD, Ayres RM, Cifuentes E, Peasey A, Stott R, Lee DL, Ruiz-Palacios G (1996). Evaluation of the WHO Nematode Egg Guidelines for Restricted and Unrestricted Irrigation. Water Sci. Technol. 22(10-11):277-283.

Blumenthal UJ, Mara DD, Peasey A, Ruiz-Palacios G, Stott R (2000). Guideline, for Microbiological Quality of Treated Wastewater used in Agriculture. Recommendation for Revising WHO Guidelines. Bull. World Health Org. (WHO) 78(9):1104-1116.

Blumenthal UJ, Mara DD, Peasey A, Ruiz-Placios G, Stott R (2001). Reducing the Health Risk of using Wastewater in Agriculture. Recommended Changes to WHO Guidelines http://www.rauf.org/1-3/26-29.html.

Dalsgard A (2001). Health aspects of the Reuse of wastewater In Agriculture and Aquaculture in Vietnam. In Liga RS, van der Hoek W, and Ranwaka M. (Eds), Water Management, Environment and Human Health Aspects. Proceedings of a Workshop in Hanoi, Vietnam 14th March 2001. IWMI Working paper 30, Colombo, Sri Lanka pp. 26-27.

Feanstra S, Hussain R, van der Hoek W (2000). Health Risks of Irrigation with Untreated Urban Wastewater in the Southern Punjab, Pakistan. IWMI Report No. 107. Institute of Public Health, Lahore. Pakistan Program. IWMI Lahore Pakistan.

Ensink JHJ, van der Hoek W, Matsuno Y, Munir S, Aslam MR (2002). Use of untreated wastewater in peri-urban agriculture in Pakistan: Risks and opportunities. Research Report 64. Colombo, Sri Lanka. Int. Water Manag. Inst. p. 12.

Pescod MD (1992). Wastwater Treatment and Use in Agriculture. Food and Agricultural Organisation (FAO) Irrigation Drainage Paper 47. Rome, Italy.

Ray HH, Bashir BA (1999). Irrigated Agriculture; In Adebayo AA and Tukur AL. (Eds), Adamawa State in Maps. Paraclete, Yola pp. 41-43.

Sangodoyin AY (1992). Wastewater Application: Changes in Soil Properties Livestock Response, and Crop Yield. Environ. Manag. Health. 3(1):11-17.

Saxena R, Frost S (1992). Sewage Management in some Asian countries. Environ. Manag. Health 3(1):18-26.

Swatzbrod L (1995). Effect of Human Virus on Public Health Associated with the use of Wastewater and Sludge in Agriculture and Aquaculture. WHO, Geneva.

Taylor C, Yahner J, Jones D, Dun A (1996). Waste and Public Health. On-site Wastewater Disposal and Public Health. Purdue on-site Wastewater Disposal. Purdue Wastewater Articles on-line, Purdue University. Pipeline Summ. 7:3. www.onsiteconsortium.org/files/Wastewaterreuse.

Van der Hoek W (2001). Reuse of Wastewater - A Global Perspectives. In Liqa RS, van der Hoek W, and Ranawaka M (Eds), Wastewater Reuse in Agriculture in Vietnam: Water Management, Environment and Human Health Aspects. Proceedings of a Workshop held in Hanoi, Vietnam 14 March, 2001. IWMI Working Paper 30, Colombo Sri Lanka. pp. 4-5.

van der Hoek W, Mehmoud UIH, Ensink JHJ, Feenstra S, Raschid LS, Munir S, Aslam R, Ali N, Hussain R, Matsuno Y (2002). Urban Wastewater: A Valuable Resource for Agriculture. A Case Study from Haroonabad, Pakistan. IWWMI Research Report 63 http://www.iwmi.org.

Wikipedia (http://www.en.wikipedia.org/wiki/Jimeta. Downloaded on Wednesday 22nd August, 2012.

World Health Organization (WHO) (1996). Analysis of Wastewater for use in Agriculture: A Laboratory Manual of Parasitological Techniques. WHO Geneva. http://www.who.int/docstore/watersanitation_health/labmanual/ch5.htm654.2%Helminth% 20eggs.

17

Influence of suspended matters on iron and manganese presence in the Okpara Water Dam (Benin, West Africa)

Tomètin A. S. Lyde[1], Mama Daouda[1,2], Sagbo Etienne[1], Fatombi K. Jacques[3], Aminou W. Taofiki[3] and Bawa L. Moctar[4]

[1]Laboratoire de Chimie Inorganique et de l'Environnement (LACIE), Université d'Abomey-Calavi, 01 BP 526, Cotonou (Bénin) République du Bénin.
[2]Laboratoire d'Hydrologie Appliquée (LHA) Université d'Abomey-Calavi, République du Bénin.
[3]Laboratoire d'Expertise et de Recherche en Chimie de l'Eau et de l'Environnement (LERCEE) Faculté des Sciences et Techniques/ Université d'Abomey-Calavi, 01 BP 526, Cotonou, Bénin.
[4]Laboratoire de Chimie de l'Eau, Faculté Des Sciences, Université de Lomé, BP 1515, Lomé, Togo.

Iron and manganese, which their concentrations are seasonally high in the Okpara Water Dam, have a high mobility links to suspended matters (SM). The presence of the water plants constitutes SM retention support within the water column. The anthropoid disturbances caused by the fishermen activities constitute a factor of sediments renewing in the suspension. Solid (>0.45 µm)-liquid (< 0.45 µm) fractionation has been carried out and enabled us to observe that SM are responsible for iron content for more than 10 to 98% and for Manganese content about 23 to 93% filtration removes from 10 to 98% of the colloidal iron which is related to filtrated SM. The phosphorus constitutes a combined factor of the iron mobility in the dam. We have noticed in the top water layers a high proportion (low repartition coefficient) of dissolved (and colloidals) iron respectively. The iron retained by SM is more concentrated in the middle of the water column than anywhere. The west side of the dam is identified as a manganese enrichment source. The manganese ion concentration influences positively on pH ($r = 0.57$), conductivity ($r = 0.78$), color ($r = 0.66$), and SM ($r = 0.66$) after decantation. Furthermore, the iron concentration is negatively influenced by the pH ($r = -0.52$) and positively by the TDS ($r = 0.51$) after filtration. A pre-filtration or a pre-decantation could reduce the quantity of chemicals used during water treatment.

Key words: Iron, manganese, suspended matters (SM), surface water, fractionation, mobility.

INTRODUCTION

Okpara Dam water constitutes a very significant source of supply drinking water of population of Parakou in the north of Benin. This water is polluted, that pollution can be observe through the eutrophication of the dam (Zogo, 2010). Otherwise, the concentration of iron and manganese are influenced by season. Iron, and in a lesser extent, manganese, are the most abundant metallic elements in the earth's crust. It had been shown that both iron and manganese found in water coming from the lixiviated grounds and industrial pollution. These elements

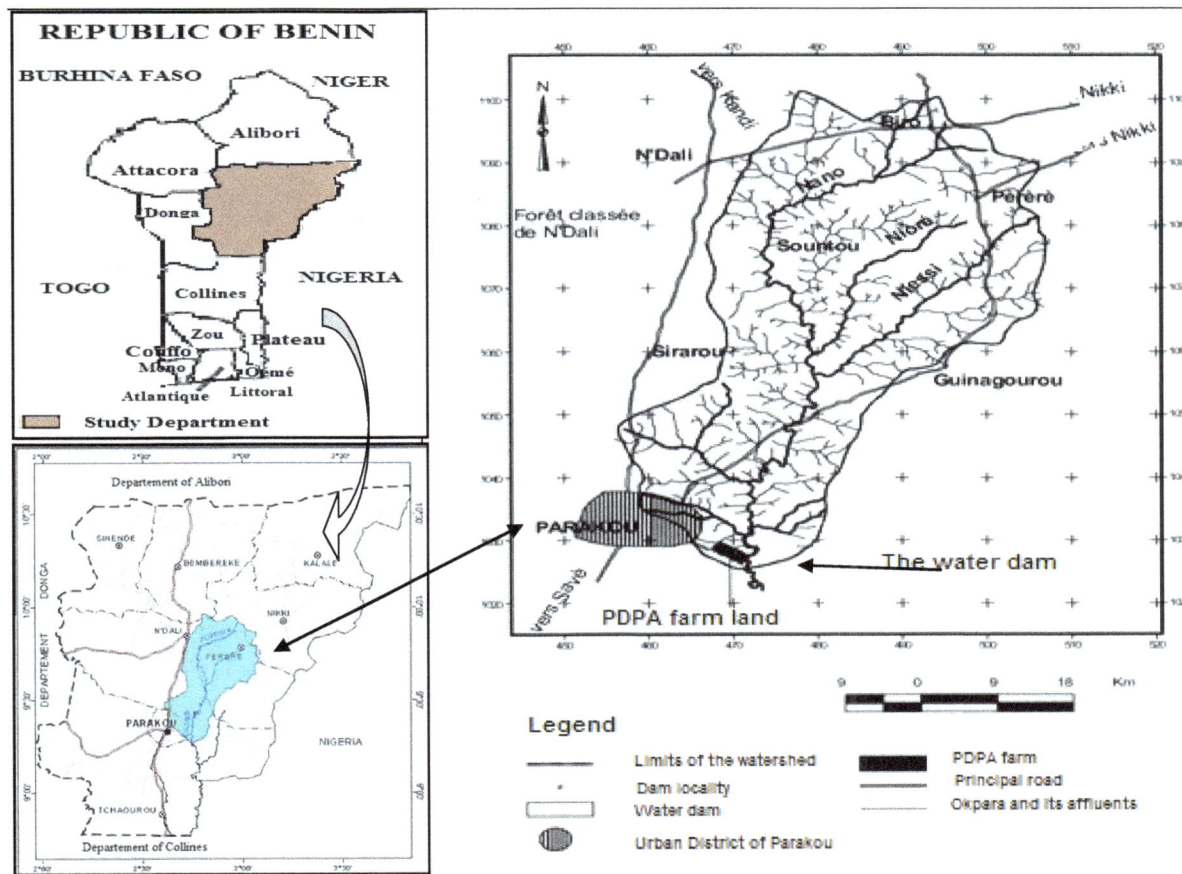

Figure 1. Localization of Benin, Parakou's town and hydrographic system in the under catchment of Okpara's Dam. Source: CENATEL (2003).

do not damage human health or environment. But they cause esthetic and organoleptic damages. Iron and manganese are found in surface or underground water. Removal of iron and manganese from Okpara raw water has been highlight through some studies (Zogo, 2010). This technique consumes a high quantity of potassium permanganate. Furthermore, the origins of iron and manganese, and their transfer in the water column remained not understood. The objective of this study consists of determining the mobility of iron and manganese along the water column across solid-liquid fractionation in raw water.

MATERIALS AND METHODS

Context

This study is focused on Okpara water dam built on an affluent of the Ouémé River called Okpara to supply the town of Parakou (located at the North-East of Benin with 450 kilometers from Cotonou) in drinking water. The dam under catchment is located at the East of the town of Parakou and has been studied in some previous papers. This pouring under catchment of reserve located between 9° 16' and 9° 58' of Northern latitude and 23° 6' and 3° 05'

of East on longitude, covers mainly the districts of Pèrèrè in the East, Nikki in the North-East, N'dali in the West, Tchaourou in the South and a small part of the Commune of Bembèrèkè in North (Figure 1).

Analysis methods and testing

Water samples were carried out from five sampling points in December 2011 (D), April 2012 (A) and in June (2012) (Figure 2).

Two samples sites (N1, N2) were more focused. Raw water samples were taken 20 cm from the surface and from 1.5 m depth to 6.5 m in 1.5 L polyvinyl chloride (PVC) plastic containers the day of the tests. The samples were identified as follows: Sampling level in the water column profiles "surfaces (S); middle (M) and bottom depth (P) "followed by the etiquette of the sampling point (N) and period" as indicated in the Table 1. Example: The water sample taken from the surface of water at the research point N1 in April is identified as follows"S-N1-A". Filtration: Samples of raw water were filtered in an erlenmeyer by using a 0.45 μm pore size Durex folded filter paper. Decantation: A portion of the raw sample water was left for decantation in anoxic condition during one to two days, and some physic-chemical parameters were determined. The difference between the total solids in suspension and the non-decantable solids in suspension gives the concentration of decantable solids in suspension. SM are responsible of water turbidity; these SM are various such as silt, clay, organic and inorganic matter in small

Figure 2. Localization of the sampling points in station on the reserve of Okpara.

Table 1. Geo-referred situation of the sampling stations on the reserve.

Sampling points	Localization	Water column depth	Period
N1	N 09° 17' 08.88" E 002°43'58.67"	2.5 m to 4 m	(A) ; (Jn)
N2	N 09° 17' 02.06" E 002° 43'58.66"	6 m to 7.5 m	(D) ; (A) ; (Jn)
N3	N 09° 17' 02.04" E 002° 43'58.57"	2 m to 3.5 m	(A)
N4	N 09° 16' 58.25 E 002° 44'00.00"	4 m to 6.5 m	(Jn)
N5	N 09° 17' 1589" E 002° 43'43.25"	0.2 m to 0.5 m	(A) ; (Jn)

particles, made up of soluble colored organic matter, plankton and other micro-organisms (American Public Health Association, 1989). Electric conductivity, the total dissolved solids (TDS) and the pH of the samples were measured using a Waterproof pH-meter. The turbidity, the color and the SM were measured by using a HACH colorimeter. The total iron and manganese content were measured; also dissolves and colloidal iron and manganese content were also determined in the raw, filtrated and decanted water samples. Iron concentration was determined by using the orthophenanthroline method and measure at 510 nm wavelength. Manganese concentration was determined according to the potassium periodate acidic oxidation method at 525 nm wavelength. A JENWAY spectrophotometer model 6305 was used for the absorbencies measurement. The iron and manganese analysis methods where

based to the standards of the French Standardization Association (AFNOR) of the water quality, Tome3 (1997).

Statistical analyses were performed on SPSS 16.0 software.

RESULTS

According to Figure 3, the suspended matters are well eliminated after filtration and decantation. This condition is favourable to the measurement of the true color. Filtration allowed more SM removal than decantation. The pH of these raw water samples reveals an acid

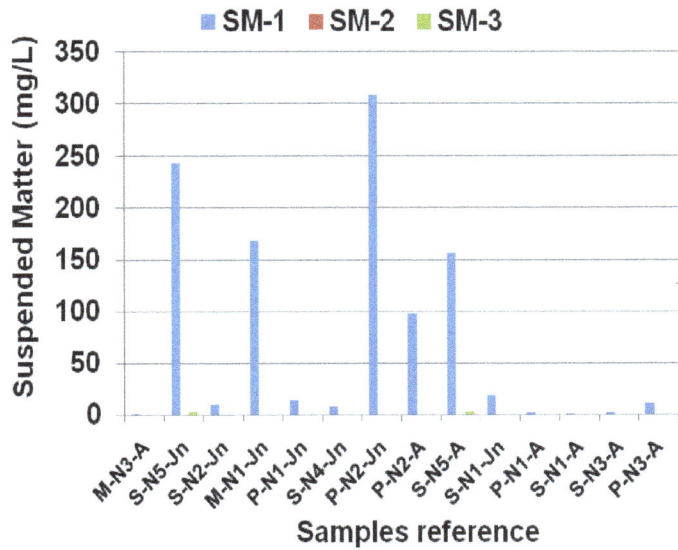

Figure 3. Suspended matter of each sample before filtration (SM-1), after filtration (SM-2) and after decantation (SM-3).

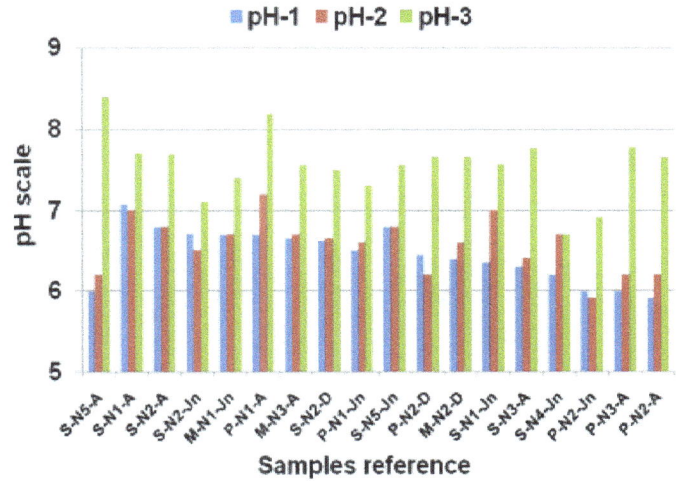

Figure 4. Samples pH values before filtration (pH-1), after filtration (pH-2) and after decantation (pH-3).

character higher from the surface to the bottom of water column (Figure 4). For these waters, there is an average value of all the points which is about 6.51 for raw waters, 6.62 after filtration and 7.56 after one to two days of decantation in anoxic condition. Thus there is an average variation of one unit of pH. All the collected water samples have pH increasing during their storage to a basic pH ranging between 7 and 8.5. The pH is one of the factors (other that the potential rédox) which has a strong influence on the behavior of oligo-elements in the external environment. For example, the reduction in the pH of one unit can lead to an increase of more than one order of magnitude in the concentration of certain metals like aluminum, beryllium (Edmunds and Smedley, 1996). The Metals and metalloids concentrations in the oxy-anions form increase in the aqueous phases when the pH increases (Bonnet, 2000).

Two types of surface water's colors can be measured. One type is the apparent color taking into account the SM (Coul-1) and the second type which is true color measured after SM removal by centrifugation or filtration (Coul-2). We observed according to Figure 5 that whatever is the sampling point, the elimination of SM is always accompanied by the reduction of the color. This fact can be explained by the inorganic matter contribution (sand, clay, etc), and the presence of organic matter and other renewed mineral matters (iron and manganese) in the suspension carried by the surface water during rainy season. The samples taken in the bottom, points NI, N2, N3 and surface in April at point N5 are less suitable for color elimination after filtration or decantation because they presented color values more than 55 color unities.

We have also observed on Figure 6, a light increase of conductivity following the SM removal. This increase

varies from 1 to 17 µS/cm except in the point (N5) for the two distinct periods April and June. Water samples from point (5) presented higher conductivity (>100 µS/cm) which decreases after filtration. We could think about a precipitation reaction of the dissolved ions or their adsorptions on eliminated SM. According to Graeme and Jameson (1999), ventilation during filtration can also lead to the flotage of the iron particles, by linking bubble to particle, which can more concentrate the particles in the flocks and give a better turbidity reduction after filtration (Béchir et al., 2007). In these conditions, waters which have high conductivity with strong rate of TDS will be unbalanced and one will observe the decrease of TDS which also involve a conductivity reduction. On the other hand the decantation process of samples from point N5 show conductivity increase with value around 485 µS/cm. Phenomena after decantation process at point N5 are the same, with less values for points N1 and N2 water samples.

Figure 8 reveals that the turbidity of raw water samples varies from 13 to 885 NTU which are results like the values found on the surface water by Hector (2006) which varies from 1.8 to 1948 NTU. The rate of turbidity elimination after filtration is 10 to 94%. December water samples have lowest turbidities and vary from 13 NTU on the surface and 80 NTU in the bottom (4 m) with the rate of turbidity elimination after filtration around 46 to 80%. For waters which turbidities are lower than 100 NTU, the turbidity elimination tends to be 100% after the decantation. The rate of turbidity elimination after decantation is a function of initial turbidity (before and after SM elimination turbidity correlation r = 0.76). In dry season where there is no movement of the water the reduction of turbidity by the sedimentation of the suspended matter (SM) must be observed. However, the presence of the water plants, combined with the daily

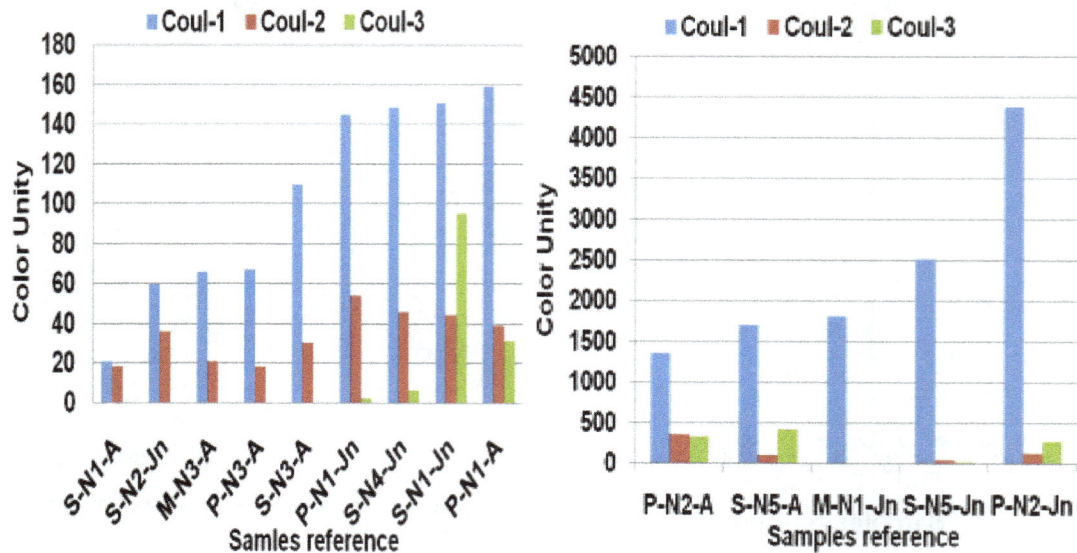

Figure 5. Samples color before filtration (Coul-1), after filtration (Coul-2) and after decantation (Coul-3).

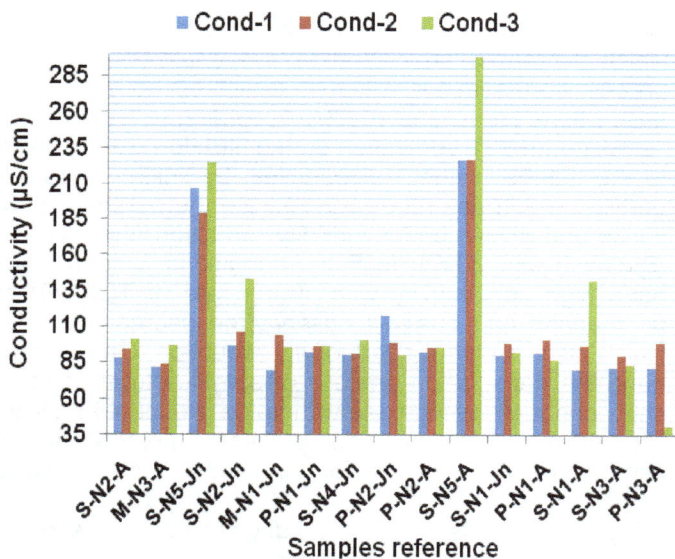

Figure 6. Samples conductivity before filtration (Cond-1), after filtration (Cond-2) and after decantation Cond-3).

impact of fishermen' activities are not favorable to the decantation process *in-situ*. The biological perturbation of the sediment is also a permanent factor of exchange of SM between sediment and water column (Santschi et al., 1997).

Figure 9 reveals that the manganese concentration decreases after filtration and this is more significant after the decantation except the point N5 where we observe the contrary phenomenon. The great values of concentrations are recorded in June (Jn) and more in the water samples taken on the surface. The water samples taken in April thus present the greatest manganese concentrations after filtration; the dissolved form of manganese has significant values and this in the middle and the bottom of the water column. Less than 50% of manganese is eliminated in this point but this is not observed with other samples. The measured manganese concentrations in the point N5 water samples are higher than the results obtained by Zogo (2010) on the waters dam whatever is the season. One measured values are about the double of Zogo results. This sampling point (N5) on Figure 2 represented a water collecting point coming from the rivers located in the districts of Eastern of Parakou town, the water treatment plant of "Soneb", the agro-pastoral farm (Figure 1). This point N5 thus represented a source of manganese which enriches the reserve water.

We deduced from the obtained iron values presented on Figure 10 that the content of particular iron varies from 5 to 78% of the total iron content. The ratio between metal fixed by SM and dissolved metal corresponding to the partition coefficient (Kd) allowing the distribution between the dissolved and colloidal form and the particular form. Weak values of Kd indicate a strong contribution of dissolved metals to their total concentration (Vignati, 2004). For analyzed water samples, Kd is very high and varies between 1 and 70 with an average value around 17. According to the work of Valérie (2009) Kd measured for metallic traces elements vary typically in range of 1 to 3 for a given metal according to the sites. Kd is often considered as a site - specific parameter at a given time (Fournier-Bidoz and Garnier-Laplace, 1994). For that, we cannot expect a sufficient reproductibleness of the partition coefficient

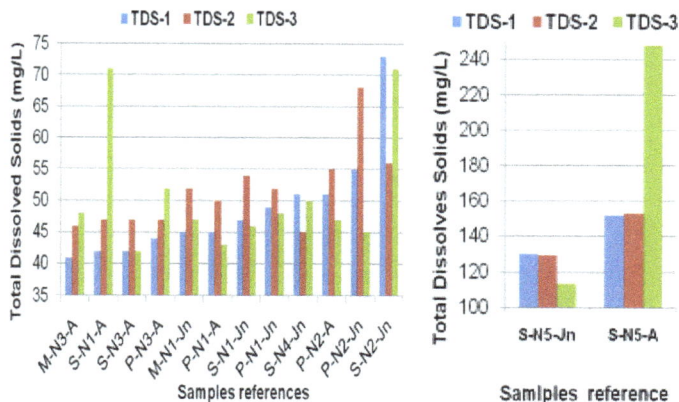

Figure 7. Samples total dissolved solids (TDS) before filtration (TDS-1), after filtration (TDS-2) and after decantation (TDS-3).

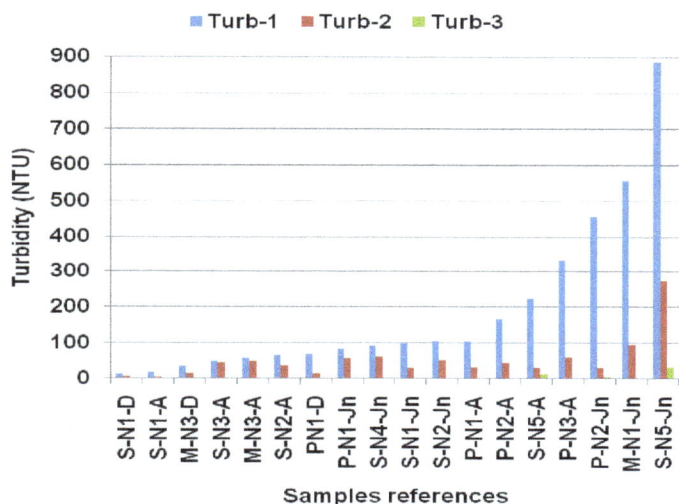

Figure 8. Samples turbidities before filtration (Color-1), after filtration (Turb-2) and after decantation (Turb-3).

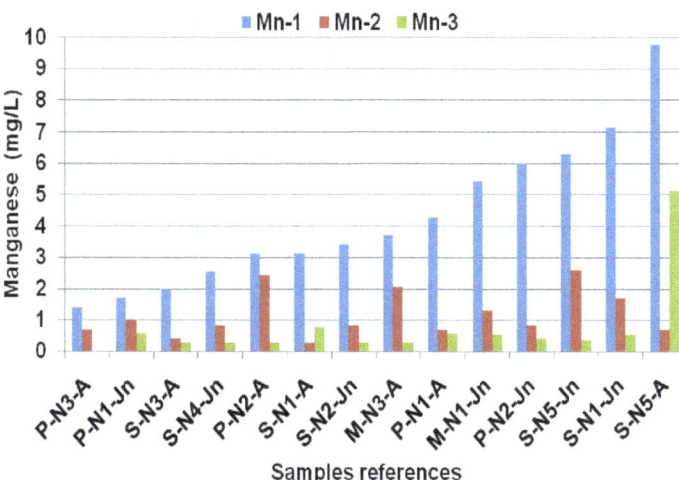

Figure 9. Manganese concentration of the samples before filtration (Mn-1), after filtration (Mn-2) and after decantation (Mn-3).

(Kd) value. The water samples taken on the surface (S) presented the weaker Kd and are less concentrated in dissolves and colloidal iron (< 5mg/L) than those of the bottom (P) from 5 to 30 mg/l whatever in December, April or June. SM contained iron in range of 53 to 99% in the water column. After filtration, all the collected water samples present lower values than 5 mg/L of iron, with an average around 0.96 mg/L in the surface layer and 2.57 mg/L in the bottom waters.

DISCUSSION

The main goal of a drinking water station production is to provide high quality water with a cheaper cost for the consumer. The various water quality parameters included such as turbidity, color, pH, SM, iron, manganese, relied on the water treatment process. To achieve the desired goal, each water treatment will ensure a good water quality and use high qualified human resources (Valentine, 2000).

Correlations are often established between turbidity, suspended matter, total solids and color. Analyzed water turbidity and to a lesser extent the color of analyzed water have show high decrease after filtration or the decantation and this is conform to result found by Zogo et al. (2011) which show that the rates of elimination of turbidity are higher at pH 6.50 and 40 mg/L of aluminum sulphate on the level of water of Okpara. After intense rains, there is SM increase with a good decantation. SM is strongly concentrated at the bottom of the water column from 2 to 310 than surfaces from 1 to 20. Simple filtration does not allow a correct elimination of the color and turbidity in the case of Okpara water in any season and any sampling point. The elimination of the color and turbidity by simple filtration is weak, probably because of the presence high content of iron and manganese seasonally. This weak turbidity elimination could be due to the TDS because (Figure 7), the turbidity of raw water can also be reinforced by the presence of inorganic solids like metallic oxides and hydroxides (iron or manganese) and biological organisms like the seaweed, the zooplankton and the filamentous bacteria or in cluster (Foley, 1980). The manganese ion concentration influences positively on pH (r = 0.57), conductivity (r = 0.78), color (r = 0.66), and SM (r = 0.66) after decantation. After decantation dissolved manganese form is strongly positively correlated (r = 0.72) with TDS-2 and, in contrast of results before filtration (r = 0.07) and after decantation (r = -0.15)
.This observation point out the particular forms of manganese in which concern their natural removal from water samples with the environment physic-chemical condition afferent. Furthermore, the iron concentration is influenced by the pH and in a less proportion by the TDS.

The average measured pH observed is 6.51. Zogo et al. (2011) has used the same pH value for pre-chlorination followed coagulation-flocculation to reach 50

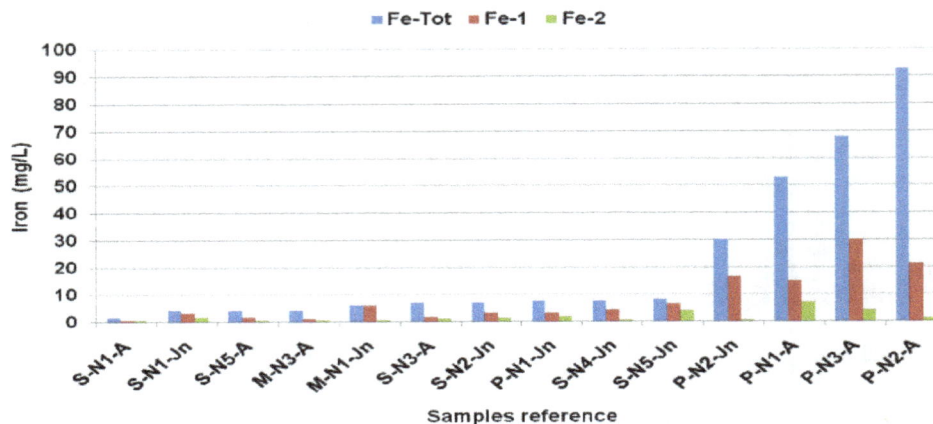

Figure 10. Total iron concentration (Fe-Tot); dissolved and colloidal iron before filtration (Fe-1) and after filtration (Fe-2).

to 95% of iron removal or 20 to 45% of manganese. The initial iron concentration variable was between 8.00 and 35.00 mg/L and an initial manganese concentration in range of 1.00 to 4.00 mg/L for experiences carried out by Zogo. The fact that we have observed a little increase in the pH after filtration and decantation, can be explained by an oxidation or a reduction of organic matters rate in the presence of the humic acids. Kedziorek and Bourg (1996) also showed that the humic substance presence could increase the pH; cation metals react then like anions (Bonnet, 2000). According to Soetaert et al. (2007) reoxidation by oxygen and the reduction of manganese and iron oxides tends to increase the pH. In addition, certain geochemical processes as the dissolution of calcite involve a consumption of hydronium ions (H_3O^+) and thus increase the pH. This can also justify in certain cases the increase of TDS after decantation.

In the reducing medium, the iron and manganese oxides are reduced and dissolved. This followed by a departure of the heavy metals from the different compartments of the sediment such as the organic matter, clay and especially the sulphides to the water column (Tack et al., 1996). This phenomenon appears observable in the deep layer of water column where Zogo (2010) observed high contents of organic matter. Filtration has eliminates 10 to 98% of colloidal iron which is related to SM filtrated. There would thus exist in water samples an important adsorbed iron. Because of their small size, the colloidal particles have specific surfaces (report/ratio surfaces/mass) significant (> 10 $m^2.g^{-1}$) and can thus represent adsorbent or absorbent form for the chemical elements in general and the metallic elements in particular (Citeau, 2004). According to IUPAC (1997), the colloids are molecules or polymolecular particles dispersed in a medium and having, at least in a direction, a dimension ranging between 1 nm and 1 μm. if the lower limit is considered without ambiguity around 1 nm (Lead et al., 1997),

determination of the higher limit seems more complex. It can vary, according to authors, from 0.2 to 0.5 μm (Singhal et al., 2006), and can even reach 1 μm (Lead and Wilkinson, 2006). Then a part of the colloids could be eliminated with the SM.

After filtration, the content of iron dissolved and colloidal in water varied from 0.41 to 7.10 mg/L. Physical treatment like filtration has thus a beneficial effect in terms of the reduction of chemical quantity to carry out the treatment without a pre-filtration of these water samples. The iron form in the surface layer of the Okpara Dam water in dissolved and colloidal which that is combined more with SM, and this could explain the weak rate of variation of the iron content related to SM after the filtration of surface water. The iron concentration is negatively influenced by the pH (r = -0.52) and positively by the TDS (r = 0.51) after filtration. In the absence of oxygen and nitrates, metallic oxides are the most powerful oxidants and are reduced in the anoxic area. According to Audry (2003) the redox species of metals which have more than one oxidation step in natural water have different mobility, solubility, toxicities and reactivity. According to Boust et al. (1999), reduced iron and manganese will be found in dissolved form Fe (II) and Mn (II) in porous water. In this form, they can either (i) diffuse towards the oxic layers or re-oxidize into oxides or (ii) precipitate, if the products of solubility are reached, with the chemical species produced by the degradation of the organic matter (sulphides, phosphates, carbonates). Manganese precipitates in carbonate salts form $MnCO_3$ (rhodochrosite), iron precipitates under carbonates salts form $FeCO_3$ (siderite) and salts of monosulphurs FeS (mackinawite) (Boust et al., 1999). In fact the form of precipitated iron and manganese will be higher in the presence of phosphate ions which are mainly present in Okpara water (Zogo et al., 2011). The anoxia and acidity in the sediments are favorable with the dissolution of oxidize-hydroxylated forms. Thus, certain metals, like Fe,

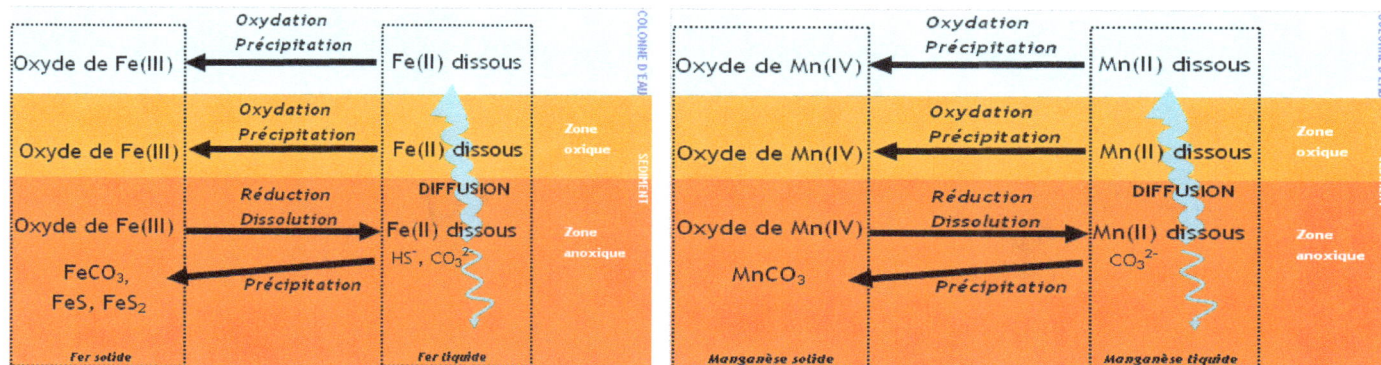

Figure 11. Behavior of iron and manganese in the sedimentary area of the rivers. (Boust et al., 1999).

Mn and Cr are more movable in their reduced form. The work archived by Zogo (2010) has show that the surface water layer in a given sampling point of the reserve of Okpara presented 10% of oxidized form (Fe III) against 90% of particular form iron with a total concentration of 2.015 mg/l. The redox conditions can increase or decrease the quantities of oxyhydroxides and sulphides adsorbed (Lions, 2004). Reducing media will support the solubilization of the metal species (oxides) and then increase or decrease the mobility of the metallic trace elements which diffuse from the sediments to the water surface layer (Blanchard, 2000).

The work achieved by Zogo (2010) concluded indeed that the splitting up related to the use of the grounds of under catchment area of the dam for intensive cotton culture and other food products accompanied by significant quantities of artificial fertilizers using are probably at the base of the imbalance observed on the level of this water ecosystem.

The work made possible to identify in the reserve of Okpara six macrophytes species and to count eighteen microphytes species Moreover, according to trophic levels' based on the chlorophyll a contents (OECD, 1982), with a concentration sometimes higher than chlorophyll 25,00 μg/l has and the level of development of the macrophytes in the stopping of Okpara, this reserve can be classified in the category of the eutrophic lakes'. The amount of ortho-phosphorus could reached 30 mg/L in the surface water then.

Phosphorus measured in the bottom water (Tomètin et al., 2013) showed that the total phosphorus could reach to 191 mg/L at station N1 and 397 mg/L at station N2. Also the sediment constitutes a source of phosphorus renewed for the water column. Then the high phosphorus amount in the water and sediment under different forms (mobility) could be control by the iron and manganese. The fraction of phosphates related to iron is very sensitive to the variations of the sediments redox potential. When the redox potential is lower than 200 mV (ESH), a fraction of Fe^{3+} available to the level of the sediments is reduced in Fe^{2+}. Gomez et al. (1999) show

that this value of potential redox is variable according to the pH of the sediments, to the presence of organic ligands and to porous water salinity. Phosphorus is thus likely to be salted out in the water column when the interface water-sediment becomes anoxic (Gomez et al., 1999) by various mechanisms of diffusion of interstitial water (Enell and Löfgren, 1988). In anoxic conditions on the surface of the sediments, it formation of iron hydroxide will thus not have there as the straight lines part of the Figure 11 indicates it and thus not of fixing of the phosphorus which is thus released in great quantity in the water column. In oxic conditions on the surface of the sediments, there are however two possible cases: the phosphorus diffuse of the sediment (anoxic) towards the interface water sediment (oxic), where it is trapped the iron hydroxide which is excellent adsorbing phosphorus in the presence of oxygen dissolved (Figure 12).

Conclusion

Iron and manganese, present in Okpara Dam water have various concentrations according to the season. Their contents and mobility are tied in the major part; with SM quantities are strongly retained more by the water plants. The colloidal SM is the major factor of iron high amount in the water middle and bottom column. The high amount of phosphate in the water column constitutes a factor of iron and manganese precipitation which migrate from the top to the water bottom. Also, these naturals' factors can be added to the anthropological activities which increase the exchange between sediments and water column. SM constituted a major vector of the mobility of iron and manganese within the water column. After decantation or SM removal from the water sample, the pH of this water varies from 6.51 to 7.62. The organic matter decomposition in the sediments and the water column maintains the water column acidity favorable to the increase of the dissolved forms which migrate towards the water surface layer under free form to be oxidized or engaged as ionic or colloid forms, tied with SM which

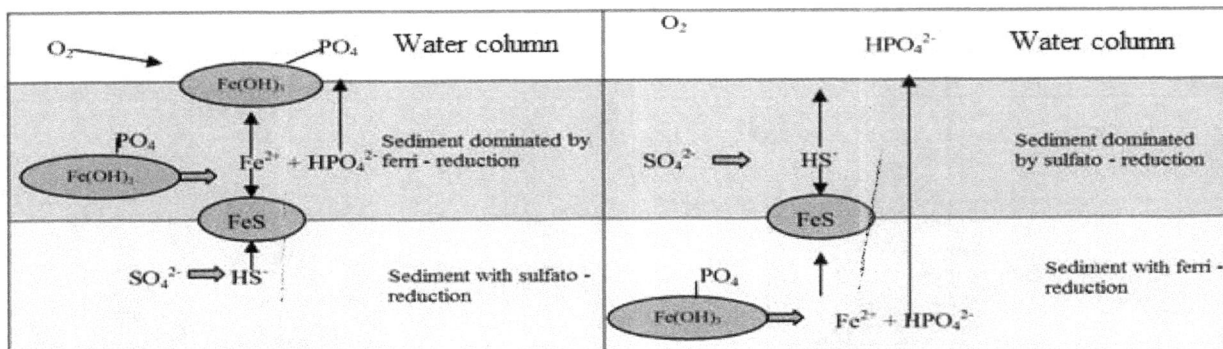

Figure 12. Diagrammatic representation of two possible mobility cases of phosphorus at water sediment interface in oxic conditions. The simple arrows represent flows of diffusion and the double arrows represent the chemical reactions (according to Matthiesen et al., 1998).

concentration increases from the surface to the bottom of the water column. Iron concentration along the water column strongly dependent on the organic matter responsible of the humic acids and the water acidic pH. Organic matter is the fundamental component of the SM. A probable source of manganese enrichment of the dam water is located at the point N5 where one river supplies water to the dam. The high concentration of manganese can be provided by total solid quantity of this river.

Physical treatment like pre-filtration has a beneficial effect in terms of chemical quantity reduction to carry out the treatment without a pre-filtration of these water samples. The fractionation and the speciation of iron and manganese in the water column and the sediments are new concept of control and treatment efficiency of Okpara Dam water.

ACKNOWLEDGEMENTS

The author extend their thanks to the Ministry of the Higher Education and Scientific Research of the Benin Republic for its financial grant, and Mrs. Chrysostome Montcho and Martial Dossou for technical assistance in the water analysis laboratory (Parakou Hydraulic General Direction) where some analyses were carried out immediately after sampling process. They acknowledge the useful comments and suggestions of the anonymous reviewers.

REFERENCES

American Public Health Association (1989). American Water Works Association/Water Pollution Control Federation. Standard methods for the examination of water and wastewater. 17th edition. Washington, DC. pp. 2-12.

Audry S (2003). Bilan géochimique du transport des éléments métalliques dans le système fluvial anthropisé Lot-Garonne-Gironde, Université de Bordeaux pp. 1-415.

Béchir BT, Khalifa R, Houda B (2007). Élimination de la turbidité par oxygénation et filtration successives des eaux de la station de Sfax (Sud de la Tunisie). Revue des sciences de l'eau, J. Water Sci. 20(4):355-365.

Blanchard C (2000). Caractérisation de la mobilité potentielle des polluants inorganiques dans les sols pollués, Institut National des Sciences Appliquées de Lyon, 241.

Bonnet C (2000). Développement de bio-essais sur sédiments et applications de l'étude, en laboratoire de la toxicité de sédiments dulçaquicoles contaminé. Thèse; Université de Metz, p. 326.

Boust D, Fischer JC, Ouddane B, Petit F, Wartel M (1999). Fer et manganèse: Réactivités et recyclages. Rapport Seine Aval. Ifremer, p. 40.

CENATEL (2003). Banques de données intégrées. Ministère de l'Agriculture de l'Elevage et de la Pêche (MAEP), Bénin.

Centre d'Expertise en Analyse Environnementale du Québec Détermination des solides totaux et des solides totaux volatils: Méthode gravimétrique. Édition : 2009-01-19 Révision : 2010-11-05 (1).

Citeau L (2004). Etude des colloïdes naturels présents dans les eaux gravitaires de sols contaminés: Relation entre nature des colloïdes et réactivité vis-à-vis des métaux (Zn, Cd, Pb, Cu). Thèse. Institut National d'Agronomie Paris-Grignon.

Edmunds WM, Smedley PL (1996). Groundwater chemistry and health: In: J.D. Appleton, R. Fuge and G. J.H. McCall (Eds.), Environmental geochemistry and health. Geological Society of London, special publication, 113:91-105.

Enell M, Löfgren S (1988). Phosphorus in interstitial water: Methods and dynamics, Hydrobiologia, 170:103-132.

Foley PD (1980). Experience with direct filtration at Ontario's Lake Huron treatment plant. J. Am. Water Works Assoc. 72:162.

Fournier-Bidoz V, Garnier-Laplace J (1994). Etude bibliographique sur les échanges entre l'eau, les matières en suspension et les sédiments des principaux radionucléides rejetés par les centrales nucléaires, Rapport IRSN SERE 94/073 (P).

Gomez E, Durillon C, Rofes G, Picot B (1999). Phosphate adsorption and release from sediments of brackish lagoons : pH, O_2 and loading influence. Water Res. 33(10):2437-2447.

Graeme J, Jameson J (1999). Hydrophobicity and flock density in induced-air flotation for water treatment. Coll. Surf. A Physicochem. Eng. Asp, 151 :269-281.

Hector RHDL (2006). Supervision et diagnostic des procédés de production d'eau potable. Thèse, Université de Toulouse; p. 163.

IUPAC (1997). Compendium of Chemical Terminology. The Gold Book, 2nd Edition, Blackwell Science, 1997 [ISBN 0865426848].

Kedziorek MAM, Bourg ACM (1996). Acidification and solubilization of heavy metal from single and dual-component model solids. Appl. Geochem. 11:299-304.

Lead J, Davison W, Hamilton-Taylor J, Buffle J (1997). Characterizing Colloidal Material in Natural Waters. Aquat. Geochem. 3(3):213-232.

Lead JR, Wilkinson KJ (2006). Aquatic colloids and nanoparticles: Current knowledge and future trends. Environ. Chem. 3:159–171.

Lions J (2004). Etude hydrogéochimique de la mobilité de polluants inorganiques dans des sédiments de curage mis en dépôt: expérimentations, étude in situ et modélisations. Ecole Nationale Supérieure des Mines de Paris, 260 pp.

Matthiesen H, Emeis KC, Jensen BT (1998). Evidence for phosphate release from sediment in the Gotland deep lake during oxic bottom water conditions. Meyniana 50:175-190.

OCDE (1982). Eutrophisation des eaux. Méthodes de surveillance, et d'évaluation de lutte. Organisation de Coopération et de développement Economiques, Paris, p. 164.

Santschi PH, Lenhart JJ, Honeyman BD (1997). Heterogeneous processes affecting trace contaminant distribution in estuaries: The role of natural organic matter. Mar. Chem. 58:99-125.

Singhal R, Preetha J, Karpe R, Tirumalesh K, Kumar SC, Hedge AG (2006). The use of ultra filtration in trace metal speciation studies in sea water. Environ. Int. 32(2):224-228.

Soetaert K, Hofmann AF, Middelburg JJ, Meysman FJR, Greenwood J (2007). The effect of biogeochemical processes on pH. Mar. Chem. 105:30-51.

Tack FM, Calewaert OWJJ, Verloo MG (1996). Metal solubility as a function of pH in contaminated dredged sediment affected by oxidation. Environ. Pollut. 103:199-208.

Tomètin LAS, Daouda M, Zogo ND, Boukari O, Bawa LM (2013). Eutrophication, sediment Phosphorus fractionation and short-term mobility study in the surface and under profile sediment of a water dam (Okpara Dam, Benin, West Africa). J. Appl. Sci. Environ. Manage. 17(4):517-526.

Valentin N (2000). Construction d'un capteur logiciel pour le contrôle automatique du procédé de coagulation en traitement d'eau potable. Thèse de doctorat, UTC/Lyonnaise des Eaux/CNRS, 2000.

Valérie D (2009). Transferts et mobilité des éléments traces métalliques dans la colonne sédimentaire Des hydro-systèmes continentaux. Thèse, Académie d'Aix-Marseille Université de Provence P. 304.

Vignati D (2004). Trace Metal Partitioning in Freshwater as a Function of Environmental Variables and its Implications for Metal Bioavailability. University of Geneve, p. 272.

Zogo D, Bawa LM, Soclo HH, Atchekpe D (2011). Influence of pre-oxidation with potassium permanganate on the efficiency of iron and manganese removal from surface water by coagulation-flocculation using aluminum sulphate: Case of the Okpara Dam in the Republic of Benin. J. Environ. Chem. Ecotoxicol. 3(1):1-8, January 2011.

Zogo D (2010). Etude de l'élimination du fer et du manganèse lors de la potabilisation de l'eau d'une retenue en cours d'eutrophisation : Cas de la retenue d'eau de l'Okpara a Parakou au Bénin. Thèse Université d'Abomey-Calavi; p. 205.

18

Water quality assessment of a wastewater treatment plant in a Ghanaian Beverage Industry

Emmanuel Okoh Agyemang[1] , Esi Awuah[2], Lawrence Darkwah[3], Richard Arthur[1] and Gabriel Osei[4]

[1]Energy Systems Engineering Department, Koforidua Polytechnic, P. O. Box KF981, Koforidua, Ghana.
[2]Civil Engineering Department, Kwame Nkrumah University of Science and Technology (KNUST) Private Mail Bag, Kumasi, Ghana.
[3]Chemical Engineering Department, Kwame Nkrumah University of Science and Technology (KNUST) Private Mail Bag, Kumasi, Ghana.
[4]Mechanical Engineering Department, Koforidua Polytechnic, P. O. Box KF981, Koforidua, Ghana.

The research is aimed at assessing the performance of the wastewater (excluding sewage) treatment plant in a Ghanaian beverage industry. Sixteen (16) water quality parameters were analyzed by collecting influent and effluent wastewater samples of the treatment plant for a year and their average values were compared with EPA (Ghana) guidelines for beverage industries discharging into water bodies. Most of the effluent wastewater pollutant content met the set guidelines, while others were unacceptable. However, the ability of the wastewater treatment plant to effectively deal with key pollutant such as BOD (93%), Ammonia (82%) and COD (82%) suggests that the treatment plant is efficient. In order to improve on the final effluent quality, sand filters may be introduced after the Sequential Batch Reactor II before final discharge into the environment.

Key words: EPA guidelines, de-sludging, environment, sand filters, wastewater.

INTRODUCTION

The importance of water in all facets of life cannot be over emphasized. It is vital for consumption, health and dignity. It is a fundamental resource for human development, especially residential area location. The development of any city has practically taken place near some source of water supply (Rangwala et al., 2007). The ever increasing levels of pollutants and complexity of effluents from municipality and industry, demand effective technologies to reduce pollutants to the desired levels. The use of current wastewater treatment technologies for such reclamation is progressively failing to meet required treatment levels. Advanced wastewater treatment technologies are essential for the treatment of industrial wastewater to protect public health and to meet water quality criteria for the aquatic environment and for water recycling and reuse (Agyemang, 2010).

The protection of receiving waters is essential to prevent eutrophication and oxygen depletion in order to sustain fish and other aquatic life. Discharge of untreated effluent wastewaters into water bodies may put at risk riparian communities that depend on these waters for domestic and personal use (Tchobanologous et al., 2003). Though treated wastewater may not comply with drinking water standards, contacts with water carrying high pathogenic loads may potentially lead to the transmission of enteric infections (Kamala and Kanth Rao, 2002).

The wastewater treatment plant discharges close to 500 m³/day of treated effluent to an open channel that leads to the Atonsu stream. Downstream are a number of riparian communities that rely on the stream for bathing, cooking, drinking, irrigation and cleaning. The main objective of the study is to conduct an assessment on the quality of the treated wastewater effluent that is discharged into the Atonsu stream. This objective was achieved through physical, chemical and biological analysis of the effluents and the standards compared with EPA (Ghana).

METHODOLOGY

Wastewater sources

Wastewater (excluding blackwater and urine) generated in the plant is channeled into a central concrete drain that leads to the pre-treatment tank of the treatment plant. Sources of wastewater generated in the plant includes wastewater generated from cleaning the production floors, washing of process equipment (mixing tanks, storage vessels, holding tanks, etc.), washing and rinsing of beverage bottles, washing and cleaning laboratory floors and equipment and wastewater generated from the kitchen.

Wastewater treatment plant

The wastewater treatment plant is the batch type and consists of four main components; the pre-treatment tank, balancing equalizing and neutralization basin, the sequential batch reactor I and sequential batch reactor II. Figure 1 presents the layout of the wastewater treatment process at the beverage plant.

Pre-treatment tank

The pre-treatment tank is an underground rectangular concrete tank with the capacity of 280 m³ and consists of three main sections. The first section removes wastewater constituents that are likely to cause operational problems during the treatment process. This includes screening for the removal of debris, crown corks, straws, rags, grit and flotation for the removal of small quantities of oil and grease. The second section is the neutralization basin and is equipped with an automatic pH meter that corrects the pH of the wastewater by dosing sulphuric acid when it is necessary. It also contains mechanical agitators which continuously stir the wastewater to ensure a uniform pH within the chamber. The third section is the neutralized tank and is equipped with submersible pumps to automatically pump the wastewater when it gets to a set maximum limit.

Balancing neutralizing and equalizing basin

The balancing, neutralization and equalization basin receives preliminary treated wastewater from the pre-treatment tank and is automatically controlled. It is circular in shape, has a capacity of 780 m³ and is made of stainless steel. The basin serves to neutralize the wastewater pumped from the pre-treatment tank and

to receive and balance shocks such as high pH and temperature levels. The basin has a retention time of 8 h. From the basin water is pumped to the sequential batch reactor I for subsequent treatment to be effected.

Sequential batch reactor I

The sequential batch reactor I is circular in shape, has a capacity of 780 m³ and is made of stainless steel. The reactor contains bacteria employed to feed on the high concentration of organic and inorganic compounds present in the wastewater. The bacteria convert the colloidal and dissolved carbonaceous organic matter into various gases and cell tissue. The resulting cell tissue is removed from the treated wastewater by gravity settling. The reactor is automatically operated and desludging is done periodically. Wastewater is retained for 8 h.

Sequential batch reactor II

From the sequential batch reactor I, wastewater is pumped to the sequential batch reactor II for further treatment to be effected. The sequential batch reactor II is also circular in shape, has a capacity of 780 m³ and is made of stainless steel. The second reactor is employed to further breakdown organic and inorganic compounds present in the wastewater by the use of bacteria. It is automatically operated and desludging is done periodically. Treated effluent wastewater is finally discharged into the Atonsu stream.

Wastewater sampling and analysis

Sampling was done monthly starting from August 2009 to July 2010. Twenty-four (24) samples representing twelve (12) influent and twelve (12) effluent samples were analysed. Temperature of the samples was measured in-situ. Parameters that were analyzed include BOD, COD, turbidity, colour, pH, temperature, total dissolved solids, total suspended solids, conductivity, coliforms, nutrients and trace metals. Table 1 presents the methods and instruments used for the water quality analysis.

RESULTS AND DISCUSSION

Wastewater characteristics

The mean value of each water quality parameter considered for both influent and effluent wastewater samples have been computed and tabulated (Table 2) as well as the standard deviation and standard errors of 95% confidence interval.

Temperature

The temperature of the influent wastewater to the treatment plant ranged from 47 to 50°C and with a mean of 48.2°C. The effluent temperature ranged from 28 to 30°C. The drop in the effluent temperature could be due to heat losses by convection to the atmosphere and conduction to the walls of the receiving treatment tanks. A drop in temperature is paramount to aiding bacterial

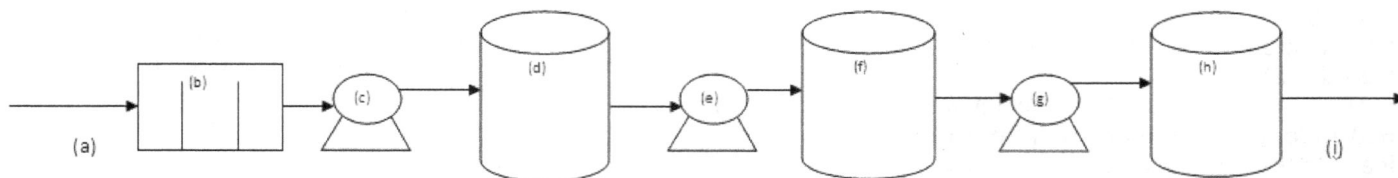

Figure 1. Layout of wastewater treatment process at the Beverage Plant. (a) Influent wastewater (b) Pre-treatment tank (c) Centrifugal pump (d) Balancing equalizing neutralizing basin (e) Centrifugal pump (f) Sequential batch reactor I (g) Centrifugal pump (h) Sequential batch reactor II (i) Effluent wastewater.

Table 1. Methods and instruments used for water quality analysis.

Parameter	Method used	Instrument used
BOD	Winkler modification	-
COD	Closed Tube method	-
Turbidity	APHA Standard method (USEPA)	HACH Model 2100P Turbidimeter
	Cyberscan PC 300 Series	Cyberscan PC 300 Series
Total dissolved solids	Cyberscan PC 300 Series	Cyberscan PC 300 Series
Total suspended solids	Gravimetric method	-
Ammonia-Nitrogen	Titrimetric method	Micro Kjeldhal method
Sulphates	-	HACH Type DREL/2010 Spectrophotometer
Trace metals	-	A.A.S 220 model
Temperature	-	Thermometer
pH	-	Cyberscan PC 300 series pH meter
Colour	-	Nesselerizer
Coliforms	Membrane Filtration method	Membrane filter
DO	-	Oximeter

activities in the treatment tanks. The mean effluent temperature of 29°C was below the EPA Ghana guideline of 30°C.

pH

All the influent wastewater samples analyzed were alkaline. The mean pH value was 11.3 and was in the range of 10.8 to 11.6. The mean pH values of the effluent wastewater ranged from 7.9 to 8.9 and were all within EPA Ghana guideline range of 6 to 9. The decrease in the pH value of the effluent wastewater indicates that some form of treatment had been achieved. The decrease in the effluent pH value could be attributed to the dosing of sulphuric acid to the influent wastewater at the pre-treatment section of the treatment process, in order for biological processes to be effected.

Conductivity

Generally conductivity of water is determined to ascertain the ability of the waters to conduct electrical current. The mean influent conductivity value ranged between 1750 and 1999 μS/cm and was 1750 μS/cm. The high influent conductivity values may be attributed to the high concentration of dissolved ions present in the wastewater during the bottles washing stage of the bottle preparation process. Mean effluent conductivity was 842.8 μS/cm in a range from 923 to 756 μS/cm. Even though the drop in conductivity shows some amount of ion removal, the conductivity levels of the effluents wastewater were unsatisfactory compared to EPA (Ghana) guideline value of 750 μS/cm.

Turbidity

Turbidity, a measure of the light transmitting properties of wastewater, is a test used to indicate the quality of wastewater discharges with respect to colloidal and residual suspended matter. High levels of turbidity in industrial effluents contribute large amounts of suspended solids to receiving waters. The mean influent turbidity value was in the range of 39 and 57 NTU and

Table 2. Wastewater characteristics.

Parameter	Mean Influent	Mean Effluent	EPA Ghana (2000)
Temperature (°C)	48.2±0.7	28.3±0.3	30
TDS (mg/l)	862.2±56.1	839.8±59.3	<1000
Conductivity (µS/cm)	1750.1±100.6	842.8±58.8	750
Colour (TCU)	77.8±36.0	100±41.3	100
TSS (mg/l)	87.7±27.8	176.7±114.3	<50
Turbidity (mg/l)	46.8±3.8	94.8±67.8	75
DO (mg/l)	3.6±0.6	5.7±0.6	<1
pH	11.3±0.2	8.5±0.3	6-9
BOD (mg/l)	1116.8±192.7	49.8±32.9	<50
COD (mg/l)	3114.2±252.7	569.2±115.9	<250
NH_3-N (mg/l)	11.5±1.1	2.1±0.3	1.0
Sulphate (mg/l)	60.6±24.2	177.7±17.7	250
Cadmium (mg/l)	0.0±0.0	0.0±0.0	<0.02
Copper (mg/l)	0.0±0.0	0.00±0.0	1
Lead (mg/l)	0.0±0.0	0.0±0.0	<1
Coliforms (mg/l)	5466.7±1952.5	15850±6377.1	400

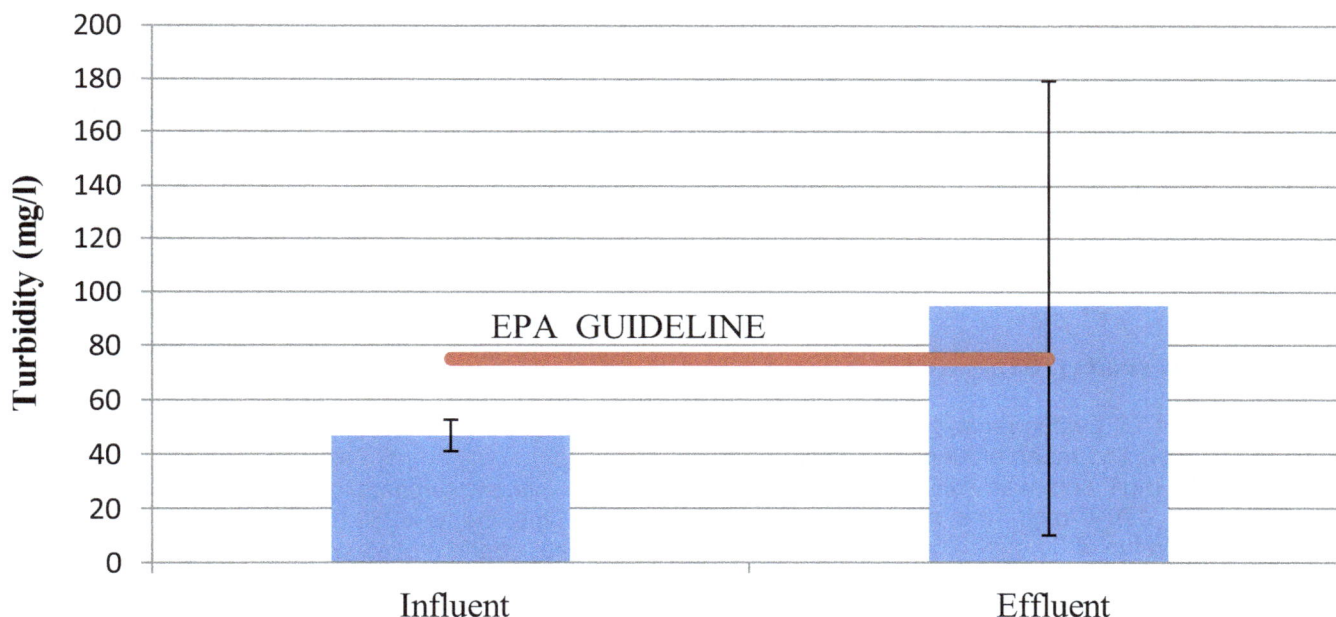

Figure 2. Influent and effluent turbidity and EPA guideline.

was 46.8 NTU. The final effluent turbidity value was in the range of 32 and 225 NTU. The mean effluent turbidity value of 94.8 NTU was above the EPA Ghana guideline value of 75 NTU. The mean effluent turbidity value could be attributed to incomplete sludge settlement during the sedimentation stage of SBR. Figure 2 is a plot of the mean influent and effluent turbidity results and the EPA Ghana guideline.

Colour

Various beverage processing activities such as production floor cleaning and bottle washing impart considerable amount of colour to water. Mean colour value for the influent wastewater ranged from 25 to 150 TCU respectively with the mean of 77.8 TCU. The mean final effluent colour was 100 TCU and ranged between 60 to 180 TCU. Mean effluent colour value was consistent with EPA guideline of 100 TCU. However, some high effluent values of colour recorded during the effluent sampling time could be attributed to incomplete sludge settlement during the sedimentation stage of sequential batch reactor.

Dissolved Oxygen (DO)

Dissolved oxygen is required for the respiration of aerobic microorganism as well as all other aerobic life forms. Mean influent DO ranged from 1.8 to 4.8 mg/l and was 3.6 mg/l. Mean effluent DO was 5.7 mg/l and ranged from 4.8 to 6.4 mg/l. The increase in the effluent DO may be attributed to the infusion of air by blowers during the wastewater treatment period. Both the influent and effluent DO values were consistent and above the EPA Ghana guideline value of 1 mg/l. Figure 3 is a plot of the average influent and effluent DO results and the EPA Ghana guidelines.

Total dissolved solids (TDS)

Total dissolved solids consist of both the organic and inorganic molecules and ions present in the true solution of the water. Mean influent TDS value ranged from 771 to 991 mg/l and was 862.2 mg/l. Mean effluent TDS value was 839.9 mg/l and ranged from 720 to 923 mg/l. It was noted that both average influent and effluent TDS results were consistent with the EPA Ghana guideline for beverage industries discharging into water bodies.

Total suspended solids (TSS)

The mean influent TSS value ranged from 44 to 144 mg/l and was 87.7 mg/l. The mean effluent TSS value ranged from 71 to 380 mg/l and was 176.7 mg/l. The mean effluent value of 176.7 mg/l was more than the EPA Ghana guideline value of 50 mg/l. The high mean effluent TSS value could be attributed to incomplete sludge settlement during the sedimentation stage of SBR. Figure 4 is a plot of the average influent and effluent TSS results and the EPA Ghana guideline.

Biochemical Oxygen demand

The influent BOD concentration of the treatment plant ranged from 700 to 1504 mg/l, with a mean of 1116.8 mg/l. The high values of BOD in the influent wastewater may be attributed to the high concentration of the organic matter content in the wastewater. The mean effluent BOD concentration was 49.8 mg/l and in the range of 19 to 120 mg/l. Figure 5 is a plot of the average influent and effluent

BOD results and the EPA Ghana guideline. The result of the average effluent signifies that the biological method is able to treat the wastewater by means of biodegradation of organic matter. It is noted that the release of excess amounts of organic matter into receiving waters could result in a significant depletion of oxygen and subsequent mortality of fishes and other oxygen dependent aquatic or marine organism. The percentage removal achieved was 93%.

Chemical Oxygen demand (COD)

The mean influent COD value ranged between 2466 mg/l to 3760 mg/l and was 3114.2 mg/l. The mean effluent COD was between 450 mg/l and 856 mg/l respectively with a value of 569.2 mg/l. Even though all the effluent COD values were low as compared to the influent values none met the EPA Ghana guideline value of 250 mg/l. The effluent values could be attributed to the presence of sulphides, sulphites, thiosulphate and chlorides that cause interferences to COD. The removal efficiency was 82%.

Ammonia-Nitrogen (NH$_3$-N)

The mean influent ammonia value ranged from 9.1 mg/l to 15 mg/l and was 11.5 mg/l. The mean effluent value was 2.1 mg/l and ranged from 1.7 mg/l to 2.8 mg/l. The initial rise in ammonia of the influent quality could be due to the presence of ammonia is a by-product of anaerobic digestion whilst the fall in the effluent values could be due to nitrification and de-nitrification processes. The percentage removal achieved was 82%. The average effluent result was above EPA Ghana set guideline.

Sulphates

Mean influent sulphate concentration ranged from 30 to 150 mg/l and was 60.6 mg/l. Also the mean effluent sulphate concentration was 117.7 mg/l and ranged from 80 to 130 mg/l. Both influent and effluent sulphate concentration results obtained during the sampling period were in the range of EPA Ghana set guideline of 250 mg/l. The increase in the sulphate concentration of the effluent wastewater could be attributed to the dosing of sulphuric acid during the wastewater pre-treatment stage in order to bring the pH down for biological activities to be effected.

Cadmium

The mean influent cadmium concentration was 0.0 mg/l and the effluent cadmium concentration 0.0 mg/l. The effluent quality is acceptable according to EPA Ghana

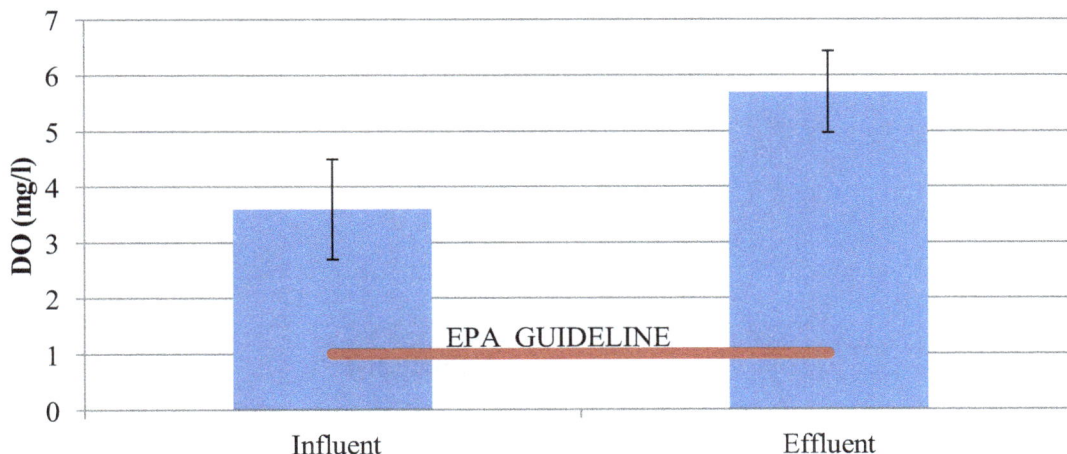

Figure 3. Influent and effluent DO and EPA guidelines.

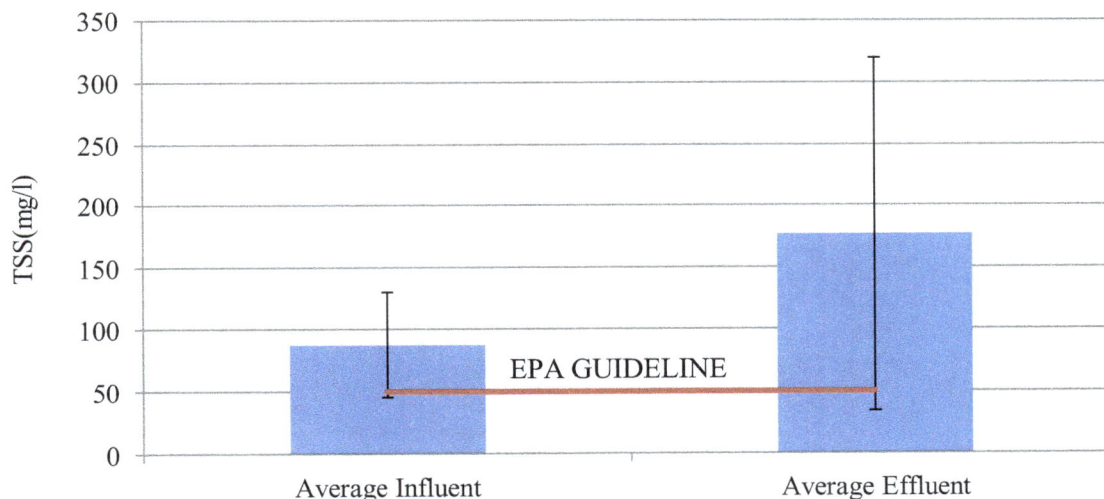

Figure 4. Influent and effluent TSS and EPA guideline.

guideline of <0.02 mg/l.

Copper

The mean influent copper concentration was 0.0 mg/l and the mean effluent was 0.0 mg/l. Both influent and effluent copper concentration were within the EPA Ghana guideline of <1 mg/l and was satisfactory.

Lead

The mean influent lead concentration was 0.0 mg/l and the mean effluent was 0.0 mg/l. Both influent and effluent

lead concentrations results obtained were below the EPA Ghana guideline of 1 mg/l.

Coliforms

The mean influent coliforms count ranged from 0.26E+04 to 1.02E+04 C/100 ml and registered an average of 0.55E+04 C/100 ml. The effluent coliforms count ranged from 0.55E+04 to 2.54E+04C/100 ml with an average of 1.66E+04 C/100 ml. The low influent coliform concentration could be due to the fact that at a high pH and temperature, most coliform group die or remain inactive. Although it was expected that the total number of effluent coliforms be reduced after treatment the reverse

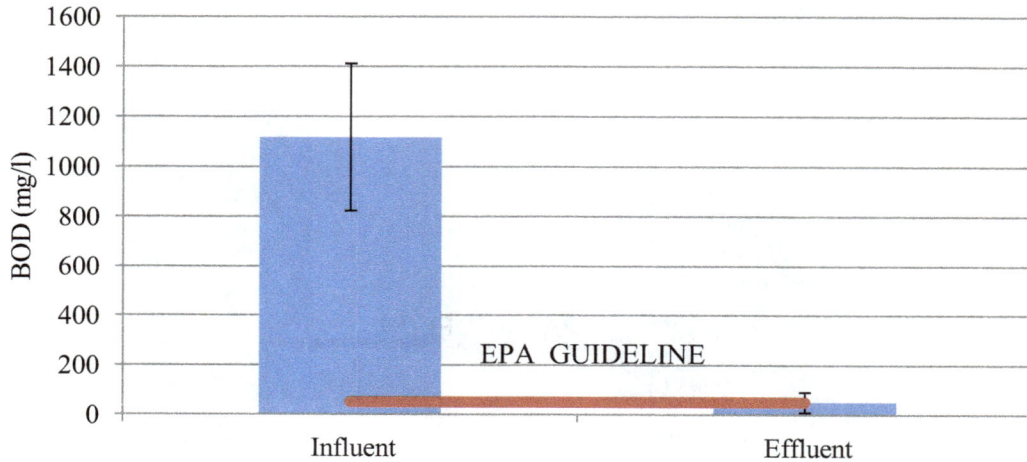

Figure 5. Influent and effluent BOD and EPA guideline.

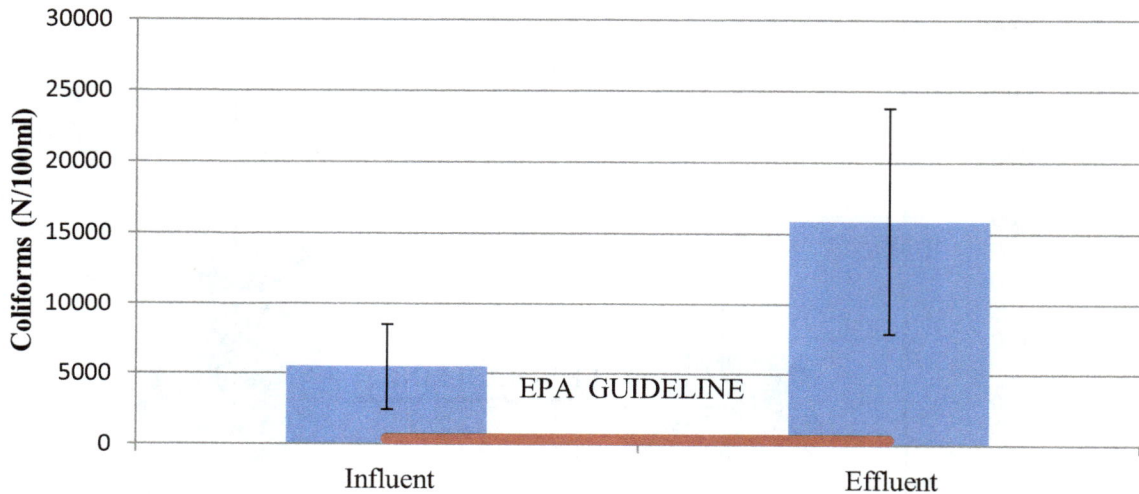

Figure 6. Influent and effluent coliforms and EPA guideline.

was observed. The increase in the effluent coliform concentration could be attributed to the high organic matter content (>20 mg/l BOD) in the treatment tank which serves as food for the bacteria to grow, proliferating rapidly in numbers, the small number of predators in the treatment tanks to devour the bacteria or on the grounds that the treatment tanks have not been desludged since its working life. Figure 6 is a plot of the average influent and effluent coliform results and the EPA Ghana guideline.

CONCLUSION AND RECOMMENDATION

The wastewater treatment plant has a high potential of removing key pollutants and could be used for better treatment of wastewater if managed properly. The removal efficiencies of key parameters such as conductivity, BOD, COD and ammonia were between 50 and 100%. The wastewater treatment plant is efficient; however parameters such as total coliforms, TSS and turbidity were unsatisfactory. By recommendation, a slow sand filter may be introduced after the sequential batch reactor II to improve the effluent wastewater quality. Tanks within the treatment units should be desludged in order to improve on the effluent wastewater quality. Consequently disinfection of the effluent wastewater may be carried out before final discharge into the Atonsu stream.

REFERENCES

Rangwala SC, Rangwala KS, Rangwala PS (2007). Water Supply and Sanitary Engineering, Environmental Engineering. 22nd edition,

Charotar Publishing House, pp. 11- 58.

Agyemang EO (2010). Water auditing of a Ghanaian beverage plant, MSc. Thesis. Kwame Nkrumah University of Science and Technology (KNUST), Kumasi, Ghana.

Tchobanologous G, Burton FL, Stensel HD (2003). Wastewater Engineering Treatment and Reuse, 4th Edition, McGraw Hill, Boston, U.S.A.

Kamala A. Kanth Rao DL (2002). Environmental Engineering: Water Supply, Sanitary Engineering and Pollution. Tata McGraw-Hill Publishing Company limited, New Delhi, pp. 48-57.

EPA Ghana (2000). General Environmental Quality Standards (Ghana), Regulations 2000, pp. 8-13.

19

Flow system, physical properties and heavy metals concentration of groundwater: A case study of an area within a municipal landfill site

Adebisi, N. O.[1], Oluwafemi, O. S.[2], Songca, S. P.[3] and Haruna, I.[4]

[1]Department of Earth Sciences, Faculty of Science, Olabisi Onabanjo University, Ago Iwoye, Ogun State, Nigeria.
[2]Department of Chemistry, Cape-Peninsula University of Technology, P.O.Box 652, Capetown, 8000, Western Cape, South Africa.
[3]Department of Chemistry, Walter Sisulu University, Private Bag X1, Mthatha, 5117, South Africa.
[4]Ikorodu Local Government, Ikorodu, Lagos State, Nigeria.

Groundwater within the Olususun landfill site in Lagos Metropolis was evaluated. Previous research on quality parameters of groundwater in the area made use of equipment of low detection capacity for heavy metals concentrations in water. Also, subsurface flow and significant attenuation of leachate due to horizontal distance between wells and landfill site are yet to be technically elucidated. In the present investigation, priority was given to heavy metals as small quantities may build up in human systems to become a significant health hazard. Then analysis was done with Inductively Coupled Plasma Mass Spectrometry (ICP-MS), while Geographical Information Systems (GIS) technique was used for spatial data analysis and management to illustrate localised flow of groundwater. Digital subsurface model of data from 20 drinking-water wells showed that flow directions are north-south, north-west and south-east. The two extremes of the pH for the groundwater are 4.04 and 8.05, indicating slightly acidic to weakly basic water. Total Dissolved Solids (TDS) are positively-strongly correlated with electrical conductivity (EC) in a line of fit TDS = 29.71 EC - 47.9. From the ICP-MS results, Fe concentrations at locations 1, 3 and 4, and Pb concentrations at locations 1, 5, 7, 8, 14 and 16 did not conform to international human-health benchmarks. Generally, the longer the horizontal distance between a well and the landfill site, the lesser its potential for groundwater contamination. This study better clarifies heavy metals concentrations in water, with GIS for satisfactorily display of positional and attribute for groundwater flow in the area.

Key words: Heavy metals, groundwater, landfill, Geographical Information Systems (GIS), Inductively Coupled Plasma Mass Spectrometry (ICP-MS).

INTRODUCTION

Heavy metals occur naturally and artificially in a groundwater system with large variations in concentration. Landfill is a common and the cheapest method for organized solid waste management in many parts of the world. However, heavy metal contamination of groundwater does arise from landfill source. In very small amounts, heavy metals such as Cobalt (Co), Copper (Cu), Chromium (Cr), Manganese (Mn) and Nickel (Ni) are required by man to support life, while in larger amounts, Mercury (Hg), Cadmium (Cd), Lead (Pb) and Chromium (Cr) are toxic. They may build up in human systems if consumed in contaminated water to become a significant health hazard. In this category are Manganese, Mercury, Lead and Arsenic which are carcinogenic, affecting among others, the central nervous system. Hg, Pb, Cd and Cu affect the kidneys or liver, while Ni, Cd, Cu and Cr affect skin, bones, or teeth.

The Ojota area of Lagos has the largest municipal landfill situated at Olusosun, while groundwater is the major source of water supply for people living at Oregun, Ikosi-Ketu, Ojota garage and Ojota-Ogudu, around the landfill site. Lagos metropolis is highly industrialized with rapidity in urban population. For this reason, probable contamination of groundwater from both domestic and industrial wastes in the area has been of major concern to researchers in environmental and health. Previous work revealed that the Olusosun landfill shows a measurable impact of leachate outflows on groundwater quality at elevated levels of anions (NO_3^-, Cl^- and SO_4^{2-}) in the groundwater body and heavy metals (Cr^{3+}, Cd and Cu) attenuation following no definite pattern (Longe and Enekwechi, 2007).

From other parts of Lagos, Longe and Balogun (2009) examined the level of groundwater contamination near a municipal landfill site at Alimosho where the mean concentrations of NO^{3-}, SO_4^{2-} and Cr^{3+} did not conform to the stipulated World Health Organisation (WHO) potable water standards and the Nigerian Standard for Drinking Water Quality (NSDWQ). Momodu and Anyakora (2010) assessed heavy metal contamination of groundwater in middle class neighbourhood of Lagos. None of the samples analysed contained Al^{3+} in concentrations above the WHO and NSDWQ maximum contamination level, with the exception of [Cd] and [Pb^{2+}], which are above the tolerable contamination level. This population is prone to a significant risk due to the toxicity of these metals and the fact that for many, hand dug wells and bore holes are the only sources of their water supply.

The impact of leachate percolation on ground water quality in the vicinity of an unlined municipal solid waste landfill site at Igando area of Lagos metropolis has been studied by Aderemi et al. (2011). From the study, total dissolved solids (TDS), electrical conductivity and [Na^+] exceeding the WHO's tolerance limits for drinking water with 75% of the samples exceeding WHO's limits for pH and ΣFe. In addition, high population of *Enterobacteriaceae* ranging from $4.0 \times 10^3 \pm 0$ to $1.0575 \times 10^6 \pm 162,705$ cfu/ml was also measured in the groundwater samples.

The objective of this paper is to develop a subsurface flow model, and analyze the horizontal distance influence of landfill site from wells on heavy metals concentration in the groundwater in the shallow aquifer overlying active the landfill. Even though, the vulnerability of groundwater to landfill operations in Lagos has been revealed, no previous work has employed a GIS to describe the subsurface flow below the active Olusosun landfill. Also, the most common instrumental method used to determine heavy metals concentrations in groundwater is

mainly atomic absorption spectrometry (AAS). Compared to AAS technique, Inductively Coupled Plasma Mass Spectrometry (ICP-MS) method is characteristic of high sensitivity and accuracy (Vladimir et al., 2007). ICP-MS is a type of mass spectrometry which is capable of detecting metals and several non-metals at concentrations as low as one part in 10^{12} (part per trillion). This is achieved by ionizing the sample with inductively coupled plasma and then using a mass spectrometer to separate and quantify those ions. This database technology will go a long way to explain subsurface flow and significant contamination or otherwise of the heavy metals chemistry taking cognisance of drinking water quality standards as recommended by FEPA (1991) and WHO (2003).

Study area

The Olusosun landfill is situated exactly on a geographical cross wire at latitude N6° 35' 40.4"and longitude E3° 22' 44.3" as shown in Figure 1, 48 m above sea level. It covers about 42.7 ha of land with estimated residual life span of 20 years. Figure 2 is a photograph of part of the landfill site showing areas of refuse dump and leachate release. LAWMA report (2007) indicates that the dump site receives approximately 40% of the total waste deposits from Lagos metropolis and has been active since Friday 19th November, 1992.

In various discussions which pertain to the physical setting of Lagos, vegetation and soil types have been implicated to influence the spatial pattern of the landform. Abegunde (1976) showed how the relief and drainage patterns of Lagos generally reflect the coastal location. The coastal lowlands which dominate the Lagos landscape form part of a wider stretch of the coastal zone of southern Nigeria (Figure 3). The mode of landform evaluation in Lagos has been largely influenced physico-tropical climatic characteristics, which include; rainfall amount, intensity and distribution character of vegetation. All of which have been implicated in the dynamics of coastal landform processes in the area.

Geologically, Lagos area is within the Dahomey Miogeosynclinal Basin, Southwestern Nigeria. It is underlain by sedimentary deposits comprising silt, clay, peat or coal and sands of various sizes, and composition with a high degree of lithologic variability (Figure 4). From hydrogra-phical study, Onwuka (1990) identified two broad geological formations in Lagos area. These are Ilaro Formation and Coastal Plain Sands in the seacoast of Lagos. The Quarternary Formation of the Coastal Plain Sands is more wide spread over the study area.

MATERIALS AND METHODS

Field materials, measurements and sampling

Well depth and water level measurements were obtained using

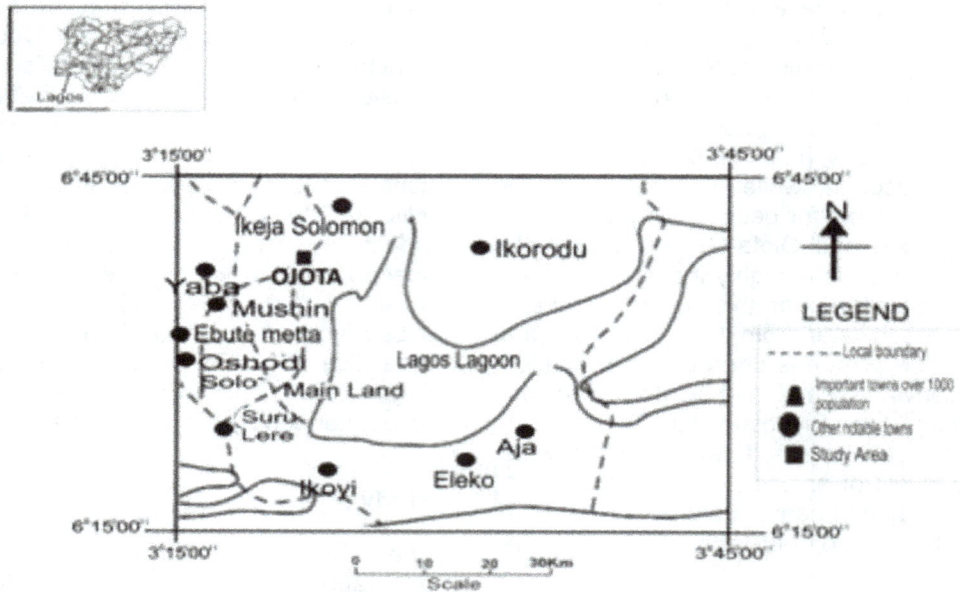

Figure 1. Map of Lagos metropolis showing location of the study area.

Figure 2. Photograph of part of the Olususun landfill site showing areas of refuse dump and leachate release.

string, beaker, hook and sinker with measuring tape. Geographical locations of wells and elevation were captured on the field using Global Positioning System (GPS). During the reconnaissance survey, twenty wells comprising 11 tube wells (boreholes) and 9 large diameter wells (hand dug wells) were earmarked for sampling and designated L1 to L20. Wells numbers 15 and 16 were earmarked for control sampling because of the distance from the land fill site. The wells were between less than 200 and above 3000 m within and around the landfill site. The leachate sample was designated SLE. Two of the wells at 3010.2 m and 2963.2 m were also earmarked for control sampling.

Twenty raw water samples were collected from the wellhead in 20, 500 ml laboratory certified polyethylene, which has been earlier

rinsed 3 times with the well water to be sampled and acidified. Maximum holding periods were designated for each physical parameter, while the samples were preserved by keeping it cool in order to slow chemical and biochemical reactions. The leachate accumulating at the base of the landfill was sampled from one location within the landfill site.

Vector mapping of water flow

A vector map was created from information in a-two-grid vector using *Surfer 8*. The vector was drawn at each grid node based on static water level and elevation information. This was with a view to emphasize on the groundwater flow direction within the study area where arrow symbols will indicate the "downhill" direction and their lengths will be indicative of the magnitude, or steepness, of the slope enhancing the flow direction of groundwater and groundwater potential zones (Figure 5).

Analytical methodology

Equipments employed to measure physical parameters are pH meter (For measuring hydrogen ion concentration) and Horiba U-10 multiprobe meter (For Total Dissolved Solids (TDS), Electrical Conductivity, EC) and Salinity. The concentrations of trace elements were determined using an Inductively Coupled Plasma Atomic Absorption Mass Spectrophotometer (ICP-MS). The internal standard of the ICP-MS also serves as the diluent consists primarily of deionized water, with nitric or hydrochloric acid, and Indium. 5 ml of the internal standard was added to a test tube along with 10 to 500 µl of the water sample. This mixture was then vortexed before it was pipetted and analyzed.

PRESENTATION OF RESULTS

The consistency of groundwater datasets helped in the explanation of subsurface flow analysis model in the study area. Figure 3 shows a steady state flow, where well head is independent of depth. This is important in order to generate a raster for groundwater flow vector. The standard output raster generated represents the groundwater volume balance residual, which measures the difference between the flow of water into and out of every cell within the 3 km radius of the Olususun landfill site. Since the flow calculations are performed through each of the cell walls independently (flow is governed by the differences between adjacent cells), it is possible that more (or less) water may flow into a location than out of it, resulting in a positive (or negative) volume balance residual.

Unlike surface water which flows from areas of high to low elevation on the terrain, groundwater flows from a point of higher pressure gradient to a point of lower pressure gradient. The magnitude of pressure gradient in the subsurface is indicated by the length of the arrow, whereby areas of higher pressure gradients have longer arrows than areas of lower pressure gradients. The flow directions are north-south, north-west and south-east. Areas of high groundwater potentials are where the gradient is enhancing the flow direction of groundwater.

Results of physical properties of the investigated groundwater are summarized in Table 1 compared with the Federal Environmental Protection Agency (FEPA, 1991) and World Health Organisation (WHO, 2003) standards. The decimal logarithm of the reciprocal of the hydrogen ion activity of a water sample is defined as its pH. The pH value varies with temperature. Electrical conductivity (EC) in this context measures the ability of groundwater to conduct an electric current. This measurement is indispensable as it strongly relates to the Total Dissolved Solids (TDS) in the water, which in turn

reflects groundwater catchment geology as it finds application in water classification (Ela, 2007).

Figure 6 is a regression model for the line of best fit TDS = 29.71 EC - 47.9 in which the values for slope, TDS-intercept and EC of the groundwater are defined. Strong positive correlation (r = 0.85) exists between TDS and EC of the groundwater with high coefficient of determination (r^2 = 0.72). The studied water samples at locations 5, 8 and 19 have elevated SL values in areas of shallow water tables, though still lower than the amount that could adversely affect human health.

Table 2 shows values of heavy metals concentration compared with drinking water standards of FEPA (1991) and WHO (2003), which provides the suitability picture of the groundwater for domestic and other purposes in the study area. Zinc (Zn) is one of the important trace elements that play a vital role in the physiological and metabolic process of man. Nevertheless, at higher concentration, zinc can be toxic to the man. Copper (Cu) is also an essential element for human beings, excess of which can be toxic, causing hypertension, pathological changes in brain tissues and specific diseases of the bone (Iyengar et al., 1988). In this study, the maximum concentration of Cu recorded is 0.02 mg/l which is within the recommended limits of drinking water.

The most common source of Iron (Fe) and Manganese (Mn) in the local groundwater around the study area is the leachate from the municipal landfill site. Fe in groundwater occurs mainly in form of ferric hydroxide. It is found in significant concentration in some sample of the studied groundwater only at locations 1, 3 and 4. The shortage of Fe causes anemia, while its prolonged consumption in water may lead to a liver disease called haermosiderosis (Iyengar et al., 1988). The main source of Cd in groundwater is industrial activities such as electroplating, pigments, plastic, stabilizes and battery manufacturing. Small quantities of Cd can cause adverse changes in the arteries of human kidney (Iyengar et al., 1988). It replaces zinc biochemically and causes high blood pressures, kidney damage and etc. It interferes with enzymes and causes a painful disease called Itai-itai. Cd concentrations in groundwater of the study area are mostly below the detectable level, except at location 14 where a concentration of 0.01 mg/l is recorded.

Lead (Pb) is an undesirable trace metal and a serious cumulative body poison, which can inhibit several key enzymes involved in the overall process of haemo-synthesis whereby metabolic intermediate accumulates. Pb concentrations above the safe limit were measured in groundwater samples at locations 1, 5, 7, 8, 14 and 16. Measured values of heavy metals in groundwater are presented in Table 2.

From the descriptive statistics (Table 3) Pb has the maximum average concentration of 0.07 mg/l compared to other heavy metals, while Cd has the minimum standard deviation of 0.003. C_U is the most variable heavy metal, while Fe is the least variable trace element

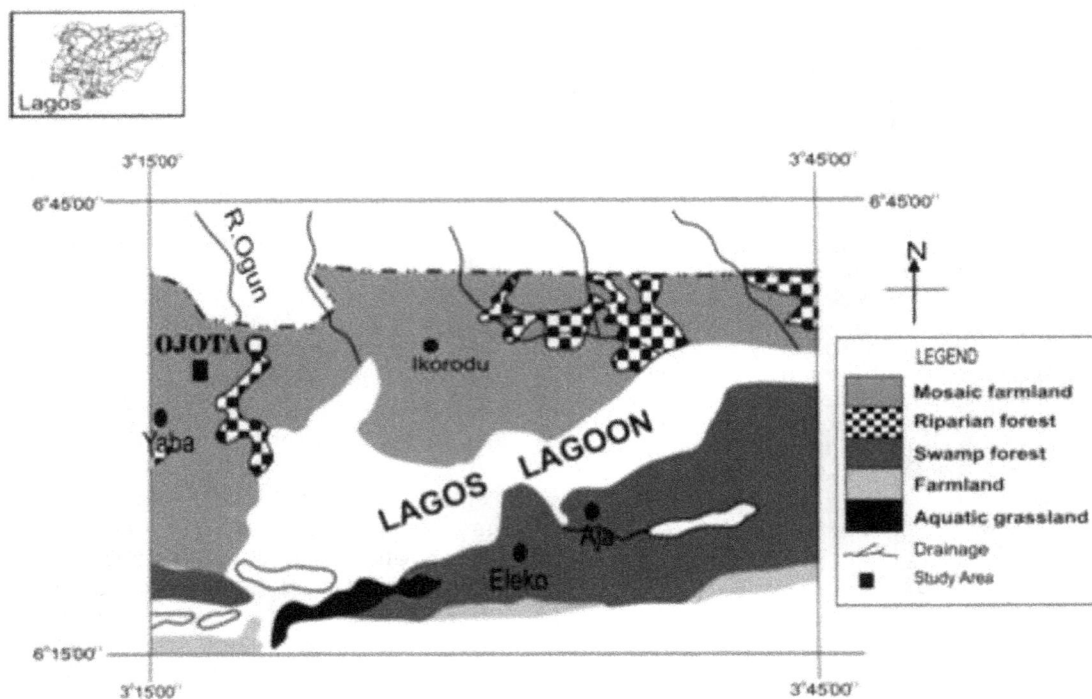

Figure 3. Map showing the major rivers and vegetation of Lagos.

Table 1. Physical properties of the studied groundwater.

Sample S/N	pH	E.C. (µS/cm)	TDS (mg/l)	Salinity (mg/l)
FEPA (1991)	6 – 9	1.00	200	250
WHO (2003)	6.5 - 9.2	1.00	500	250
LE	8.05	13.2	464	40
L1	5.87	1.28	10	10
L2	5.36	0.75	11	2
L3	5.61	0.94	17	6
L4	6.8	5.70	17	253
L5	6.58	4.80	0	180
L6	5.25	1.25	0	34
L7	5.96	1.72	6	30
L8	6.76	4.64	0	165
L9	5.73	1.98	4	90
L10	5.15	0.18	1	50
L11	7.25	2.40	17	10
L12	6.48	1.80	4	63
L13	4.04	1.66	16	47
L14	5.96	1.01	0	32
L15(CTR)	4.16	1.71	6	52
L16(CTR)	4.48	0.95	6	32
L17	4.25	2.03	4	59
L18	6.57	4.06	0	169
L19	4.44	0.81	0	16
L20	4.83	0.81	5	33

NS: Not specified; CTR: control; LE: leachate.

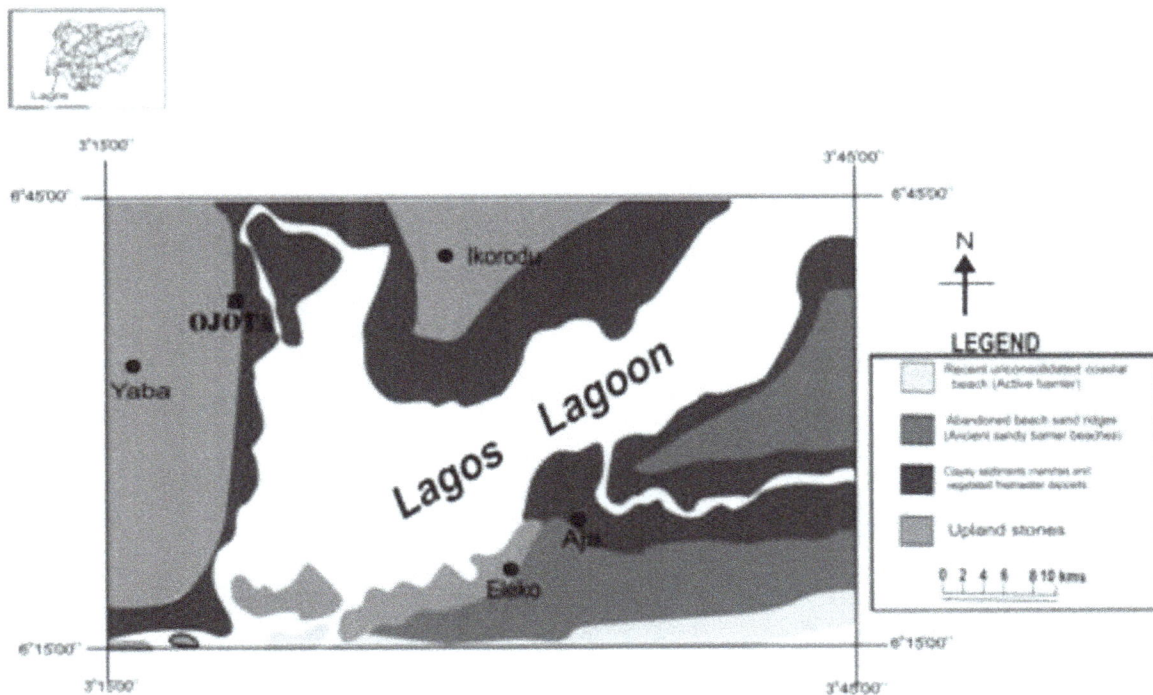

Figure 4. Map showing the surface geology of Lagos.

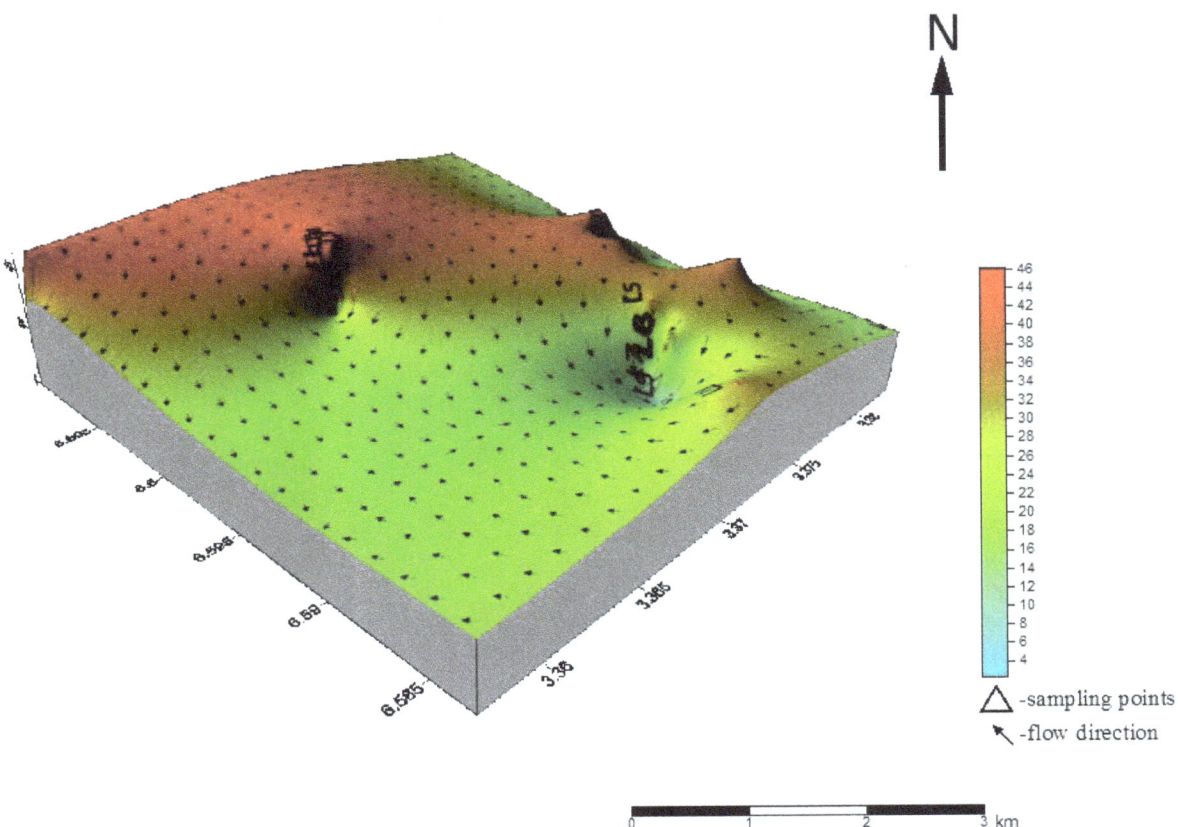

Figure 5. Digital terrain model of the study area and groundwater flow direction.

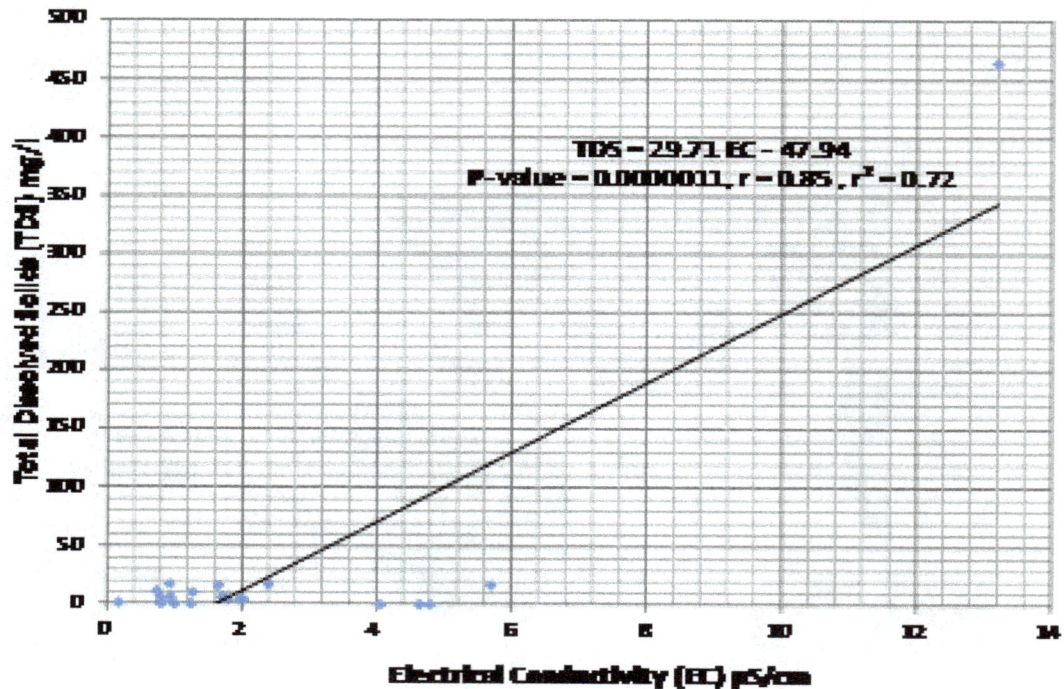

Figure 6. Regression plot of Total Dissolved Solids against electrical conductivity of groundwater.

Table 2. Heavy metals concentration of the studied groundwater.

Sample S/N	Distance (m)	Zn (mg/l)	Cu (mg/l)	Mn (mg/l)	Fe (mg/l)	Cd (mg/l)	Pb (mg/l)	Cr (mg/l)
FEPA	--	3.0	1.00	0.20	0.30	NS	0.01	0.1
WHO	--	1.5	0.50	0.50	0.03	0.002	0.01	0.1
LE	0.	0.0	0.01	0.00	0.06	0.01	*0.20*	0.0
L1	269.2	0.0	0.01	0.01	0.05	0.00	*0.09*	0.0
L2	414.4	0.0	0.00	0.01	0.02	0.00	0.01	0.0
L3	145.0	0.0	0.00	0.00	*0.36*	0.00	0.01	0.0
L4	121.0	0.0	0.00	0.00	*0.64*	0.01	*0.20*	0.0
L5	735.6	0.0	0.00	0.01	0.00	0.00	*0.92*	0.0
L6	900.8	0.0	0.00	0.00	0.00	0.00	0.00	0.0
L7	977.0	0.0	0.00	0.00	0.00	0.00	*0.06*	0.0
L8	888.8	0.0	0.00	0.01	0.00	0.00	*0.04*	0.0
L9	1061.0	0.0	0.00	0.01	0.00	0.00	0.00	0.0
L10	687.63	0.0	0.01	0.00	0.00	0.00	0.00	0.0
L11	737.6	0.0	0.01	0.04	0.01	0.00	0.00	0.0
L12	1539.6	0.0	0.01	0.02	0.01	0.00	0.00	0.0
L13	1609.6	0.0	0.01	0.02	0.00	0.00	0.00	0.0
L14	1603.6	0.0	0.01	0.00	0.00	0.01	*0.02*	0.0
L15(CTR)	3010.2	0.0	0.01	0.01	0.01	0.00	0.01	0.0
L16(CTR)	2963.2	0.0	0.01	0.00	0.00	0.00	*0.04*	0.0
L17	1446.4	0.0	0.01	0.02	0.00	0.00	0.00	0.0
L18	1471.4	0.0	0.01	0.02	0.00	0.00	0.00	0.0
L19	643.6	0.0	0.02	0.00	0.00	0.00	0.00	0.0
L20	740.8	0.0	0.01	0.00	0.00	0.00	0.00	0.0

NS: Not specified; CTR: control; LE: leachate.

Figure 7 shows line plots of groundwater heavy metals concentration and well horizontal distances from the Olususun landfill site. Among the measured heavy metals, only Pb and Fe show elevated concentration of 0.6 and 0.9 mg/l respectively, at a distance less than 1000 m. It follows that wells need be located at least 1500 m away from the landfill site.

DISCUSSION

Aquifer flow system

The well head raster generated so far is reasonable with respect to transmissivity, porosity, and aquifer thickness. Considering the elevation information, the directions at which arrows are pointing are in the direction of pressure release where wells are concentrated. The hydrogeological condition of the landfill site is consistent with the regional hydrogeological setting of Lagos (Longe et al., 1987). The subsurface geology of the landfill area comprises a lateritic cover of variable thickness averaging 4 m, which overlies an alternating sequence of sand and clay deposit. The water-bearing zone consisting of loose, medium to coarse sand with an average thickness of 10.4 m is directly below it.

Physical properties

The physical properties of groundwater are significant factors for the occurrence of many trace elements. For instance, pH below 7 is a significant factor in the occurrence of many cationic metals, such as Al, Fe, Mn, and Ni; these metals, as well as Cu, Pb, and Zn, adsorb more strongly to aquifer materials as pH increases (Ayotte et al., 2011). Chemically, water dissociates according to the equilibrium

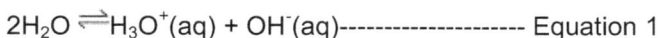

$$2H_2O \rightleftharpoons H_3O^+(aq) + OH^-(aq)\text{-------------------- Equation 1}$$

with a dissociation constant, K_w defined as

$$K_w = [H^+][OH^-]\text{---------------------------------- Equation 2}$$

Where $[H^+]$ is the concentration of the aquated hydronium ion, and $[OH^-]$ represents the concentration of the hydroxide ion. K_w has a value of about 10^{-14} at 25°C.
The two extremes that describe the pH of studied water samples are 4.04 and 8.05, indicating slightly acidic to weakly basic groundwater. Tested samples of groundwater in the study area have TDS values which placed them in the class of fresh waters. Raw water samples from wells in the study area have TDS values ranging from 0 to 17 mg/l, which is far below the recommended limits of FEPA (1991) and WHO (2003).
The trend of specific electrical conductance with dissolved solids is such that as the number of charged ions in the water (TDS) increase so does the EC. Also about 72% in the variation of the TDS was associated with the EC of the groundwater. On the basis of the p-value which is far less than the predetermined significance level α (which is often 0.05 or 0.01), it can be concluded that the TDS is highly dependent on the EC of the studied groundwater. Therefore, it is possible to estimate TDS of the water measured values of EC.
Salinity (SL) is a major water quality limitation on the environmental values of groundwater. Human action such as waste disposal among others has been implicated to cause excessive salinity in groundwater. Salinity has limited the use of groundwater sources as they are too salty for human consumption.

Heavy metals concentration and quality standards

In addition to landfill site contribution, the geologic composition of aquifers and aquifer geochemistry are among the major factors affecting trace-element occurrence (Katz et al., 2007). Zn concentration in the groundwater is below the detectable level. This can be due to its restricted mobility from geological formation which underlies the dump site. At concentrations found in most of the sampled groundwaters, Fe and Mn are at concentrations below the aesthetic objective, and therefore, not considered a health risk. This study reveals that the concentration of Pb is below the detectable level in most of the groundwater samples.

Variation in heavy metals concentration

Trace metals that occur in the studied groundwater are not of concentrations large enough to constitute a significant health impact. However, the order of decrease in concentration is Cu > Mn > Cd > Pb > Fe.

Landfill site and well distance

Locating a well in a safe place takes careful planning and consideration of factors such as horizontal distance from pollution source. The horizontal distance between a well and landfill site is one of the most crucial safety factors to consider when studying groundwater contamination. The minimum distance between a well and the landfill site can lead to contamination, but longer distances would better protect the well. For wells located at a distance less than 1000 m, the landfill site can have a major impact on them.

CONCLUSIONS AND RECOMMENDATIONS

The Olusoun landfill was considered a critical site for groundwater pollution as it comprises heavy domestic

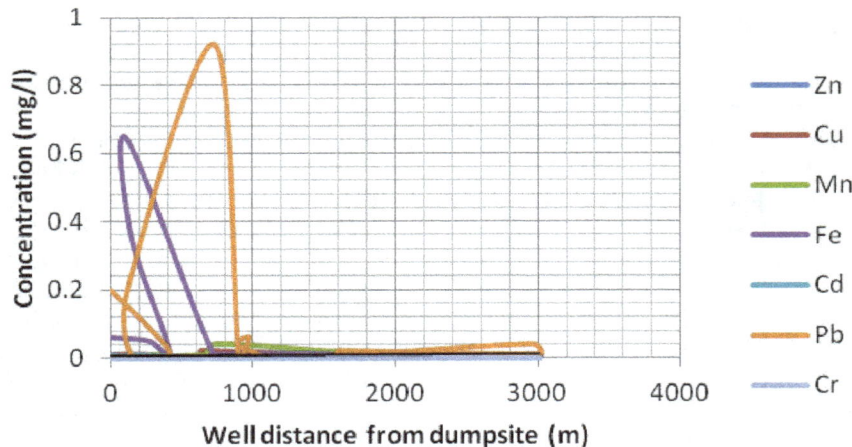

Figure 7. Well distances and heavy metals concentration of groundwater.

and industrial wastes. Groundwater flows from points of higher pressure gradient to a point of lower pressure gradient. The flow directions are north-south, north-west and south-east. Areas of high groundwater potentials are where the gradient is enhancing the flow direction of groundwater. The two extremes that describe the pH of studied water samples are 4.04 and 8.05, indicating slightly acidic to weakly basic groundwater. The regression model for the line of best fit TDS = 29.71 EC - 47.9 shows that it is possible to estimate TDS of the water measured values of EC, and about 72% in the variation of the TDS was associated with The EC of the groundwater.

There exist a moderate number of heavy metal pollutants exceeding the guideline level for water supply. Significant concentration of Fe in water samples are only at locations 1, 3 and 4. Cd concentrations in groundwater of the study area are mostly below the detectable level, except at location 14 where a concentration of 0.01 mg/l is recorded. Pb concentrations above the safe limit were measured in groundwater samples at locations 1, 5, 7, 8, 14 and 16. Among the measured heavy metals, only Pb and Fe show elevated concentration of 0.6 and 0.9 mg/l respectively, at a distance less than 1000 m.

This study concludes that wells for domestic needs be located at least, 1500 m away from the landfill site. For wells located at a distance less than 1000 m, the landfill site can have a major impact on the waters from them. In order to authenticate these findings, more detailed study on groundwater quality in and around the landfill site will be necessary. In addition groundwater pollution monitoring programme for quality status in the vicinity of the landfill is suggested.

Ayotte JD, Gronberg AM, Apodaca EL (2011). Trace elements and radon in groundwater across the United States, 1992-2003. US Geol. Surv. Sci. Inventory Report pp. 0011-5059.

Ela WP (2007). Introduction to Environmental Engineering and Science, Prentice Hall, 3rd ed.

FEPA (1991). SI. 8 National EnvironmentalProtection (Effluence limitations) regulations. In Odiete 1991 Environmental Physiology of Animal and Pollution, Published by Diversity resources Ltd., Lagos, Nigeria pp. 157-219.

Iyengar GV, Gopar-Ayengar AR (1988). Human health and trace elements including Effects of high-altitude populations. Ambio. 17(1):31-35.

Katz BG, Crandall CA, Metz A, McBride WS, Berndt MP (2007). Chemical characteristics, water sources and pathways, and age distribution of groundwater in the contributing recharge area of a public-supply well near Tampa, Florida, 2002–05. US Geol. Surv. Sci. Invest. Rep. 2007-5139. <http://pubs.usgs.gov/sir/2007/5139/pdf/sir2007-5139.pdf>.

LAWMA (2007). Waste dumped by landfilled for the year 2007. http://www.lawma.org/DataBank/Waste%20data%202007.pdf

Longe E O, Malomo S, Olorunniwo M A, (1987). Hydrology of Lagos metropolis. J. Afr. Earth Sci. 6(3): 163-174.

Longe EO, Enekwechi LO (2007). Investigation on potential groundwater impacts and influence of local hydrogeology on natural attenuation of leachate at a Municipal Landfill. Int. J. Environ. Sci. Tech. 4(1):133-140.

Longe EO, Balogun MR (2010). Groundwater Quality Assessment near a Municipal Landfill, Lagos Nigeria. Res. J. Appl. Sci. Eng. Technol. 2(1):39-44.

Momodu MA, Anyakora CA (2010). Heavy metal contamination of groundwater: The Surulere case study. Res. J. Environ. Earth Sci. 2(1):39-43.

WHO (2003). Guideline for drinking water quality. Geneva, (WHO/SDE/WSH 03.04).

Vladimir NE, Douglas ER, Jian Zheng OFXD, Masatoshi Y (2007). Rapid fingerprinting of [239]Pu and [240]Pu in environmental samples with high U levels using on-line ion chromatography coupled with high-sensitivity quadrupole ICP-MS detection. J. Anal. At. Spectrom. 22 (9):1131–1137. doi:10.1039/b704901c.

REFERENCES

APHA (1992). Standards methods for examination of water and wastewater. 18[th]edn., American Public Health Association.

20

Study on the effects of vegetation density in reducing bed shear stress on the downstream slope of earthen embankment

H. M. Rasel[1] , M. R. Hasan[2] and S. C. Das[3]

[1]Department of Civil Engineering, Rajshahi University of Engineering and Technology, Rajshahi, Bangladesh.
[2]Institute of Environmental Science, University of Rajshahi, Bangladesh.
[3]Bangladesh Water Development Board (BWDB), Dhaka, Bangladesh.

River bank, coastal erosion and embankment failures are endemic and recurrent natural hazards, causing loss of lands and livelihoods along major rivers and coastlines. Overflow is the most damaging factor for earthen embankments on the downstream (d/s) slope due to the erosion of surface material which is caused by the high bed shear stress and supercritical flow over the entire d/s slope. Vegetation planted on d/s slope may have a considerable effect in reducing bed shear stress which makes erosion. However, the influence of vegetation for flow resistance as well as bed shear stress control has been studied a little. Therefore, the aim of this research is to introduce the use of natural resources like vetiver grass for controlling bed shear stress which actually causes embankment erosion starts from the high flow region that is, in supercritical flow state. A small-scale experiment is performed with four different densities of vegetal cover considering different discharges on d/s slope. Vegetative barrier affect the mean flow rate by induced drag force against hydraulic forces. Hydraulic characteristics and mechanism of vegetation in open channel during overflow are studied through flume experiment. Results demonstrated that vegetation markedly reduced the effective bed shear stresses which are the influencing factor for causing erosion on the d/s slope, toe and bed of embankment. Results also revealed a significant reduction of damage due to erosion and risk of failure from overtopping flow can be achieved by providing dense vegetation cover on the d/s slope of earthen embankment.

Key words: Embankment, downstream slope, drag force, bed shear stress, vegetation density.

INTRODUCTION

Bangladesh with its repeated cycle of floods, cyclones, and storm surges, has proved to be one of the most disaster-prone areas of the world. River bank and coastal erosion, and embankment failures happen continuously throughout Bangladesh. These are endemic and recurrent natural hazards, causing loss of lands and livelihoods along major rivers and coastlines. There are several traditional embankment protection works are available such as CC blocks, Geobags etc. From a strictly economic point of view, the cost for mitigating these problems is high. In addition, the State budget for such works is never sufficient which confines rigid structural protection measures to the most acute sections, never to the full length of the river bank or coastline and embankment, this bandage approach compounds the problem.Not only that, hard engineering structure makes the scenic environment unpleasant and helps only to transfer the problem to another place, to the opposite site or downstream, which aggravates the problem rather than reducing (Grimshaw, 2008). Scientists on river management and restoration claim that the river not only includes nonliving substances (flow, sediment) but also organism, so the vegetation in the watercourse should be included in the river dynamic system, at the same time, the vegetation will cause flow resistance, rise of water level and reduction of discharge capacity (Wu, 2008). Powledge et al. (1989a) observed that the initial erosion process begins anywhere within the supercritical flow region, especially at a point of slope discontinuity that is, from crest and toe of downstream slope of embankment.

Bangladesh Water Development Board (BWDB) has been constructed nearly 13000 km of embankments (over 4000 km of coastal embankments along the coastline surrounding the Bay of Bengal and offshore islands, nearly 4600 km of embankments along the bank of big rivers and nearly 4500 km of low-lying embankments along the small rivers, hoars and canals). A large number of sea dike or river embankment failures have been initiated from damages induced by wave or flow overtopping (Islam, 2000). For earth coastal dikes or river embankments, overtopping is one of the most damaging factors for the downstream slope. Eventually a failure of the downstream slope may lead a failure of the dike. Water infiltrates into dike crest, downstream slope and reduces the shear resistance of the soil (Hanson and Temple, 2001). Vegetation planted on the downstream slope may have a considerable effect in reducing overtopping erosion. Vegetative barriers increase the surface roughness and slow the flow rate. It dissipates energy and protects the slope from erosion. In a number of tropical countries Vetiver grass (Vetiveria zizanioides) is well-known as bioengineering. Recently, Vetiver grass has been planted on downstream slope as sea dike or river embankment protection.

An eco-friendly Vetiver grass (V. zizanioides) has been used successfully over 120 countries for more than a century as traditional technology for riverbank stabilization and embankment erosion control (Truong and Loch, 2004). Recent research and case studies have shown that vetiver grass can be used as a cheap method to protect shorelines or embankments. Mature vetiver is extremely resistant to washouts from high velocity flow due to its extraordinary root depth and strength. The stiff shoots and strong roots can keep the plant stand steadily in water with 0.6 to 0.8 m deep and 3.5 m/s velocity of water flow (Ke et al., 2003). On this subject, Bangladesh is in advantageous position as it has been abundant supply of Vetiver grass. But Vetiver grass technology (VGT) is very new in Bangladesh and using Vetiver as a living barrier against erosion remains untouched excepting

as few small-scale trials in some foreign aided projects such as earthen embankment project of BWDB supported by World Bank. Moreover, its application has been based on experience rather than hydraulic principles (Das and Tanaka, 2009). Basically, the upstream side of earthen sea dikes or embankments are armored with stones or concrete blocks where there any exist of important infrastructures while the downstream slope often covered with grass. Very little is known about the strength and stability of downstream slope of sea dikes or river embankments covered with grass during overflow. To minimize the impact of natural disasters as well as to achieve the aim of agricultural production, sustainable and cost-effective maintenance of those embankments is a sine qua non for Bangladesh. However, the understanding of the processes and properties between flows and Vetiver grasses and the flow characteristics of d/s slope of sea dikes or river embankments covered with grass during overflow are still limited. Therefore, the main aim of this study is to introduce the use of natural resources like vegetation to protect the downstream slope of embankments or sea dikes during overflow and to diminish the bed shear stress which makes embankment erosion starts from the high flow region that is, anywhere within the supercritical flow state.

MATERIALS AND METHODS

Experimental set up and procedure

A small-scale laboratory flume experiment is performed to investigate the effects of vegetation (commercially available polypropylene Hechimaron vegetation model with 5 and 20% blocking that is, 95 and 80% of water passes through the model respectively which called the porosity of the vegetation model) roughness on the behaviour of d/s bed shear stress. Note that, four different densities of vegetal cover considering different discharges on the d/s slope are used. Firstly, row vegetation (2D type) and then grid type 3D vegetation is used. Afterward the experiments are conducted with staggered type 3D vegetation. All these experiments are conducted with maintaining a ratio of 0.25 (5 cm spacing case) and 0.75 (15 cm spacing case) with 20% uniform blocking, where, ratio = width of spacing within the vegetation rows over width of vegetation rows, here 5 and 15 cm spacing and 20 cm fixed vegetation width is used. Finally, the test is conducted with all over vegetation on d/s slope and bed with 5 and 20% blocking respectively, to evaluate the comparative results with no or WOV case in case of different vegetation effects as well as to investigate the effects of vegetation density for different blocking effect for controlling the bed shear stress on the d/s slope of embankment. A model of wooden embankment was constructed and placed at the middle of the flume which separated the flume along its length, forming main stream on one side (upstream) and the floodplain on the other side (downstream). The size of the embankment is 0.25 m in height, 0.25 m in crest width and 1.5 m in length, with slopes 3H:1V and 2H:1V in the upstream and downstream sides, respectively as shown in Figure 1. The details test program with four different types of vegetation arrangements are shown in the Figure 2.

To understand the processes and properties of hydraulics of

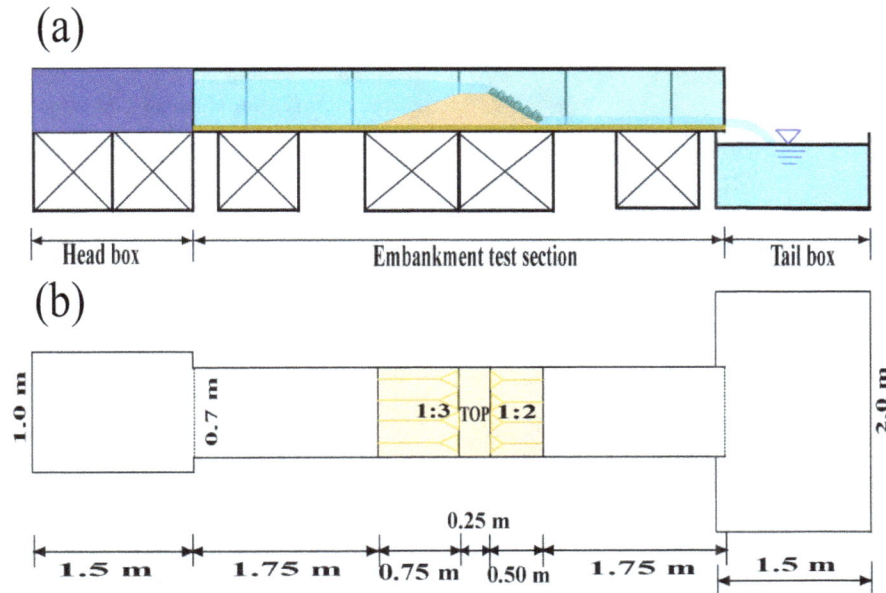

Figure 1. Profile of testing facility (a) front view and (b) top view.

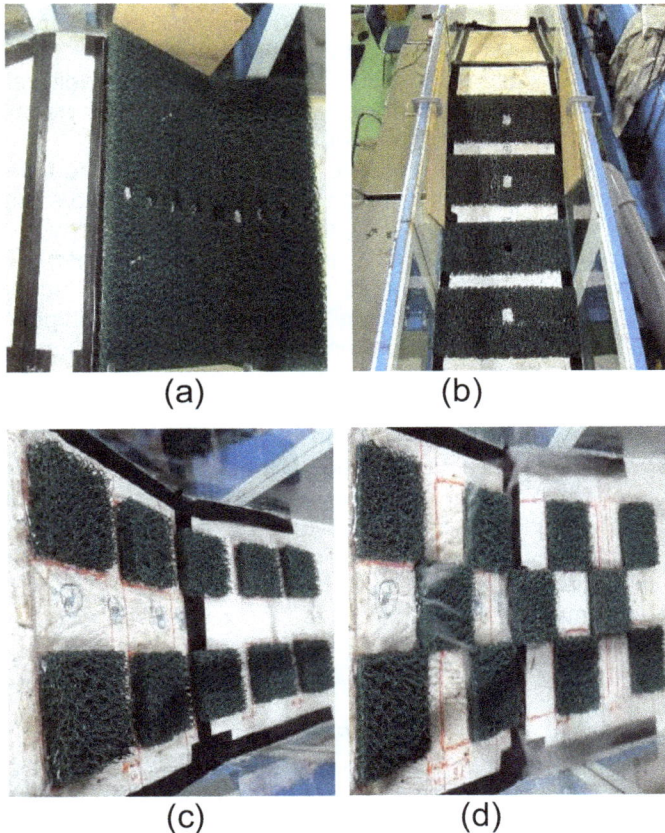

Figure 2. Experimental setup top view- (a) all over vegetation case, (b) row vegetation (2D type), (c) grid type 3D vegetation and, (d) staggered type 3D vegetation placed on the d/s slope and bed in the flume.

overtopping flow on the downstream slope of the embankment with the vetiver hedges, a small-scale laboratory flume experiments with an embankment using emerged vegetal cover were carried out with three different unit discharges as 0.018, 0.013 and 0.010 m³/s was taken in our experiments. The longitudinal flow velocities and water depths were measured along the centre line of the flume at 26 points with an interval of 0.10 m up to upstream (u/s) crest of the model and later 0.05 m interval from u/s crest to the d/s end of the flume until steady flow condition was established. A scale factor (ratio of a variable in the model that is, height of embankment model 0.25 m to the corresponding value of its prototype height of 4 m usually used for earthen embankment) 0.0625 was kept constant throughout the tests. Usually earthen embankments are made with 4 m height with slopes 3H:1V and 2H:1V in the upstream and downstream sides, respectively. In the tests the velocities and the water depths were measured for situations with and without vegetation in the flume. The height of the vegetation model was kept constant, 0.05 m and width and length was kept same as the d/s side of the embankment model considering emergent flow conditions.

Data collection

The discharges were measured with an electromagnetic flow meter (model: MK -515/ 8510-XX, paddle flow sensor, Georg Fischer Signet LLC, USA). An electromagnetic velocity meter (type of main amplifier: VM-2000; type of sensor: VMT2-200-04P, Kenek Company, Ltd.) was used to measure the flow velocities at the centerline of the channel and the model. The water surface elevation was measured at the same locations as the velocity profiles by the point gauges (with accuracy up to 0.1 mm), fixed and mounted on a movable sliding carriage.

Drag force measurement

A two-axis load cell (streamwise (X) and transverse (Y) directions,

(a)

(b)

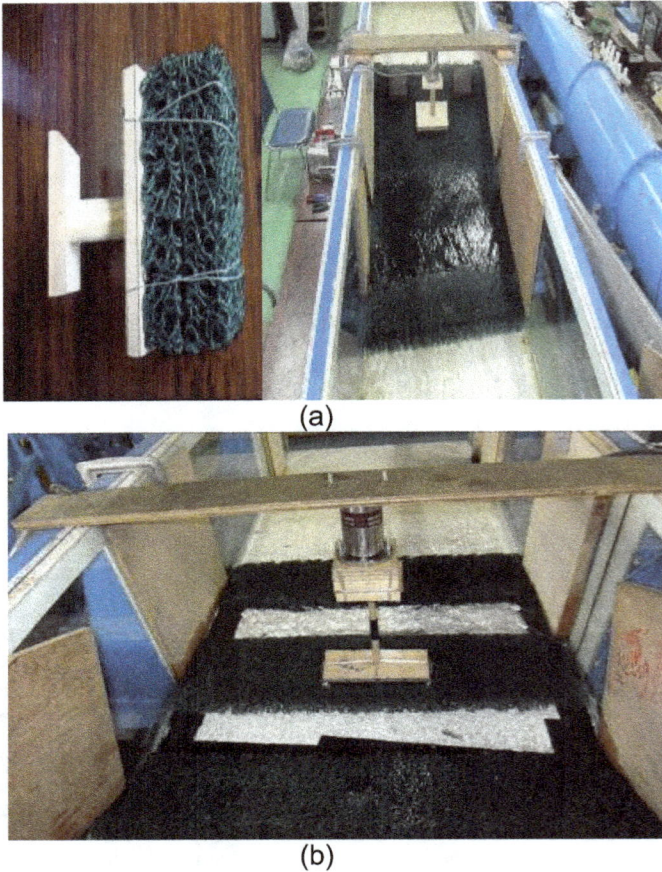

Figure 3. (a) Drag force measurement setup for all over vegetation on d/s slope and bed, and (b) Drag force measurement apparatus and setup for 2D type or row vegetation case.

type LB-60, SSK Co., Ltd.) with a resolution of 1/1000 that can measure a maximum load of 10.0 N was used to measure the drag force on the vegetative roughness model (Figure 3).

Calculation of effective shear stress

During the experiments the steady uniform overflow condition was established at the upstream side of the model and during overtopping flow became unsteady and non-uniform. But for simplicity we consider steady non-uniform flow for calculation of effective shear stress. For a control volume of unit area along the slope the balance of horizontal momentum in this case can be expressed as total shear stress which is equal to the sum of the bed shear stress and the equivalent shear stress due to the vegetation drag:

$$\tau_w = \tau_b + \tau_v \tag{1}$$

where, τ_b= bed shear stress transfer to the soil; τ_w= stream wise component of the weight of water mass and τ_v=resistance due to the drag around the vegetation. The drag force was measured directly by a two-axis load cell apparatus as mention above.

The streamwise weight component of the weight of water mass per unit bed area is expressed as,

$$\tau_w = \rho g h i_e (1 - \lambda) \tag{2}$$

where, λ = area concentration due to vegetation. Based on the Equation (2) the effective bed shear stress which directly transfers to the soil was calculated by deducting the drag forces induced by the vegetation for different discharges.

RESULTS AND DISCUSSION

Overtopping flow over earthen embankment in without vegetation (WOV) condition shows that the flow was static at the beginning upstream, accelerating sub-critical flow state over a portion of the embankment crest; through critical flow on the crest and supercritical flow over the remainder of the crest; and supercritical flow on the downstream slope and extend to the further downward same as observed in the previous study (Powledge et al., 1989b). Whereas the flow characteristics with vegetation (WV) condition shows that, the flow was static at the beginning of upstream, accelerating subcritical flow state both on the embankment crest and the downstream slope.

It is clear that the initial erosion process begin anywhere within supercritical flow region due to high bed shear stress, from the point of slope discontinuity that is, from d/s slope to d/s toe region especially for embankment. The effective bed shear stress (EBSS) which actually transfers to the soil surface is well below the shear stress developed in the case of WOV because of the induced drag force due to vegetation.

Figures 4 and 5 illustrate the comparison of bed shear stresses behind vegetation rows (BVR) along d/s slope and bed between row type (2D type) and grid type 3D vegetation on d/s slope and bed of embankment. The flow velocity behind the vegetation rows is much higher in 2D type row vegetation whereas, in grid type 3D roughness flow accelerates between the vegetation rows and therefore velocity is not much significant behind the rows, results the reduction of bed shear stress is attained on just behind the rows due to grid type 3D vegetation. Where, the reduction of bed shear stress from 28 to 40% is attained on the d/s slope. Vegetation also reduced the bed shear stress from 45 to 54% on the d/s bed. In comparison of the BSS behind the vegetation rows (BVR) in between 2D type row vegetation and grid type 3D vegetation, it is seen that the maximum reduction was 40% at d/s slope and 54% on bed, both for the grid type 3D roughness and for 0.25 ratios of rows with 20% blocking.

Herein the following Figures 6 and 7 shows the comparison of bed shear stresses along the d/s slope and bed in between grid type 3D and staggered type 3D vegetation placed both on d/s slope and bed. For grid type 3D roughness, flow accelerates within vegetation rows and increased the flow velocity results in increasing of bed shear stress. The effects of grid type 3D vegetation shows that bed shear stress increased on the

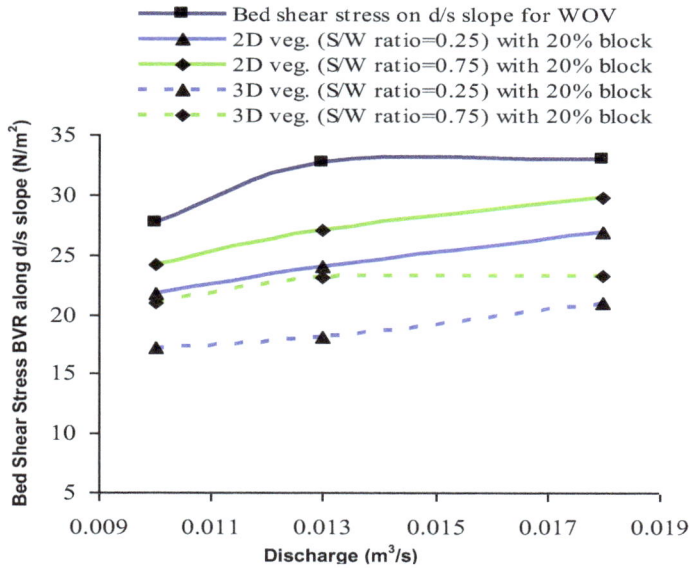

Figure 4. Comparison of effective bed shear stress behind the vegetation rows (BVR) along the d/s slope in between 2D and 3D type vegetation effect.

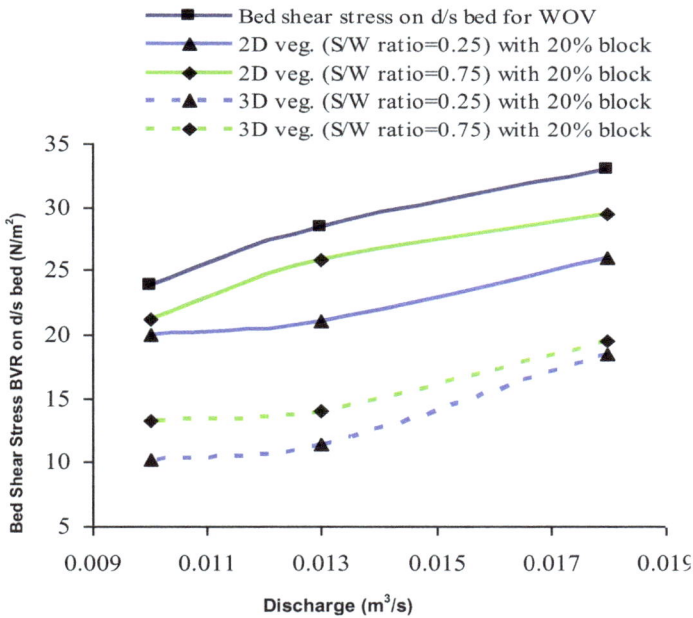

Figure 6. Comparison of effective bed shear stress along d/s slope in between grid-type and staggered type 3D vegetation effect.

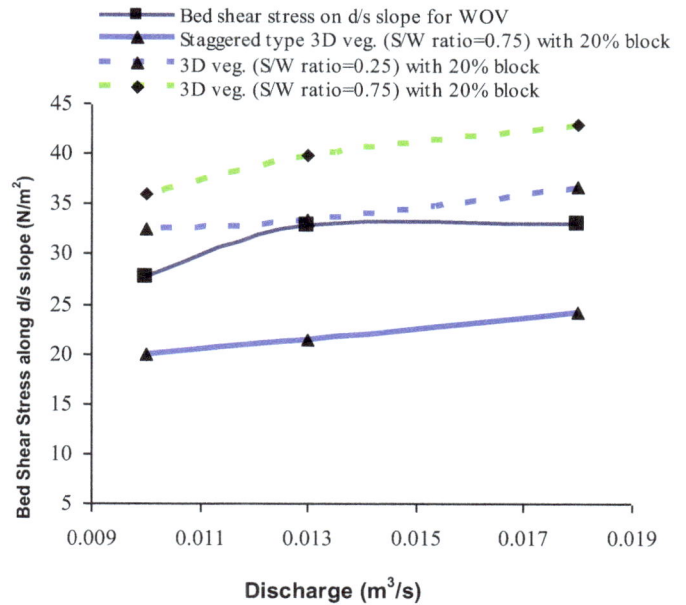

Figure 5. Comparison of effective bed shear stress behind the vegetation rows (BVR) on the d/s bed in between 2D and 3D type vegetation effect.

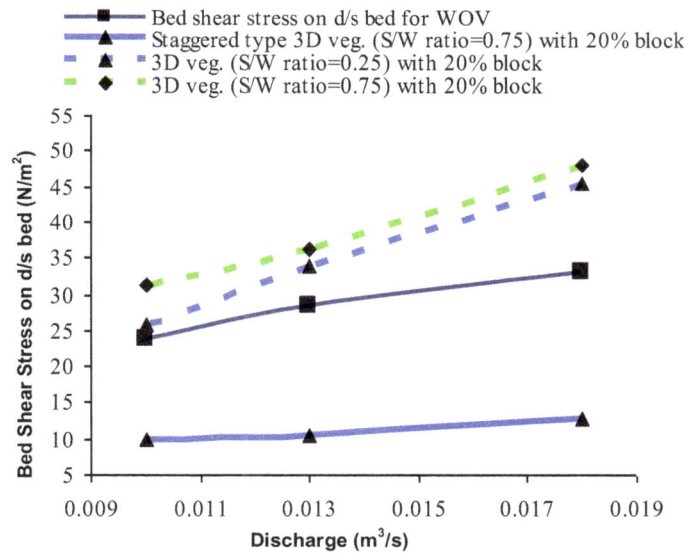

Figure 7. Comparison of effective bed shear stress on the d/s bed in between grid-type and staggered type 3D vegetation effect.

d/s slope as well as bed due to the acceleration of flow in between the vegetation rows. Bed shear stress increased from 10 to 27% on d/s slope and 22 to 35% on bed for 5 and 15 cm spacing rows (0.25 and 0.75 ratios) respectively.

On the other hand, in staggered type 3D roughness the flow could not accelerate directly between the vegetation rows because flow retard by the alternate rows in this type and decreased the flow velocity that results decreased the effective bed shear stress on both d/s slope as well as bed of embankment. The reduction of effective bed shear stress is found out 30 and 61% on the d/s slope and bed respectively for staggered type 3D roughness.

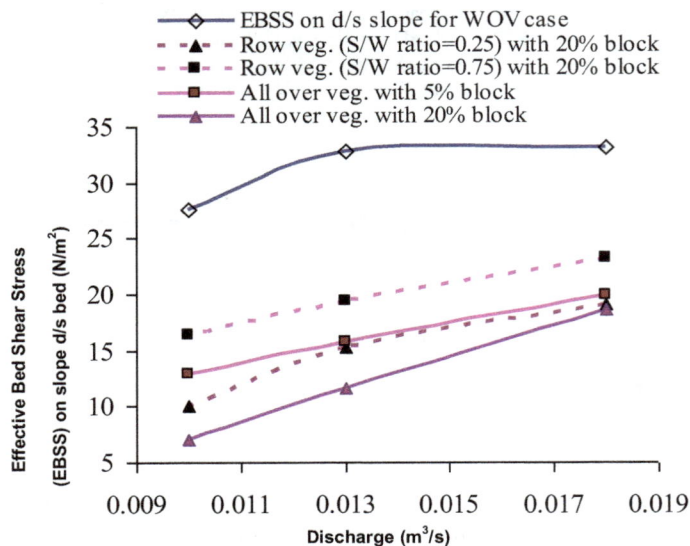

Figure 8. Comparison of effective bed shear stress along the d/s slope in between different row type or 2D veg. and all over vegetation effect on both slope and d/s bed.

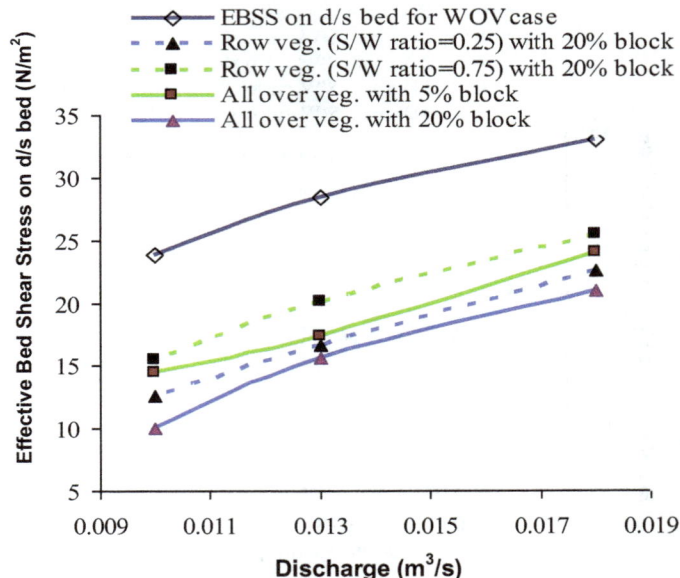

Figure 9. Comparison of effective bed shear stress on the d/s bed in between different row type or 2D vegetation and all over vegetation effect on both slope and d/s bed.

The Figures 8 and 9 illustrate the comparison of effective bed shear stresses for row type 2D vegetation with different spacing and all over vegetation on the d/s slope and bed. It is found that, in row case, maximum reduction of bed shear stress is attained for 5 cm spacing rows (for 0.25 ratio) with 20% blocking, on the other hand, maximum reduction of bed shear stress is achieved due to 20% blocking for all over vegetation case.

Comparative effects of row type 2D and all over vegetation on d/s slope show that, for all over vegetation the maximum reduction of EBSS is established as 61% on d/s slope and 48% on bed, both for 20% blocking case. Similar effects are found on the d/s bed. Vegetation reduced the bed shear stress about 47% for 20% blocking however, the bed shear stress reduction was 35% on the d/s bed due to low blocking case taken as 5% blocking in compared with 2D row type vegetation effect.

Conclusions

A cost effective and eco-friendly solution for stabilization of earthen embankment is presented in this paper. Vegetation markedly reduced the effective bed shear stress on the surface that demonstrated a significant reduction of damage and risk of failure from overtopping flows can be achieved by providing vegetation cover on the d/s slope of the embankment.

Based on the results of this investigation, the following conclusions can be drawn:

1. Bed shear stresses behind the vegetation rows (BVR)

are much lower in block type 3D roughness whereas, higher values occurred in 2D type row vegetation.

2. BSS in between grid type 3D vegetation and staggered type 3D vegetation, it is seen that BSS is increased in 3D type vegetation due to the acceleration of flow within the rows whereas, BSS decreased on both d/s slope and bed of embankment as the flow retard directly by the alternative rows in staggered type.

3. It is observed that the maximum reduction of effective bed shear stresses is obtained for 20% blocking all over vegetation on the d/s slope and bed of the embankment compared to all the cases.

4. Spacing of roughness element is also the influencing factor in reducing bed shear stress.

5. It is concluded that vegetation can be the effective and innovative solution for the stabilization of river bank or coastal embankments.

REFERENCES

Das SC, Tanaka N (2009). The effectiveness of vetiver grass in controlling water borne erosion in Bangladesh. Proceedings of the 11th International Summer Symposium, JSCE, 11 September, Tokyo, Japan, pp. 69-72.

Grimshaw D (2008). Vetiver System Applications, Technical Reference Manual (second edition) published by The Vetiver Network International, US, pp. 1-10.

Hanson JG, Temple MD (2001). Evaluation mechanics of embankments erosion during overtopping. Proceedings of the Seven Federal Interagency Sedimentation Conference, March 25 to 29, Reno, Nevada, USA, pp. 24-30.

Islam MN (2000). Embankment erosion control: Towards cheap and simple practical solutions for Bangladesh, Proceedings of the Second

International Conference on Vetiver, Office of the Royal Development Projects Board, Bangkok, pp. 307-321.

Ke CC, Feng ZY, Wu XJ, Tu FG (2003). Design principles and engineering samples of applying Vetiver eco-engineering technology for landslide control and slope stabilization of riverbank. Proceedings of the Third International Conference on Vetiver, Guangzhou, China, October, pp. 349-357.

Powledge GR, Ralston DC, Miller P, Chen YH, Clopper PE, Temple DM (1989a). Mechanics of Overflow Erosion on Embankments, I: Research Activities. J. Hydraul. Eng. ASCE 115(8):1040-1055.

Powledge GR, Ralston DC, Miller P, Chen YH, Clopper PE, Temple DM (1989b). Mechanics of Overflow Erosion on Embankments, II: Hydraulic and Design Considerations. J. Hydraul. Eng. ASCE 115(8):1056-1075.

Truong PNV, Loch R (2004). Vetiver system for erosion and sediment control. In Paper Presented at 13th International Soil Conservation Organization Conference, Brisbane, 4-7 July (Paper No. 247), pp:1-6.

Wu F-S (2008). Characteristics of flow resistance in open channels with non-submerged rigid vegetation. J. Hydrodyn. 20(2): 239-245.

21

Contamination of groundwater due to underground coal gasification

R. P. Verma[1,2], R. Mandal[1], S. K. Chaulya[1], P. K. Singh[2], A. K. Singh[1] and G. M. Prasad[1]

[1]CSIR-Central Institute of Mining and Fuel Research, Barwa Road, Dhanbad-826015, India.
[2]Department of Environmental Science and Engineering, Indian School of Mines, Dhanbad-826004, India.

Underground coal gasification (UCG) generates potential groundwater pollution because it changes local hydrogeological parameters. Groundwater pollution is caused by diffusion and penetration of contaminants generated by UCG processes towards surrounding strata and possible leaching of residue by natural groundwater flow after gasification. A large number of hazardous water-borne contaminants were identified during different UCG operations conducted so far, and in some locations long-term groundwater contaminations were observed. Organic pollutants were detected after UCG process are phenols, benzene with its derivatives, polycyclic aromatic hydrocarbons (PAHs), heterocycles etc. and inorganic pollutants includes ammonia, mercury, zinc, sulphates, cyanides, heavy metals etc. Adsorption function of coal and surrounding strata makes a significant contribution to decrease the contaminants over time and distance from the burn cavity. Possible pollution control measures regarding UCG include identifying unsuitable zone, sitting a hydraulic barrier and pumping contaminated water out for surface disposal. The paper enumerates major pollutants identified in different UCG sites and its mitigation measures during gasification processes, and groundwater remediation after gasification.

Key words: Underground coal gasification, groundwater contamination, water influx.

INTRODUCTION

Underground coal gasification (UCG) is a procedure for extracting the synthesis gas (syngas) from *in-situ* underground coal seams that could not be economically extracted by conventional mining methods. UCG process produces gas suitable for high-efficiency power generation, chemical feedstock, fuel etc. The first available environmental data on UCG came from later United States trials, mainly regarding the Hanna and Hoe Creek UCG trails (Cooke and Oliver, 1983), at this site groundwater contamination monitoring was conducted pre- and post-UCG process. The results showed that UCG at shallow depths can pose a significant risk to groundwater in adjacent strata. UCG process, involves air/oxygen and steam pumped into the underground coal seam through an injection well. The introduction of an oxidizing gas produces heat, which partially combusts the coal *in-situ* and creates the syngas product, primarily composed of hydrogen (H_2), carbon monoxide (CO), carbon dioxide (CO_2) and methane (CH_4) (Stephens et al., 1985). The syngas is extracted from the UCG burn cavity by a production well, which brings the gas product to the surface for industrial uses. A schematic diagram of UCG process is shown in Figure 1. UCG is partly environmental friendly due to non-discharge of tailing, decrease of sulfur emission, and non-discharge of ash, mercury and tar. In UCG process, physico-chemical interaction changes natural stress in the surrounding rock mass, which influences contaminants formation in the UCG reactor as well as inducing potential subsidence, and pollution of groundwater, surface water and atmospheric quality. Subsidence creates a hazard for any surface infrastructure that presents above the UCG zone and may create detrimental changes in surface or groundwater hydrology above the cavity (Torres et al., 2014). UCG cavity is a source of both gaseous and liquid pollutants. The risk of groundwater pollution from UCG depends on whether the contaminants can migrate beyond the immediate reaction zone to more sensitive groundwater areas. Transport of aqueous phase contaminants depends on permeability of *in-situ* rocks, geological setting of gasification reactor and hydrogeology of the adjacent area.

For the commercialization of UCG technology environmental impact studies, especially of groundwater pollution prevention and control is necessary. The paper identifies possibility of underground water pollution due to UCG and analyzes the fate of contaminants (organic and inorganic) as well as formulates strategies for groundwater pollution control.

GROUNDWATER CONTAMINATION BY UNDERGROUND COAL GASIFICATION

Contamination of ground water is considered as a serious environmental threat. Degradation of ground water can result from any of the following three sources (Ahern and Frazier, 1982):

(i) Organic contaminants in tars produced during carbonization or gasification;
(ii) Inorganic salts or trace elements in the leachate from the ash; and
(iii) Changes in flow patterns or rates resulting from subsidence or interconnection of aquifers by fracturing.

Groundwater contamination around UCG reactor is caused by dispersion and penetration of pyrolysis products of coal with migration of groundwater and escaped gases. During the gasification process, air or oxygen is injected with high pressure equal to or greater than the surroundings hydrostatic pressure (P_h). Some of the gas products are therefore lost to the surrounding permeable media and overlying strata, asa result of cracks in the overburden, as shown in Figure 2. In UCG

reactor temperature is very high and due to this, solvent power of water increases, and density and viscosity of water decreases. After gasification coal ash is left in cavity and due to low pressure of rector groundwater begins to enter into the gasifier. It may contain some higher molecular weight organic substances that are produced during pyrolysis of coal seam. The more volatile the product, the farther it is transported out into the surrounding coal strata before condensing or dissolving in groundwater. As the cavity cools and fills with water, the residual ash is leached, leading to increase in pH and in the concentration of many inorganic species. During this period, thermally driven convection currents transport some of the non-volatile inorganic contaminants from the ash into the surrounding formation. After that the concentrations of many contaminants will continue to change as a result of adsorption on the coal and strata or reactions among different species.

Due to hydraulic gradients, migration of groundwater will occur through coal seams and burned out areas which lie below the water table. This may cause soluble components sorb on ash or char to be leached out and transported away from the gasification site. An increase in dissolved organic material could result from partial dissolution of coal tars formed during gasification (Phillips and Muela, 1977).

Subsidence of the overburden above the UCG burn cavity also can cause groundwater contamination problems. An example of this phenomenon was found at the Hoe Creek UCG test site in the United States where aquifers cross connection occurred during gasification operations. The problem was transmission of pollutants generated in the burn zone through fractures caused by subsiding overburden into overlying aquifers. However, this should be avoided during the phase of choosing the study site (Burton et al., 2004). Initially at the time of gasification when the temperature of the gasification cavity is high, most of the returning water is vaporized and may be vented to the surface via the processing wells.

WATER INFLUX DURING UNDERGROUND COAL GASIFICATION PROCESS

Groundwater plays an important role in gas formation process during UCG. The major sources of water in the cavity areas are as follows:

(i) Water already available in the coal;
(ii) Water influx by permeable coal beds: It is the major source of water influx into the cavity; and
(iii) Water influx by overburden: It is the second major source of water into the cavity. Water influx by overburden is mainly two types, firstly when UCG process starts, water influx happen due to permeability of

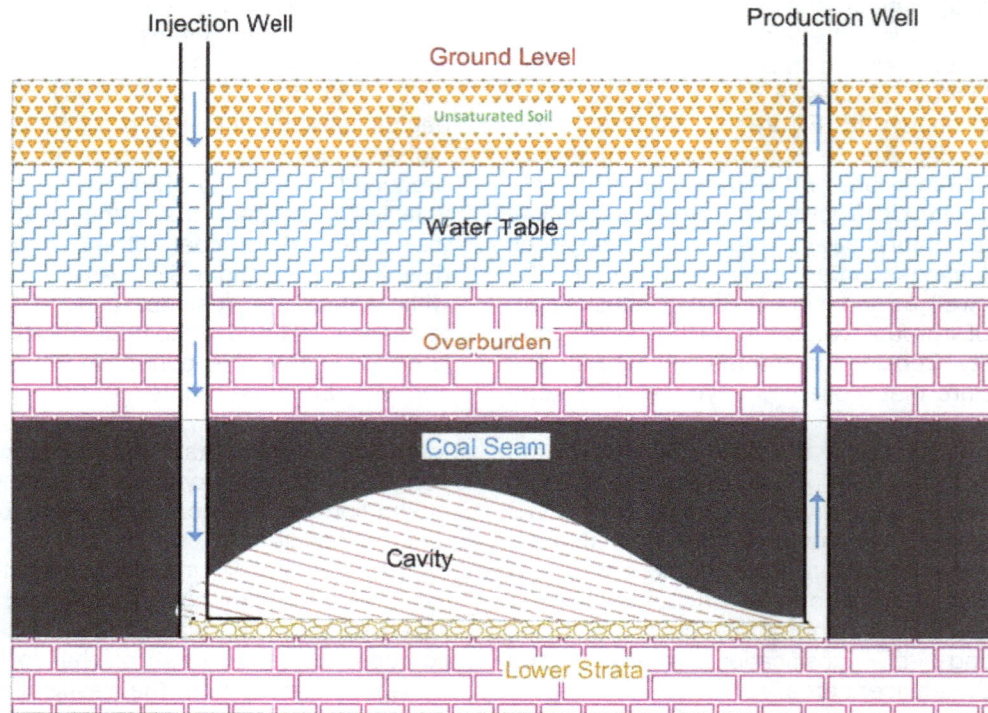

Figure 1. Schematic diagram of UCG process.

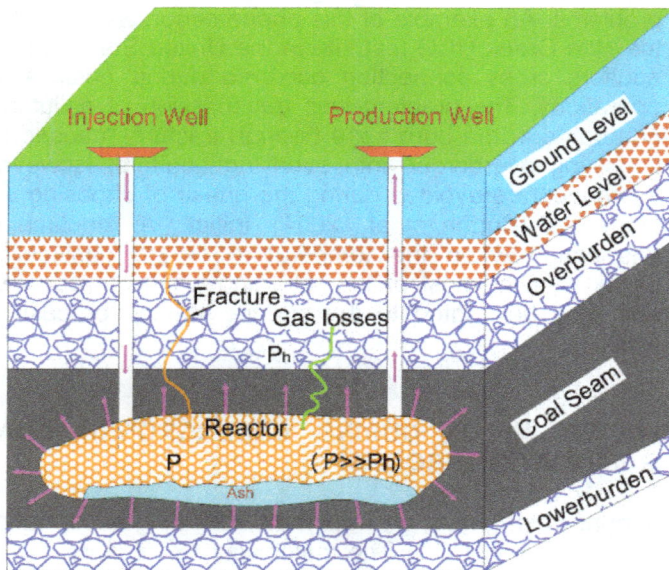

Figure 2. Fracture developed due to collapse of strata and gas escaped from the fracture.

rock overburden as shown in Figure 3, and secondly as coal burns away during UCG and overburden is exposed to intense heating. This causes extensive spalling of rock as it dries. Rock is exposed to thermal stresses and steam pressure (Sury et al., 2004).

Role of water in UCG process depends on whether the water influx is into the hot UCG cavity or downstream into the carbonized link zone. Water influx into the cavity can participate in gasification reaction and downstream into the carbonized link can participate directly in water-gas shift reaction which exchanges carbon monoxide for hydrogen, and indirectly in the methanation reaction; furthermore, it cools the product gas and thus decreases the potential for surface heat exchange to recover the sensible heat of the product gas but, at the same time, reduces high temperature corrosion of the production well and surface piping system. The amount of water flowing in is measured and controlled by the ratio of water to gasified coal (W/W). When the water influx ratio is higher than 0.5, a normal gasification process ceases at temperatures below 800°C. Therefore, a controlled groundwater influx is very important for enhanced-hydrogen gas preparation in UCG field tests by using the stored heat energy underground (Shu-qin et al., 2008).

Important reactions of underground coal gasification

The controlled gasification of UCG process involves burning of coal with air or other oxidants with complex series of reactions including carbonization, distillation, oxidation, reduction, pyrolysis, water-gas shift conversion, water-gas reaction, methanation and Boudouard reactions (Haider, 2012). The followings are

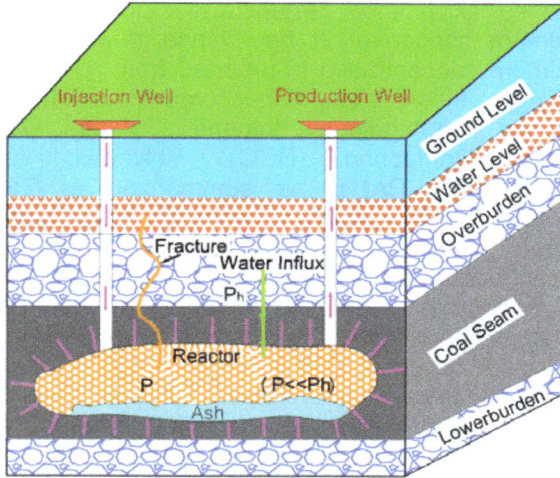

Figure 3. Water influx into the burned cavity.

significant reactions involved during UCG process:

(i) Oxidation reaction: $C + O_2 \rightarrow CO_2 + 393.8$ MJ/kmol
(ii) Partial oxidation: $2C + O_2 \rightarrow 2CO_2 + 231.4$ MJ/kmol
(iii) Oxidation: $2CO + O_2 \rightarrow 2CO_2 + 571.2$ MJ/kmol
(iv) Reduction/ Boudouard reaction:$C + CO_2 \rightarrow 2CO - 162.4$ MJ/kmol
(v) Hydrogenous water-gas reaction: $C + H_2O$ (g) $\rightarrow CO + H_2 - 131.5$ MJ/kmol
(vi) Shift conversion: $CO + H_2O \rightarrow CO2 + H_2 + 42.3$ kJ/mol
(vii) Methanation:$CO + 3H_2 \rightarrow CH_4 + H_2O + 206.0$ kJ/mol
(viii) Hydrogenation gasification: $C + 2H_2 \rightarrow CH_4 + 87.5$ kJ/mol.

Effect of aquifer interconnection

Aquifer interconnection due to UCG process is investigated in Hoe Creek II and Hoe Creek III experiment in Wyoming (Stone et al., 1982). Aquifer interconnection would be happened when two coal seams are gasified in a column and a strata lies between them. After gasification of both coal seams (gasification process started at first lower coal seam) and the strata may be collapsed, depending on the strata strain and aquifers interconnect to each other. So, due to aquifer interconnection possibilities of groundwater contamination increases.

GOVERNING EQUATION OF GROUNDWATER FLOW AND TRANSPORT

Ground water flow equation

The two-dimensional equation of continuity for flow in one aquifer to multi-aquifer systems is represented by the following equation (Contractor and Shreiber, 1987):

$$\frac{\partial}{\partial x}(K_x B_I \frac{\partial}{\partial x}(h_I)) + \frac{\partial}{\partial y}(K_y B_I \frac{\partial}{\partial y}(h_I)) + r + Q_I + \frac{K_{Ia}}{B_{Ia}}(h_{I-1}-h_I) + \frac{K_{Ib}}{B_{Ib}}(h_{I+1}-h_I) = S_I \frac{\partial}{\partial x}(h_I) \tag{1}$$

Where, K_x, K_y are the permeabilities in x, y directions, h_{I-1}, h_I, h_{I+1} are piezometric heads in Aquifers I-1, I, I+1; B_I is the saturated thickness of Aquifer I; r is the recharge into the aquifer; Q_I is the pumping rate from aquifer I; K_{Ia}, K_{Ib} are the permeabilities of aquitards above and below aquifer I; B_{Ia}, B_{Ib} are the saturated thicknesses of aquitards above and below aquifer I; and S_I is the storage coefficient of aquifer I. For derivation of equation some essential assumptions has taken into consideration, such as water is homogeneous with constant density, Darcy's law is valid, the Dupuit approximation is assumed to be valid, one dimensional vertical flow is assumed to occur in the aquitards without storativity effects, the off-diagonal terms of the conductivity tensor (K_{xy}, K_{yx}) are equal to zero; and x and y being the principal axes.

Water quality model

Two-dimensional equation for mass transport is described below:

$$\frac{\partial}{\partial t}(RC) = \frac{\partial}{\partial x}(D_{xx}\frac{\partial C}{\partial x}+D_{xy}\frac{\partial C}{\partial y} - V_x C) + \frac{\partial}{\partial y}(D_{yx}\frac{\partial C}{\partial x} + D_{yy}\frac{\partial C}{\partial y} - V_y C + \mu C + \gamma + \sum \eta C + F_{I-1} + F_{I+1} \tag{2}$$

Where, C is concentration; R is the retardation factor; D_{xx}, D_{xy}, D_{yx}, D_{yy} are dispersion coefficients; V_x, V_y are seepage velocity components in x, y direction; μ is decay coefficient; γ is the recharge parameter; $\sum \eta C$ is the effect of pumping summed over the number of wells, η being the discharge constant; F_{I-1} is the interaction effect with the aquifer above I; F_{I+1} is the interaction effect with the aquifer below I.

Now, F_{I-1} = interaction of aquifer I with aquifer I-1 above it.

$$F_{I-1} = \frac{K_{Ia}}{n_I B_I}\frac{h_{I-1}-h_I}{B_{Ia}}C_{I-1}, \text{ if } h_{I-1} > h_I$$
$$= \frac{K_{Ia}}{n_I B_I}\frac{h_I-h_{I-1}}{B_{Ia}}C_I, \text{ if } h_{I-1} < h_I$$

Where n_I is the effective porosity of aquifer I and F_{I+1} = interaction of aquifer I with aquifer I+1 below it.

$$F_{I+1} = \frac{K_{Ib}}{n_I B_I}\frac{h_{I+1}-h_I}{B_{Ia}}C_{I-1}, \text{ if } h_{I+1} > h_I$$
$$= \frac{K_{Ib}}{n_I B_I}\frac{h_I-h_{I+1}}{B_{Ia}}C_I, \text{ if } h_{I+1} < h_I$$

The dispersion coefficients are related to the velocity of ground water flow and to the nature of the aquifer using

Scheidegger's equation.

$$D_{ij} = \alpha_{ijmn} \frac{V_m V_n}{V} \qquad (3)$$

Where, α_{ijmn} = the dispersivity of the aquifer (dimensions of length), V_m, V_n = components of velocity in m, n direction, respectively, V = the magnitude of the velocity.

$$= [(V_m)^2 + (V_n)^2]^{1/2}$$

An isotropic aquifer dispersivity tensor defined in terms of two constants. These are the longitudinal and transverse dispersivities of the aquifer (αL and αT, respectively). These are related to longitudinal and transverse dispersion coefficients by:

$$D_L = \alpha_L v \text{ and } D_T = \alpha_T v$$

After expanding Equation (3) and substituting Scheidegger's identities, the components of the dispersion coefficient for two-dimensional flow in an isotropic aquifer are:

$$D_{XX} = D_L \frac{(v_X)^2}{v^2} + D_T \frac{(v_y)^2}{v^2}$$

$$D_{yy} = D_T \frac{(v_X)^2}{v^2} + D_L \frac{(v_y)^2}{v^2}$$

$$D_{xy} = D_{yx} = (D_L - D_T) \frac{v_X v_y}{v^2}$$

These essential assumptions were made in the derivation of equation are that the solute moves in a saturated, medium is porous, the solute concentration does not affect the density and viscosity of the water in the aquifer, and the dispersion coefficients are assumed to be proportional to the velocity, hence can vary with time from node to node.

TYPES OF GROUNDWATER CONTAMINANT

Organic contaminants

In laboratory and field studies, it has been observed that sorption of contaminants on aquifer substrate is an important mechanism that acts to decrease the concentration of contaminants in groundwater over time. However, certain contaminants are not sorbed sufficiently to alleviate concern for their transport in groundwater aquifers. The results (laboratory and field studies) specify that after a time period organic contaminants concentration are decreased and composition are changed. Phenols and low molecular weight aromatic hydrocarbons persist in solution while less soluble components such as three, four, and five-ring aromatic hydrocarbons are removed by sorption (Stuermer et al., 1982). The major organic groundwater contamination

identified as phenols. Phenol is a constituent whose maximum concentration has varied from site to site. Maximum concentration is reported from range of 20 to 450 mg/L in six sites. Other organic pollutants were benzene, naphthalene, toluene, xylene, Dissolved Organic Compound (DOC), Polycyclic Aromatic Hydrocarbons (PAHs) as summarized in Table 1. According to data reported from six sites benzene is another organic substance of particular concern for its concentrations occurring at many of the sites and its designation as a human carcinogen. But benzene contamination is generally confined to within 9 m of the gasification cavities in the affected aquifers. As expected, volatile organics in the process tar and water, alone, showed the presence of over 250 different organic components (Campbell et al., 1979). The organic materials present in the groundwater were analyzed by two different procedures. In one procedure the dissolved organic carbon was fractionated into two broad groupings of hydrophilic and hydrophobic species. Each of these was then further fractionated into groups of acidic, basic and neutral compounds. In second procedure gas chromatography- mass spectroscopy (GC-MS) analysis was used to identify and quantify the concentrations of many of the individual organic compounds present (Campbell et al., 1979).

So, based on general solubility characteristics of these potential water pollutants, phenols pose the greatest threat, while pyridines and anilines are also regarded as soluble and therefore likely to be present in the water. Quinolones, which are somewhat less soluble, may be present at trace levels. Aromatic hydrocarbons are not expected to present a significant hazard to water quality (Phillips and Muela, 1977).

The post-gasification distribution of phenolic materials is a function of distance from burn cavity and time. Inside the burn cavity zone, phenolic concentration is found low due to strong adsorption and hence retardation in movement of aqueous phenol through a sub-bituminous coal bed (Wang, 1979). Organic pollutants found in UCG process are volatile and non-volatile. Volatile aromatic materials are benzene, toluene, xylenes and naphthalene. The more volatile (lower-molecular weight) are transported farther from the burn boundary. The groundwater from several wells of Hoe Creek I sampled after gasification was analyzed using GC-MS.

Inorganic contaminants

Coal is primarily consists of organic materials such as carbon, hydrogen, oxygen, nitrogen and sulfur. Inorganic material present is due to layers of clay, carbonate or mineral matter such as pyrite which are washed into the swamps along with plant material during initial stages of coalification. After combustion of coal, inorganic material is found mainly to be associated with ash, which can

Table 1. Baseline and maximum reported organic contaminates concentrations in six sites (Ahern and Frazier, 1982).

Organic constituents	Hoe Creek I		Hoe Creek II		Hanna I		Hanna II		Fairfield		Tenn. Colony	
	Base	Max.	Base	Max.	Base	Max.	Base	Max.	Base	Max.	Base	Max.
Phenols(mg/L)	0.001	450	-	45	45	-	-	270	0.1	20	0.005	100
Benzene (µg/L)	4	192	-	-	-	-	-	607	-	-	-	-
Naphthalene (µg/L)	5	0	-	-	-	-	-	640	-	-	-	-
Toluene (µg/L)	11	68	-	-	-	-	-	2200	-	-	-	-
Xylene (µg/L)	1	10	-	-	-	-	-	1000	-	-	-	-
DOC (mg/L)	4	230	-	-	-	-	-	-	-	-	2.5	790

comprise from 3 to 20% of the original coal volume (Lang, 1982). Ash in the burn cavity was identified as a source of most of the inorganic constituents which moved with groundwater after gasification. Calcium, sulfate and hydroxide were the major inorganics leached out of ash. Hydroxide concentrations were lower in leachate from lignite ash than from bituminous and subbituminous ash (Humenick and Lang, 1980). Volatile inorganic species exhibit increasing concentration during UCG due to movement of volatile species out of the burn cavity. Ammonia produces during pyrolysis are detected at high concentrations in surrounding strata.

Several studies have identified changes in inorganic substances in groundwater due to the gasification process. Table 2 provides data for Tennessee Colony site in Texas which are representative of increase in inorganic groundwater constituents at other UCG sites. Soluble ash components are seen to increase total dissolved solid (TDS) concentrations in cavity water (Table 3). The materials include a wide array of ionic species, mainly calcium, sodium, sulphate and bicarbonate. There are, however, many other inorganic substances leached into the groundwater which are of interest, even though they are present in smaller amounts. These include calcium, aluminum, mercury, magnesium, sulphate, manganese, ammonia, arsenic, boron, iron, zinc, selenium, hydroxide and some radioactive materials such as uranium (Ahern and Frazier, 1982).

Field data and laboratory ash leaching experiments indicate that inorganic contaminates tend to increase due to ash leaching. Different results at different sites are probably due to coal and ash composition, gasifier temperatures, sampling techniques and natural water quality. Only one parameter, pH showed very large variations among investigators. Ash from Texas lignite showed little change in pH. Again, these differences could be accounted for by the inherent differences between coal and lignite, or may be site specific. Other groundwater contaminant grows during UCG process are total dissolved solids (TDS) and pH in different UCG sites (Table 3).

Surface water contaminations

Syngas produced by UCG contains a component of liquid or vaporized water which is removed from gas before the gas is used for power generation and chemical feedstock. This water contains residual hydrocarbons, benzenes, phenols and polycyclic aromatic hydrocarbons (PAHs) which are fully treatable. Potential pollution of surface water in UCG is extremely low, and the common pollutants are phenols, ammonia, chemical oxygen demand (COD), pH, conductivity and sulphides (Sury et al., 2004). The surface water can be affected by groundwater pumping and drilling operations, but the water pumped out to the surface may containphenol (Green, 1999).

MIGRATION OF GROUNDWATER CONTAMINANTS

Migration with natural groundwater flow

Groundwater pollution due to UCG is a geo-environmental problem. Initial UCG process requires water for gasification of coal. Once the UCG process started cavity gradually increase. During burning process, cavity pressure is controlled for water influx and gas losses. Upon completion of burning, water present in aquifer begins to seep back into the burn cavity which

Table 2. Baseline and maximum reported inorganic contaminates concentrations in six sites (Ahern and Frazier, 1982).

Type of constituents	Hoe Creek I		Hoe Creek II		Hanna I		Hanna II		Fairfield		Tenn. Colony	
	Base	Max.	Base	Max.	Base	Max.	Base	Max.	Base	Max.	Base	Max.
Cations(mg/L)												
Ammonium	0.6	72	-	-	2.3	15	-	-	1.0	100	0.7	16.3
Boron	0.09	0.5	-	0.8	0.06	1.5	0.06	0.3	-	0.3	0.03	2.2
Calcium	36	220	-	-	15	32	-	6	20	8	8	94
Iron	0.01	37	-	-	0.04	8	0.03	0.1	-	200	0.9	92
Lead	0.001	0.041	-	-	0.03	0.03	-	-	-	-	ND	ND
Magnesium	10	60	-	-	9	15	-	-	5	15	2	22
Manganese	-	-	-	-	0.06	0.06	-	-	-	0.07	0.06	1.3
Zinc	-	-	-	-	0.2	0.9	-	-	-	ND	1.8	7
Mercury	ND	ND	ND	ND	0.001	0.0001	0.001	0.0001	ND	ND	ND	ND
Anions (mg/L)												
Cyanide	0.01	290	-	-	-	-	-	-	-	940	ND	0.008
Sulfate	150	1230	-	-	400	1600	-	-	4	1150	1	625
Thiocyanate	-	-	-	-	-	-	-	-	-	ND	ND	0.3

ND-Not Determined.

Table 3. Baseline and maximum reported inorganic contaminates concentrations in six sites (Ahern and Frazier, 1982).

Constituents	Hoe Creek I		Hoe Creek II		Hanna I		Hanna II		Fairfield		Tenn. Colony	
	Base	Max.	Base	Max.	Base	Max.	Base	Max.	Base	Max.	Base	Max.
TDS (mg/L)	700	3400	-	-	1700	3300	-	-	350	2300	290	1460
pH	7.5	6.3	-	-	-	-	-	-	-	7.6	8.3	5.8

contains ash and rubble. Permeability in the cavity is assumed to be several orders of magnitude greater than that in the coal seam and porosity in the cavity is assumed to approach unity (Contractor and Schreiber, 1987). During water filling stage, products of combustion contaminate ground water. After a certain time, the cavity fills up and becomes part of the confined aquifer system. Storage coefficient of the cavity is now larger than that of the coal seam. A time is reached when flow occurs out of the cavity, taking with it the contaminants to be dispersed and sorbed in the aquifer. As started before and after gasification water in the coal seam (adjacent to the cavity) flows into it, picking up pollutants from the ash and char zone, and migrates from the cavity to adjacent aquifers (Sury et al., 2004).

Groundwater restoration

After completion of UCG process many organic and inorganic contaminants left in cavity, generated after UCG process. Phenol concentration in groundwater is found in large amount because its solubility in water is high. Groundwater is function of time and distance from the reactor (Campbell et al., 1979). So, concentrations of contaminated groundwater restore as fresh water after a long time. Inorganic contaminates, including cations such as Na^+, K^+, Mg^{2+}, NH_4^+ etc. and anions such as SO_4^{2-}, HCO_3^-, Cl^- etc. also showed a large decrease in concentration over time (Nordin, 1992).

MITIGATION OF GROUNDWATER CONTAMINATION

Site selection

Appropriate site selection is the best approach for groundwater pollution control. Before starting UCG process ensure the site is well characterized and the coal seam has limited connectivity with water sources. Therefore, selecting sites with favorable hydrogeology minimizes movement of contaminants. Selection of regions where overburden is expected to deform plastically reduces the concern of shearing of strata. Shearing can result in vertical propagation fractures that allows fluid communication between the gasification zone and surrounding groundwater (Moorhouse et al., 2010).

Appropriate sealing of wells and boreholes

Before starting UCG operation ensures that wells and boreholes used in the process are adequately sealed and maintain a 'cone of depression' in the groundwater around the reactor (Sury et al., 2004). Researchers have proposed several control technologies for containment, such as a hydraulic bypass around the contaminated zone, placing adsorbent clays within the cavity, placing a grout curtain around the contaminated zone and permeable reactor barrier. The permeable reactor barrier is filled with granulated activated carbon and peat for removal of hydrocarbons (Lutynski and Suponik, 2013). These contaminants are removed by physicochemical, chemical and biological processes.

Controlling the reactor pressure

Migration of hazards elements from UCG cavity could be reduced by maintaining rector pressure (P) below the hydrostatic pressure (P_h). Installation of operational monitoring systems is used to detect gas losses and ensure that reactor pressure is maintained below hydrostatic. In this case, water flow from the surrounding aquifer in to the UCG cavity, presents transport of contaminants into adjacent aquifers. It is also found that gas escape, which is driving force for contaminant dispersal, could be substantially reduced. Therefore, during UCG operation continuously measures the reactor pressure for mitigation measures (Friedmann et al., 2006).

Pumping of contaminated water

Another concept for control would be pumping the contaminated water from the cavity and surrounding area, and treating or disposing of the water on the surface. This alternative would be effective for removal of highly mobile contaminants. These contaminants would consist of materials generated by ash leaching, such as soluble organic matter and ammonia (Cook and Oliver, 1983). Remaining material would be more insoluble and less mobile condensed organic matter around the periphery of the burn zone.

Abandonment practices

At the time of shutdown of UCG process gasification zone cool slowly and gas is continue to extract until the gasification process stops completely. In this way contaminants can be evacuated out of the gasification zone before the site is abandoned. Monitoring of groundwater contaminates for a period of time after the site is abandoned. The actual duration of monitoring will depend on the specific site (Moorhouse et al., 2010).

CONCLUSIONS

Groundwater pollution caused by UCG due to migration of water and gas dispersion to the surrounding permeable strata. The major pollutants identified in groundwater include organic and inorganic substances, TDS etc. Concentration of these pollutants decrease by increase of distance from the cavity, and after some time these decrease due to restoration of pollutant water. Migration of contaminated groundwater and restoration as well as some mitigation measures are specified for pollution free UCG operation. Major organic and inorganic substances, TDS and pH data indicate that baseline and maximum contaminated groundwater constituents vary depending on types of coal and operating parameters of UCG.

Conflict of Interest

The authors have not declared any conflict of interest.

ACKNOWLEDGMENT

Authors are thankful to Dr. Amalendu Sinha, Director, CSIR-Central Institute of Mining and Fuel Research, Dhanbad, India for giving permission to publish the paper.

REFERENCES

Ahern JJ, Frazier JA (1982). Water quality changes at underground coal gasification sites- a literature review. Laramie Energy Technology Center, U.S. Department of Energy, under contractDE-AS20-79LCO1761.Available   at, http://uwyo.coalliance.org/fedora/repository/wyu   %3A4915/y-changes-at-underground-coal-gasification-sites_718e4862c1.pdf/water-qualitychanges-atunderground-coal-gasification-sites_718e4862c1.pdf, 5 June 2014.

Burton E, Friedmann J, Upadhye R (2004). Best practices in underground coal gasification. Lawrence Livemore National Laboratory.Available   at,   https://www.purdue.edu/discoverypark /energy/assets/pdfs/cctr/BestPracticesin UCG-draft.Pdf, 12 June 2014.

Campbell JH, Wang FT, Mead SW, Busby JF (1979).Groundwater quality near and underground coal gasification experiment. J. Hydrol. 44:241-266.

Contractor DN, Schreiber JD (1987). Field applications of multi-aquifer ground water models for flow and transport to underground coal gasification   sites.   Available   at,   http://info.ngwa.org /gwol/pdf/870142817.pdf, 10 June 2014.

Cooke DS, Oliver RL (1983). Groundwater quality at the Hanna underground coal gasification experimental sites, Hanna, Wyoming: Database and Summary. U. S. Department of Energy/Associated Western Universities, Inc. under cooperative contract (DE-AC07-76ET10723). Available at,http://www.fraw.org.uk/files/extreme/cooke _oliver_1984.pdf, 15 June 2014.

Friedmann SJ, Burton E, Upadhye R (2006). LLNL capabilities in underground coal gasification, Work performed under the auspices of the U.S. Department of Energy by University of California, Lawrence Livermore National Laboratory under contract W-7405-Eng-48.

Green MB (1999). Underground coal gasification - A joint European field trial in Spain. ETSU Report No.COAL R169.DTI/Pub URN99/1093.

Haider Z (2012). Site characterization, sustainability evaluation and life cycle emissions assessment of underground coal gasification, Ph.D thesis. Availabe at,http://scholar.lib.vt.edu/t heses/available/etd-09132012-155859/unrestricted/Hyder _Z_D_2012.pdf, 12 June 2014.

Humenick MJ, Lang M (1980). The effect of lignite ash composition on ground water leachate. In Proceedings 6th Symposium on Underground Coal Conversion, V38-V44.

Lang MJ (1982). Groundwater leaching of lignite ash after underground coal gasification. Master of Science thesis, The University of Texas, Austin.   Available   at,   http://www-lib.uwyo.edu/sho wcase/files/original/ground-water-leaching-of-lignite-ash-after-underground-coal-gasification_ 5e52 d66fb7.pdf, 10 June 2014.

Lutynski M, Suponik T (2013). Hydrocarbons removal from underground coal gasification water by organic adsorbents. Physicochem. Prob. Min. Proc. 50(1):289-298.

Moorhouse J, Huot M, McCulloch M (2010). Underground coal gasification: Environmental risks and benefits. Available at, http://www.pembina.org/reports/laurusreport.pdf,8 November 2014.

Nordin SJ (1992). Review of information and data relevant to Hoe Creek underground coal gasification site restoration. Western Research Institute, The University of Wyoming Research Corporation, Laramine, Wyoming. Available at, http://deq.state.wy.us/, 11 June 2014.

Phillips NP, Muela CA (1977). In-situ coal gasification: status of technology and environmental impact. Interagency Energy-Environment Research and Development Program Report, EPA-600/7-77-045.

Shu-qin L, Yuan-yuan W, Ke Z, Ning Y (2009). Enhanced-hydrogen gas production through underground gasification of lignite. Min. Sci. Technol.19:389-394.

Stephens DR, Hill R, Wand Borg IY (1985). Status of underground coal gasification. UCRL 92068. Int. J. Miner. Process. Extract. Metallurg Rev. 1(3-4):265-296.

Stone R, Raber E, Winslow AM (1982). Effects of aquifer interconnection resulting from underground coal gasification. Ground Water 21(5):606-618.

Stuermer DH, Ng DJ, Morris CJ (1982). Organic contaminants in groundwater near an underground coal gasification site in Northeastern Wyoming. Environ. Sci. Technol. 16(9):582-587.

Sury M, White M, Kirton J, Carr P, Woodbridge R, Mostade M, Chappell R, Hartewell D, Hunt Douglas, Rendell N (2004). Review of environmental issues of underground coal gasification. WS Atkins Consultants Ltd., University of Liège Belgium, FWS Consultants Ltd., 126. Report No. COAL R272 DTI/Pub URN 04/1880, pp. 1-126.

Torres VN, Atkins AS, Singh RN (2014). Assessment of an environmental sustainability index for the underground coal gasification process by using numerical analysis. In Proceedings of the 14th Coal Operators' Conference, University of Wollongong, Australasian Institute of Mining and Metallurgy, Mine Managers Association of Australia, pp. 309-323.

Wang F (1979). The sportive property of coal. In Proceedings of the 5th Underground Coal Conversion Symposium, U.S. Dept. Energy, Rep. CONF. 790630, pp. 403-408.

22

Design of a waste management model using integrated solid waste management: A case of Bulawayo City Council

Bupe Mwanza* and Anthony Phiri

Harare Institute of Technology, P. O. Box BE277 Belvedere Harare.

The purpose of this research paper was to design a waste management model for Bulawayo City Council (BCC) based on integrated solid waste management system. The increasing solid waste generation in the city of Bulawayo is evidenced by increased number of illegal dumping of waste in the urban communities. In this paper, a model that identifies cost-effective and efficient combinations of scientific and engineering tools to manage solid waste and also incorporating the key performance indicators (KPIs) or metrics for solid waste management services has been designed. Literature on different types of waste management models, quantitative analysis of waste at Ross camp, direct observations made in the city locations and policy documents of the Zimbabwean Government (EMA acts), founded the development of this model. The design shows a clear depiction of the environmental management acts (EMA) and the municipality by-laws. Performance measurement which is very critical in waste management has been incorporated. Model design focus was on waste type which is a determinant of the waste receptacles, collection, transportation and disposal methods have therefore been depicted in the model. The model provides easy access to data for the formation and implementation of effective solid waste management policies, strategies and programs to achieve sustainable waste management.

Key words: Environment, Integrated Solid Waste Management (ISWM), solid waste.

INTRODUCTION

Solid waste streams can be appropriately characterized by their sources, by the types of waste produced, as well as by generation rates and composition but this approach might be difficult for Bulawayo City Council as it is struggling to do so. Figure 1 is a depiction of the current waste management model for Bulawayo City Council. The waste management system shows that from the waste stream, some waste end up at illegal (informal) sites and some at legal sites. The formal involves a scenario, whereby once solid waste is produced on site; it is temporarily stored in waste bins or any other suitable containers or at transfer points awaiting collection. The municipality's solid waste management section collects

the wastes from the temporary storage facilities and transports them to its specially designated dumpsite. The second stream is illegally disposed and it accounts for solid wastes that do not find their way into the municipal solid waste collection system (Figure 2). Some of the wastes are illegally buried in the ground, or burnt at source or dumped in open spaces, streams, or roadsides. Both streams handling have leakages that have direct and indirect impacts on the biophysical and social environments. The current system (Figure 1) is inadequate in ensuring an integrated solid waste management system which should take into consideration waste generation; waste handling and sorting, storage; and processing at the source, collection, sorting, processing and transformation; transfer and transport; and disposal (Edward, 2004).

Zimbabwe, in particular Bulawayo City has been affected by an ineffective waste management system

*Corresponding author. E-mail: bmwanza@hit.ac.zw, aphiri@hit.ac.zw.

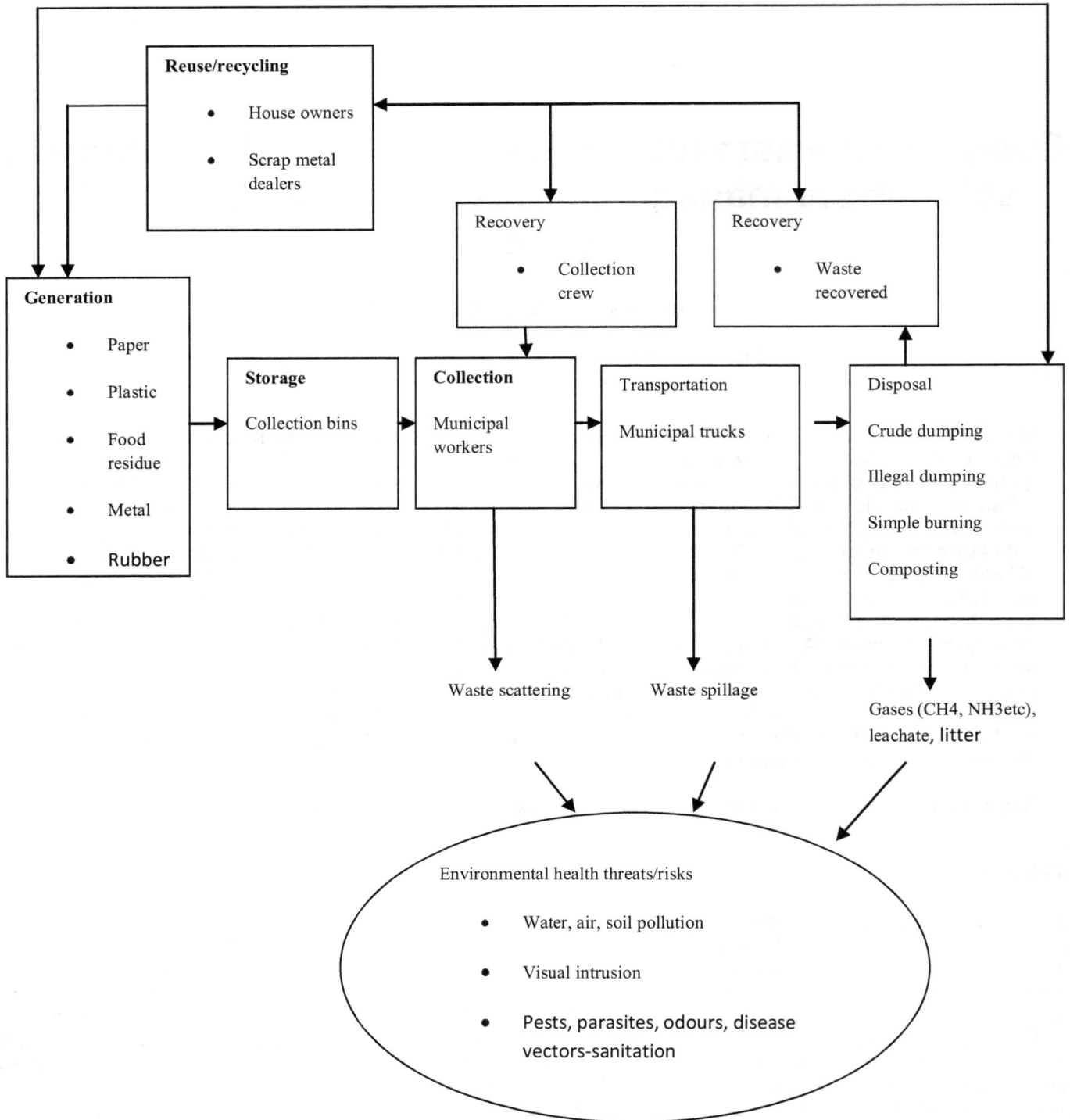

Figure 1. The Components of Solid Waste Management System in Bulawayo Municipality.

which is visible by the levels of waste which is dumped in open spaces. The research other concern was the lack of disciplin e in the way waste was dumped carelessly and the lack of commitment for waste management by the communities. The main purpose of this paper is to design a waste management model for the city councils and other organizations involved in the management of waste. The characterization of waste into different types

Figure 2. Illegal Dumpsite next to a Market (Mpopoma, 2011).

will make it easy for the waste management agents to understand the main generators of waste and how it can be stored, collected, transported and finally disposed of. The model is designed on the basis of different waste types generated from different sources of the community. The different sources considered are the different residential areas. In this study, waste characterization starts at the source to determine the type of waste receptacle, waste collection method, transportation and the best disposal method. The paper also recommends how the city council or any organization involved in waste management can improve the management of waste by including performance measures in the waste management system.

REVIEWED LITERATURE

The necessity for Waste Management cannot be over emphasized especially in modern society but it has existed for millenniums. In "Waste Management" Bilitewski et al. (1994) reports that from 9.000 to 8.000 B.C. people learnt to dispose of their waste outside their own settlement to avoid odor, wild animals and nuisances of vermin (NIMBY SYNDROME). Waste management is nowadays far more complex than it was some thousand years ago. The complexity arises not only because of the huge quantities of residuals produced by the modern society, but also because of differences in the composition of the waste. Presently various municipalities fail to supply adequate waste management service to their communities (Godfrey, 2006). This is shown by the increase in uncollected waste in high density suburbs. In various places liter is scattered and also waste accumulates in non-designated places. One of the main causes of poor service in most of the municipalities is lack of proper planning although staff, equipment and poor access to certain places are also given as reasons (Godfrey, 2006). The United Nations Development Program survey of 151 mayors during the International Colloquium, in 1997, identified insufficient solid waste collection and disposal to be among the five (5) most severe problems in cities worldwide (UNSD, 1999). This number shows the importance of well-functioning waste management systems, and the necessity of making improvements in this field. If waste is unmanaged, it becomes a source of contamination and disease (UNCHS, 1989). Proper waste management is needed to reduce health problems, water pollution risks and other environmental hazards, besides the negative aesthetic impacts. Integrated Solid Waste Management (ISWM) is a comprehensive waste prevention, recycling, composting, and disposal program (Figure 3). ISWM involves evaluating local needs and conditions, and then selecting and combining the most appropriate waste management activities for those conditions.

The approaches in the area of solid waste management are not only very capital intensive, but also difficult from the environmental and social points of view therefore, there is a need to develop, master and implement a simple, but reliable tool that will help the decision makers in the analysis process. Integrated Municipal Waste (IMW) model (Figure 4) is a tool which seems to meet all the requirements (White, 1997; Bjorklund, 1998; Eriksson, 2002; McDougall, 2001). The results of the analysis from the IMW (Figure 4) model give vast amount of information, but it is rather fragmented. Waste management is in itself a large and complex system that is difficult to survey. The system grows even more complex as one considers its links to other sectors such as manufacture, energy production, and agriculture. Based on this reviewed model and other models, the author proposes a model that is easy to understand and that will also be easy to convert into a

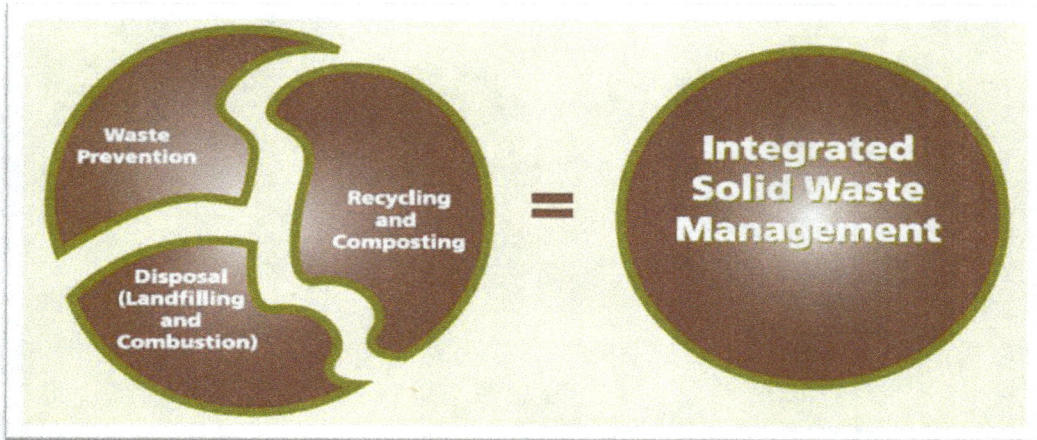

Figure 3. Integrated Solid Waste Management System (www.epa.gov/globalwarming).

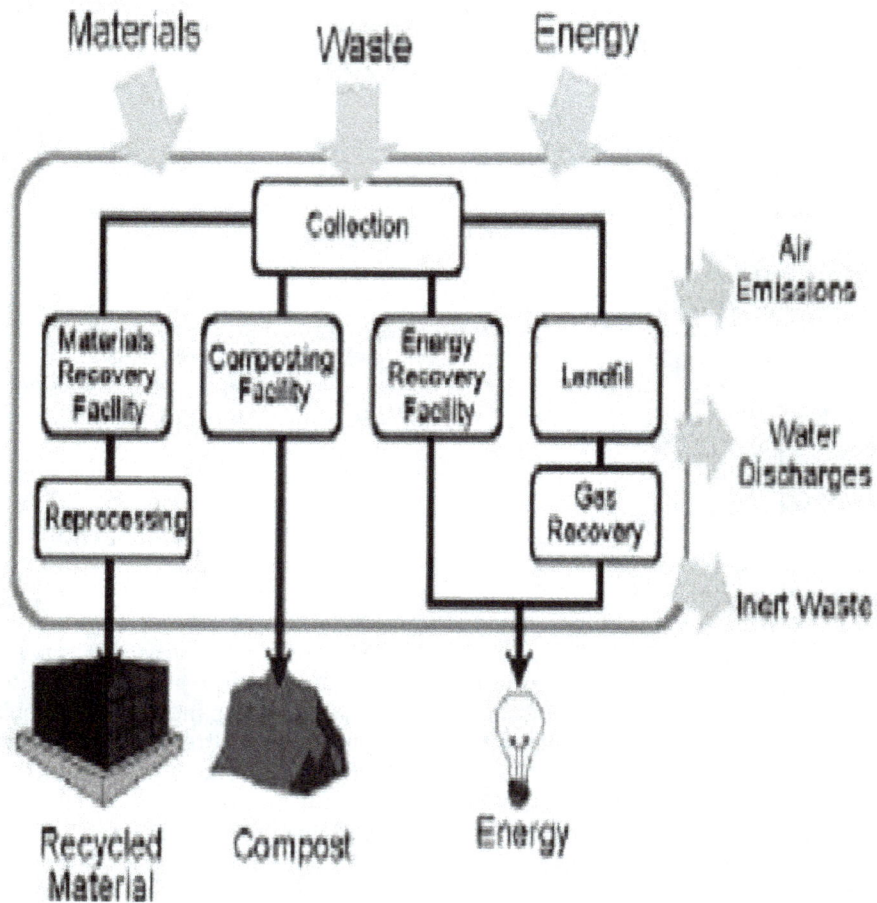

Figure 4. IWM Waste Management Model (www.plastics.ca/epic).

database that can be used by the waste management agents considering the level of technology. The mode designed in this paper gives a broad presentation of the important elements necessary in the management of waste. The major findings of this paper are that the model will enable monitoring of waste types from each location and help set targets for waste minimization.

MATERIALS AND METHODS

Quantitative analysis

An analysis of the amount and types of waste generated was conducted at Ross camp. This analysis was conducted to determine the amount of waste generated per week by a population of 483 people. Ross camp is a community that consists of 77 households and each house was given a task of accumulating their daily domestic waste for a week.

The amount of waste that was generated was analyzed in terms of the type of waste, the type of waste receptacles; separation at the source was also analyzed to determine whether the residents knew about waste separation to account for waste recycling. The amount of waste that was generated in a period of 7 days was calculated and the results are shown in Table 1.

This analysis was done to determine the amount of waste and types of waste that is generated by households as the research was more to do with domestic waste.

Data collection

The purpose of this research was to design a waste management model for Bulawayo City Council and information that was necessary for the design of the model was collected. Literature from a number of waste management models was reviewed. Information pertaining to the environmental management act concerning management of domestic waste was also reviewed. This information was reviewed to assist in the design of the new model using integrated solid waste management system.

Interviews

Informal interviews were conducted at Ross camp as it was the study area. The structured questions during the interviews focused on the following:

(i) The importance of waste management
(ii) The causes of increase in waste generation.
(iii) The methods of waste collection.
(iv) The types of waste receptacles used.
(v) The frequency of waste disposal.
(vi) Their understanding of waste separation at the source and if it is being practiced.
(vii) The effectiveness of the waste management company
(viii) Waste recycling
(ix) The role of the public in terms of waste management
(x) The impact of poor waste management.

These questions were structured as an improvement process on the waste management model of Bulawayo City Council and also to find out the views of the residents on waste management.

Observations

The researcher made tours in the different locations of Bulawayo. During those tours, it was observed that the community especially in the high density suburbs, carelessly disposed wastes anywhere depicting illegal dumping. Observations made in the city center, confirmed a different perception people have towards waste. Some people would use the bins and some did not. Coming to the low density areas, provision of waste receptacles in some households is there but the concern was on the collection regularity and time management by the city councils. During these trips, besides taking notes about the physical characteristics of the neighborhoods, photographs were taken to show and record illustrative features and interesting attributes of the areas.

RESULTS

Waste management models are getting increasing attention throughout the world. It is therefore important to understand possibilities and limitations of different models. This section covered the design of the conceptual model for Bulawayo City Council Integrated Solid Waste Management Model (Figure 5) based on the reviewed models, (Figures 1 and 3) and data that was collected from the case study at Ross camp.

Case study results

Table 1 shows the results that were obtained from the study conducted at Ross camp to determine the waste generation rate. In this study, waste was collected from the 77 houses that make up Ross camp after a period of 7 days. The rate of waste generation in Ross camp in Bulawayo is 0.00024 m^3 per capita per day (Table 1). During this study, no waste receptacles were provided for the households as the study also focused on the type of receptacles the residents used. The waste components found in Ross camp included food residues, paper, plastics, metals, glass, textiles, rubber, and wood. The results obtained from Ross camp on waste composition are shown in Table 2.

From Table 2, the sampled households reflected a total weekly waste generation of 8.06 m^3 (8060 L). This gives a mean weekly waste generation of 0.010 m^3 per household. The daily per capita waste generation is estimated at 0.00024 m^3/capita/day. The value is far below; this may be due to the fact that the study at Ross camp focused on domestic waste only.

The case study (Table 3) shows the type of waste receptacles which the households use for waste storage. According to the case study, 73% of the households use makeshift (sacks, tins, buckets, plastics), 19% use proper bins provided by the city council and 8% have no receptacles.

Residents whose households do not pay for waste collection services ended up disposing their waste in open spaces, rivers, and drainage basins (Figure 1), burning or burying it. In this study, it was observed that waste collection is done once every week for those households that pay for their waste to be collected.

The factors that influence the amount of waste generated are;

(i) The time of the year: mainly holidays of the year, that is, Christmas and New Year holidays.

Table 1. Summary of the calculation of the per capita waste generation.

Variable	Sample	Total study area
Number of households	77	770
Population	483	770*6=4620
Total weekly waste generation(m^3)	0.806	0.010*770=8.06
Weekly per capita waste generation (m^3/capita/week)	0.806 m^3/483 people= 0.00167	0.00167
Daily per capita waste generation (m^3/capita/day)	0.00167/7=0.00024	0.00024

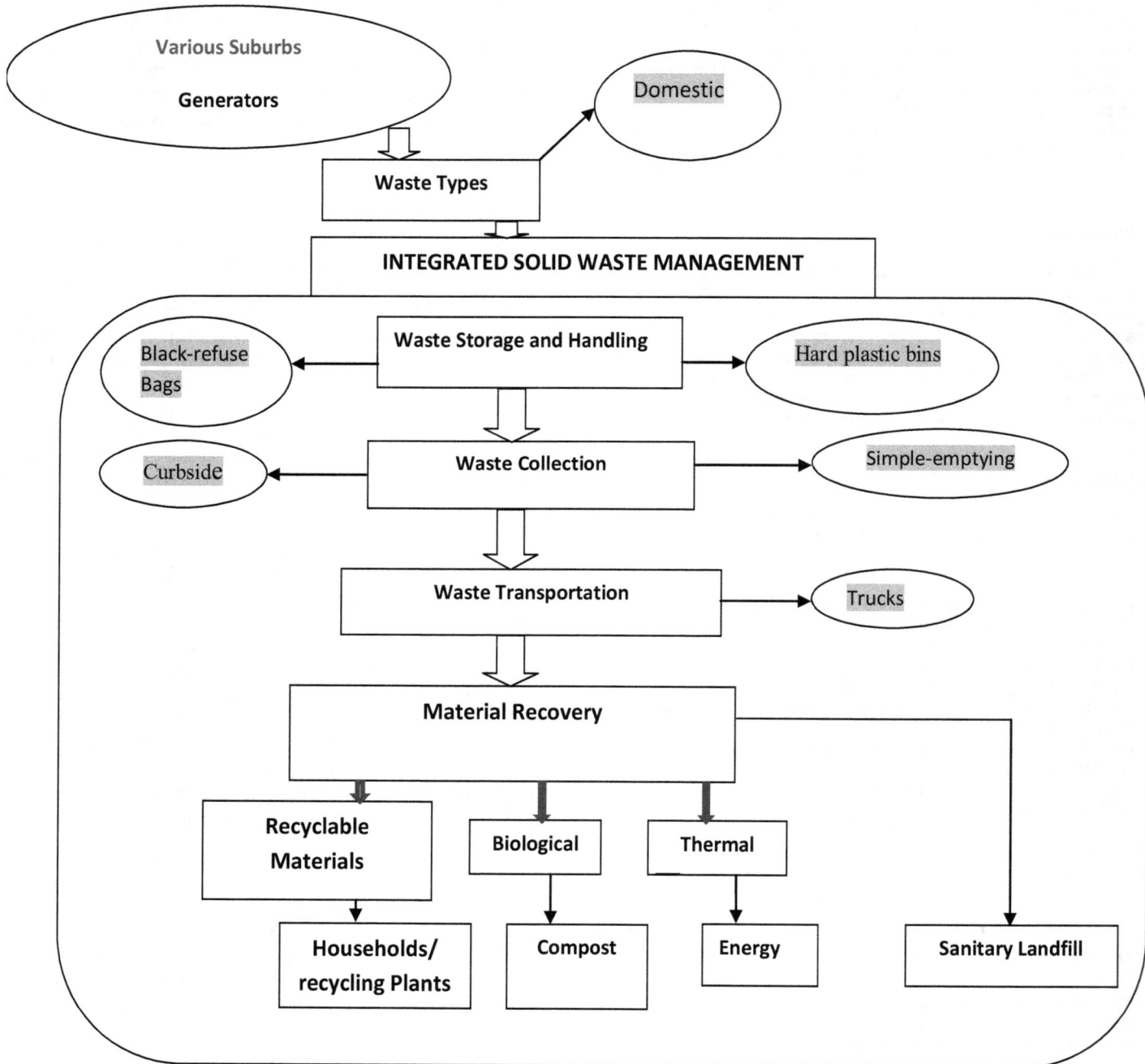

Figure 5. Conceptual Model of BCC ISWM.

Table 2. Waste Composition table.

Waste composition	%
Food remains	23
Plastic	42
Paper	30
Yard wastes	4
Glass	0.4
Metal	0.6
Total	100

Table 3. Type of Receptacles used for daily collection of waste.

Type of container	%
No bins	8
Makeshift (sacks, tins, buckets, plastics)	73
Proper bins (provided by BCC)	19

(ii) The rise in residents' earnings. (iii) Month-ends earnings.

The findings from the study also showed that waste separation at the source is not practiced by those households whose waste is collected by the waste collectors. The case is even worse for those households whose waste is not collected. From this study, waste recycling is not practiced by the residents and in any case, if waste recycling was being practiced by the waste collectors, then separation at the source would have been promoted and the residents would have full knowledge about waste recycling. Disposal of waste by residents whose waste is not collected is done on a daily basis and mostly consists of open space disposal and burning.

The information that was obtained from the case study contributed to the design of the model.

Model design

The researcher conceptualized the system based on the fact that waste is generated from different sources and this waste is of different types depending on where it is being generated. The different sources of waste in Bulawayo town are the residential areas, industries, commercial areas, institutions, farms and construction companies etc. but the focus is on residential areas. These sources generate different types of waste which can be categorized into domestic waste, industrial waste, toxic waste, biodegradable waste etc. Waste types can be handled, stored and transferred differently. Each waste type can be separated at the source if the waste receptacle is determined. Separation at the source will

bridge the gap in the current BCC model. The researcher therefore classified each waste type with a suitable waste receptacle. The different types of waste can be stored and collected differently (Table 4). Once the waste is separated at the source and stored in specific waste receptacles, it will be easy to transport waste to the designated disposal area. Other than disposal, waste can be treated by various ways to recover material. This includes energy recovery from industrial waste, compost from bio-gradable waste and recycling by both households and industries.

Bulawayo City Council as a waste management authority should comply with the regulations on waste management. The municipality has a department which manages waste and this department should comply with the acts and laws which have been put in place. The Model is designed by linking it to the environment management act and the city by-laws as compliance to the regulations. All the regulations in the environmental management act pertaining to the management of waste should be considered by Bulawayo city council otherwise it will be contravening the act.

The designer of the model included the EMA acts and the city by-laws in its waste management to show that BCC acknowledges the acts and by-laws.

This model provides a clear understanding of waste management process. The waste type characterization makes the waste recovery section easy as there will be no need for waste separation. The uniqueness of the model is that it shows;

(i) The type of waste generator.
(ii) The type of waste generated.
(iii) The type of waste receptacles.
(iv) The type of collection method.
(v) The type of transportation.
(vi) The type of disposal method pertaining to a waste type.

This model does not only reflect that the EMA is articulated, but it has included some performance measurements on the collection of waste which is part of waste management.

Strengths of the model

(i) EMA acts and the municipality by-laws are depicted in the model showing that BCC abiding by the regulations and also showing a sense of responsibility pertaining to waste management.
(ii) Performance management is very critical to waste management and for this reason the system has been designed with an aspect of performance management on waste collection.
(iii) Knowing the different types of waste is relevant in this system as the system is designed in such a way that waste type is the determinant of the waste receptacles, collection, transportation and disposal method which

Table 4. Waste Receptacle and collection method.

Source	Waste generator	Waste receptacle	Collection method
Residential	Households	Black refuse bags, Hard plastic bins	Curbside, informal, Simple-emptying
Industrial	Industries	Skips, metal bins	One-way, non-systematic
Chemical	Chemical plants	Skips, metal bins	Exchange, one way
Agricultural	Farms, Vineyards	Hard plastic bins	Simple emptying
Commercial	Restaurants, stores, hotels	Hard plastic bags, plastic bins	Simple emptying
Institutional	Schools, colleges, prisons, hospitals	Metal bins, Hard plastic bins	Simple emptying

makes it easy for learners to understand the use of the model.

(iv) Material recovery is important and this model shows the different ways in which waste can be recovered.

DISCUSSION

Waste Management acts are not fully been complied to by the city council and the community and this prompted the author to come up with a design of a model which incorporates the EMA acts as part of the driving force in waste management. The fact that waste is dumped in illegal dump sites and also the results from the interviews also contributed to the conclusion that EMA acts are not yet totally in place, and more time is needed for municipalities to be able to comply with the regulations. (EMA act is in place but no compliance)

There has been little pressure in the past for providers of solid waste management services to evaluate service quality, improve standards of performance or to justify service quality relative to its cost (Michael, 2008). This model has integrated performance measurement as an evaluation method for BCC Departments' management information systems. In practice this means establishing a reporting system on the performance of the SWM services, and routine collection of information (daily, weekly, monthly, yearly).

The results and picture (Figure 2) from the surveys show that there is no waste characterization especially in the communities and this model shows that waste characterization is the drive towards easy waste management as it draws a line between different types of waste and waste generators. This strategy of characterization of waste in terms of waste generators will help waste managers to manage the types of waste they are dealing with and how best to recover, recycle and finally dispose of it.

This model is a start to the design of a database system for the city council and other waste managers. It is a framework or skeleton for the design of a database.

RECOMMENDATIONS AND CONCLUSION

The model designed will require a design of a database

for it to be fully appreciated in its capacity for managing waste. The results show that this waste management model can be useful at a number of different levels in society. For example, it can be used by companies to support strategic decisions, by municipalities for waste management planning, and for governments for policy decisions. The researcher also recommends that BCC starts waste recycling projects by incorporating residential areas as the main source of collecting recyclable waste which is separated at the source. For this to be made possible, households should be provided with waste receptacles. BCC does not have to charge waste collection charges for those residents who separate and store the agreed amount of recyclable waste on its behalf.

ACKNOWLEDGMENTS

The authors are grateful to Eng. Phiri and Mr. N. Chirinda for offering the right guidance regarding this research work and helping to improve the paper as preciously as possible.

REFERENCES

Bilitewski B (1994). Waste Management. Berlin Springer-Verlag. pp. 10-20.

Bjorklund A (1998). Environmental Systems Analysis Waste Management with Emphasis on Substance Flow and Environmental Impact. Stockholm University of Technology, Sweden. pp. 10-21.

Edward DA (2004). Integration of municipal solid waste management in Accra (Ghana): Bioreactor treatment technology as an integral part of the management process. Lund University pp. 10-12.

Eriksson O (2002). ORWARE- A simulation tool for waste management Resources, Conservation and Recycling. Res. Gate pp. 15-40

Godfrey L (2006). Integrated waste management plans: A useful management tool for local Government or a bureaucratic burden. Biennial Waste Conference Proceedings 2006, Cape Town, South Africa.

McDougall F (2001). Integrated Solid Waste Management. A Life Cycle Inventory. Blackwell Publishing, pp. 10-15.

Michael EV (2008). Performance Measurement of the Binghamton University Waste Management System. State University of New York, USA, pp. 10-12.

Small scale waste management models for rural, remote and isolated Canadian communities in http://www.ccme.ca/assets/pdf/pn_1260_e.pdf (accessed 26/07/2011).

Solid Waste and Emergency Response (May 2002). http://www.epa.gov/global warming (accessed 22/05/2011).

UNCHS (1989). Solid Waste Management in Low-Income Housing Projects. The Scope for Community Participation. Nairobi.

UNSD (1992). Environmentally Sound Management of Solid Wastes and Sewage. Related Issues. [Online]. [United Nations Sustainable Development.]Available:http://www.un.org/esa/sustdev/agenda21.ht m [2011, June 6].

Whiteman A (2001). Strategic Planning Guide for Municipal Solid Waste Management . Integrated solid waste management in western Africa. pp. 337–353.

23

Modelling dynamics of organic carbon in water hyacinth Eichhornia Crassipes (Mart.) Solms artificial wetlands

Aloyce W. Mayo

Department of Water Resources Engineering, University of Dar es Salaam, Tanzania.

The role of water hyacinth on removal of organic carbon was investigated in free water constructed wetlands. A model incorporating the activities of suspended and biofilm biomasses was developed in order to simulate the various processes involved in the transformation and removal organic matter in the water hyacinths constructed wetland. The results show that the major processes governing the organic carbon transformation and removal in a water hyacinth constructed wetlands system are sedimentation of solids (56.5%), regeneration of organic carbon (25.5%), oxidation of organic carbon to carbon dioxide (6.4%), plant decay (4.4%) and uptake of organic carbon by heterotrophic bacteria (4.2%). The total permanent removal of organic material was 26.7% of the total influent chemical oxygen demand (COD). The COD removal efficiency of the model when the effect of biofilm was considered was 40.5%. However, in absence of biofilm activities, only 34.0% of COD was removed. This confirms the significance of the water hyacinth roots as an attachment media, which is extremely biologically active in assisting the organic carbon removal in the water hyacinth constructed wetland.

Key words: Water hyacinth, artificial wetlands, modelling, organic carbon.

INTRODUCTION

Water hyacinths have demonstrated a great potential for purification of wastewater through physical, chemical and biological mechanisms (Mayo and Kalibbala, 2007). Artificial wetlands have been used for secondary treatment (Gersberg et al., 1985; Vymazal, 2010) and for specific tertiary treatment such as removal of nitrogen (Senzia et al., 2004) and bacteria (Kalibbala et al., 2008; Mayo and Kalibbala, 2007; Vymazal, 2010). Successful case studies indicate that wetlands significantly reduce organic matter, suspended solids (SS), pathogens, heavy metals and excessive nutrients such as nitrogen, phosphorus and heavy metals from wastewater (Yi et al.,

2009; Mugasha, 1995; Mayo and Kalibbala, 2007; Mayo and Bigambo, 2005; Vymazal, 2010; Mayo et al., 2013). The deviation in chemical, biological, and physical characteristics among wetland ecosystems and complications in understanding and predicting the efficiency of such systems, have motivated the development of artificial wetland systems (Barrie, 2002). As a result, for over 50 years, natural and artificial wetlands have been engineered for wastewater treatment (Senzia, 2003) particularly for small and medium sized communities and isolated areas in Europe and the USA where over 700 artificial wetlands have been constructed

(Water 21, 2000).

Artificial wetland systems use floating or submerged aquatic plants in the treatment of industrial or domestic wastewater such as *Eichhornia crassipes* (Mayo and Kalibbala, 2007), *Phragmites mauritianus* (Senzia et al., 2004; Bigambo and Mayo, 2005) and *Typha domingensis* (Senzia, 2003; Nakibuule, 2013; Okurut, 2013). Until recently, most of the floating aquatic plant systems for wastewater treatment have been water hyacinth systems (USEPA, 1988). *E. crassipes* have been used in a variety of experimental and full-scale systems either for removing algae from oxidation pond effluents or for nutrient removal following secondary treatment in tropical regions (USEPA, 1988; Polprasert and Khatiwada, 1997). Other beneficial functions of wetlands include supplementary wildlife and human use benefits resulting from treatment wetlands. However, some researchers have also pointed potential problems of wastewater treatment in wetlands such as bioaccumulation of toxins and transmission of diseases (Knight et al., 2000; Muyodi, 2000).

Numerous studies have verified the usefulness of constructed wetlands (Barrie, 2002), and have provided a database for the development of design manuals for wastewater treatment with artificial wetlands (Reed et al., 1995). In spite of these advances on removal mechanisms of organic matter in constructed wetlands, models have failed to adequately predict performance of wetland systems. Organic carbon removal variability has tended to be influenced by a variety of factors resulting in its irregular removal pattern, which has complicated the optimization of organic carbon removal in artificial wetlands. The objectives of this paper are to determine and quantify the role of water hyacinth *E. crassipes* (Mart.) Solm for removal of organic carbon and to develop a mathematical model incorporating the activities of suspended biomass and bio-film on plant roots on removal rate of organic carbon in artificial water hyacinths wetland.

METHODOLOGY

Layout of plant and data collection

Two pilot wetland units of dimensions 7.5 m long, 3.5 m wide and 0.85 m deep were constructed adjacent to primary facultative pond at the University of Dar es Salaam (Figure 1). The location of the wetland units was at latitude 6°48'S and 39°13'E, 30 m to the north where the mean monthly air temperature of the site varies between 23 and 28°C with a mean value of 26°C (Mayo, 1989). The wetland units, which were planted with *E. crassipes,* were supplied with wastewater from the primary facultative pond at an average flow rate of 1000 L/day.

Flow was measured at the inlet and outlet of the water hyacinth unit with the aid of graduated container and a stopwatch. Samples of examination of water quality parameters were collected at the inlet and outlet of the system at 10:00 a.m once every two days. Samples were also collected along the length and depth of the wetland unit from sampling ports installed at the sides of the wetland units. Collections were done in a clean 250 ml sampling

bottles and samples were immediately taken to the water quality laboratory at the University of Dar es Salaam for examination. Samples containing settleable solids were blended with a homogenizer and preliminary dilutions were made for wastes containing high chemical oxygen demand (COD) to reduce the error in measuring small sample volumes. All physical-chemical parameters were determined in accordance to Standard Methods for the Examination of Water and Wastewater (1996). Analyses of samples were conducted within 2 h of sampling. Measurement of temperature and pH were done *in-situ* using pH meter (Metrohm pH meter, model 704). Dissolved oxygen (DO) concentrations were determined *in-situ* using a digital DO meter (YSI DO meter, model 50B). Chemical oxygen demand was measured using closed reflux method in accordance with Standard Methods for the Examination of Water and Wastewater (2012). Surfer 7.0 software was used to produce contour variation of the physical and chemical parameters with time, along the length and across the depth in the wetland unit.

RESULTS

Variation of physical chemical parameters

Concentration of dissolved oxygen (DO) in the wetland decreased from an average of 2.16 mg/l in the influent to 0.51 mg/l in the effluent. Dissolved oxygen decreased because the rate of its consumption by microorganisms for biodegradation of organic carbon was higher than its rate of production in the system. It is worth mentioning that water hyacinth plant roots generate only a small portion of the oxygen required for oxidation of organic carbon (Senzia, 2003). Unfortunately, interfacial diffusion of oxygen from the atmosphere and the production of oxygen by algae photosynthesis were suppressed by the dense plant cover, which reduced the surface gas-exchange and wind-induced turbulence.

For similar reasons, the average pH in a water hyacinth wetland decreased from 7.87 to 7.26 because of complete cover provided by the plants on the water surface resulting in prevention light penetration, consequently inhibiting algae photosynthesis activities in the wetland. Consumption of CO_2 decreased as a result of inhibition of growth of algae resulting in shift of Equation (1) towards the right. This has resulted in reduction of hydroxyl ions, thus lowering the pH of wastewater.

$$CO_2 + OH^- \Leftrightarrow HCO_3^-$$

(1)

The wetland system was 0.8°C colder at the effluent compared to the influent because of plant biomass mat, which prevented the direct solar energy from increasing the temperature of the wetland unit. This has a negative effect on the system as temperature has a significant influence on the rate of uptake of organic carbon, the interfacial gas transfer and settling rate of biological solids (Metcalf and Eddy, 1995).

Table 1 shows that COD concentration in the wetland

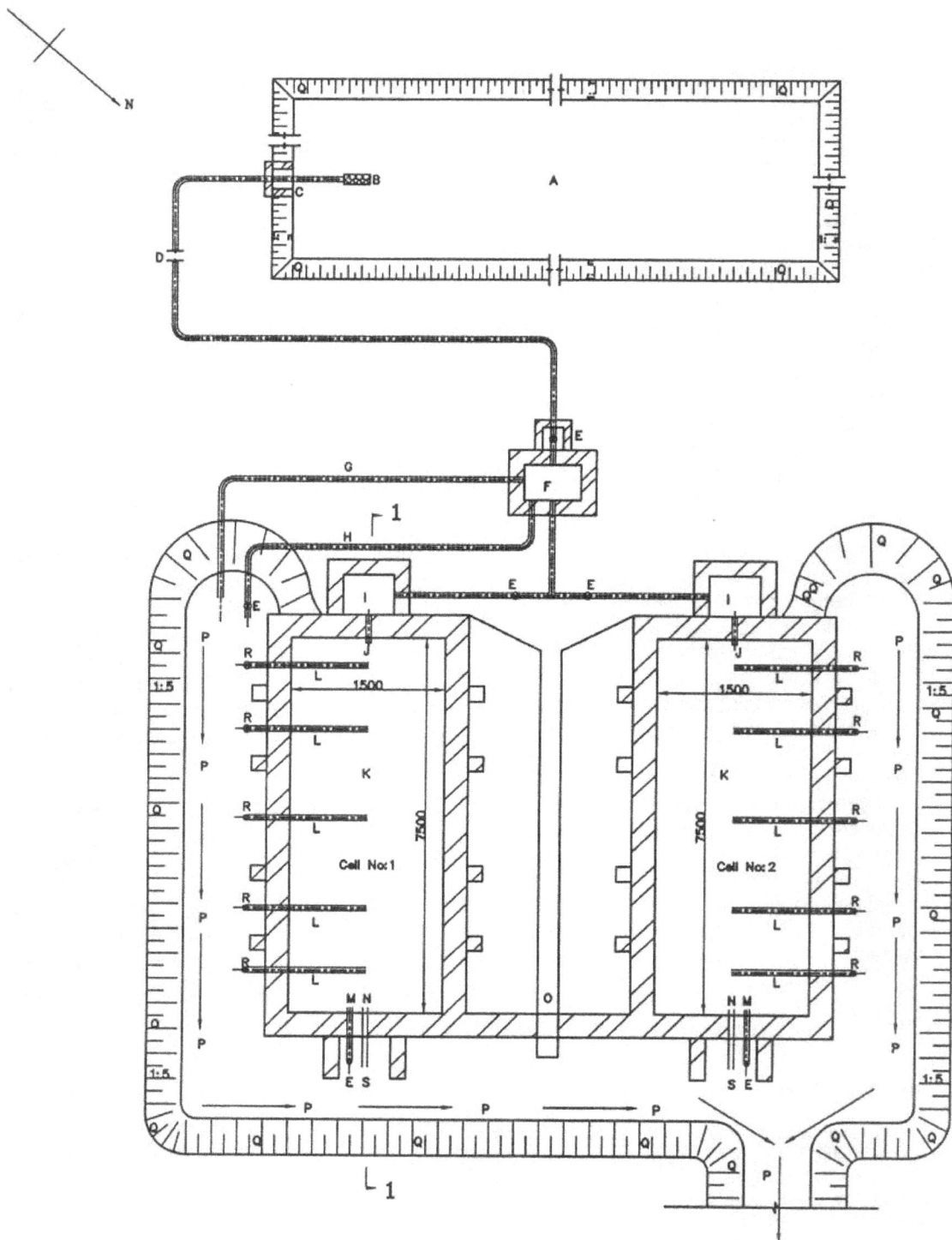

Figure 1. Layouts of the experimental wetland system.

decreased from an average of 210 mg/l in the influent to 121 mg/l in the effluent, which is equivalent to 42% removal efficiency. The variation of COD along the length and depth of the wetland unit shows that COD decreased from 215 mg l^{-1} near the influent to about 95 mg l^{-1} near the effluent (Figure 2). COD concentration decreased gradually along the length and appears to increase towards the bottom of the wetland unit. There was

Table 1. Variation of physical-chemical parameters in water hyacinth wetland.

No	Parameter	Influent		Effluent	
		Range	Mean	Range	Mean
1.	pH	7.75~8.14	7.87	7.10~7.48	7.26
2.	Dissolved oxygen (mg/l)	1.65~2.78	2.16	0.36~0.68	0.51
3.	Temperature (°C)	26.5~27.7	26.9	25.6~26.5	26.1
4.	Chemical oxygen demand (mg/l)	184~228	210	75~135	121

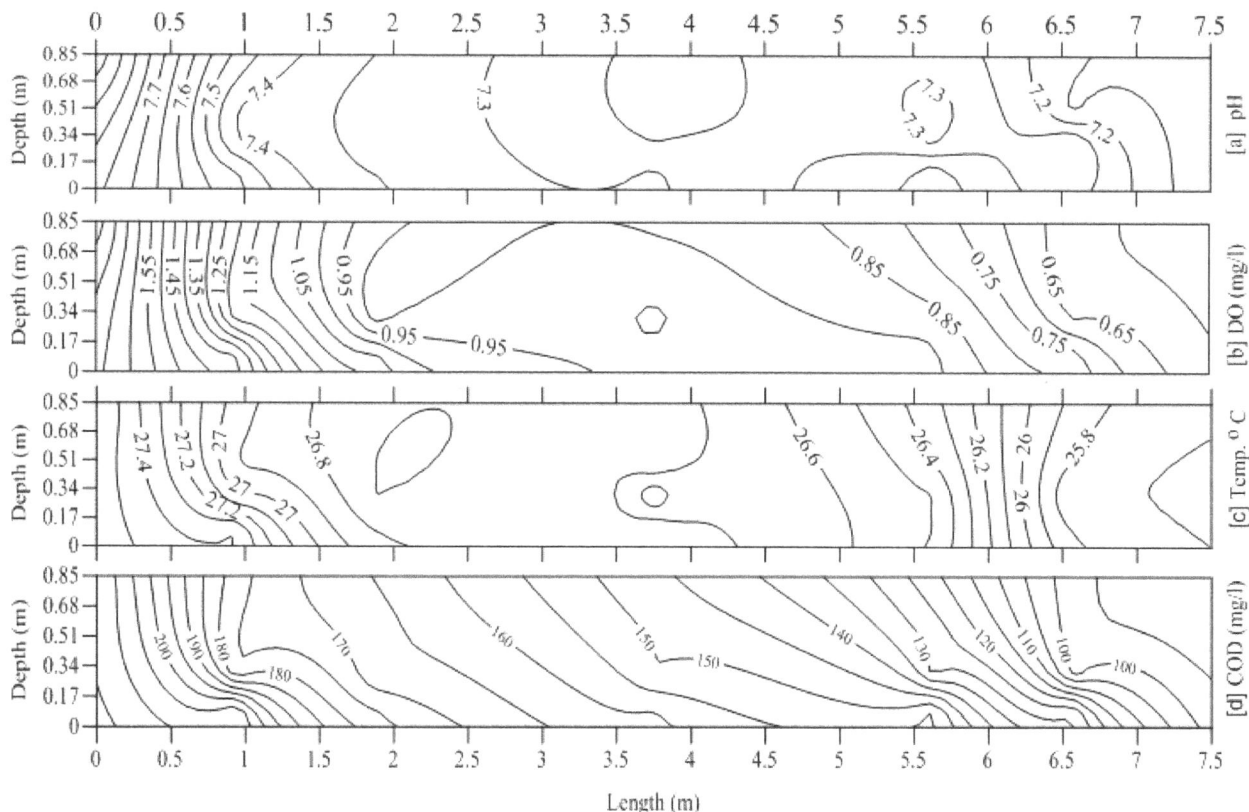

Figure 2. Variation of (a) pH (b) Dissolved oxygen (c) Temperature and (d) COD with the length and depth of the wetland.

evidence that COD was increasing with the depth of wetland unit, which is an indication of settling of organic particles to the benthic layer.

Model development

Conceptual model and organic carbon mass balance

The developed model incorporated activities of biofilm and suspended biomass on the transformation and removal mechanisms of organic carbon in the wetland unit. The model took into consideration the state variables substrate COD, COD in benthic layer and plant

organic carbon as the major forms of organic carbon in the wetland unit. The transformation mechanisms that were considered in the model include oxidation of COD to carbon dioxide (CO_2), uptake of organic carbon by heterotrophic bacteria, regeneration of organic carbon, sedimentation of organic carbon, plant decay, generation of methane (CH_4) and release of carbon dioxide from the sediment. The transformation and removal mechanisms were conducted by attached root biofilm and suspended bacteria biomass.

The conceptual model that includes suspended and biofilm biomass activities influencing the organic carbon transformations and removal were developed (Figure 3). The model illustrates the material flow in and out of the

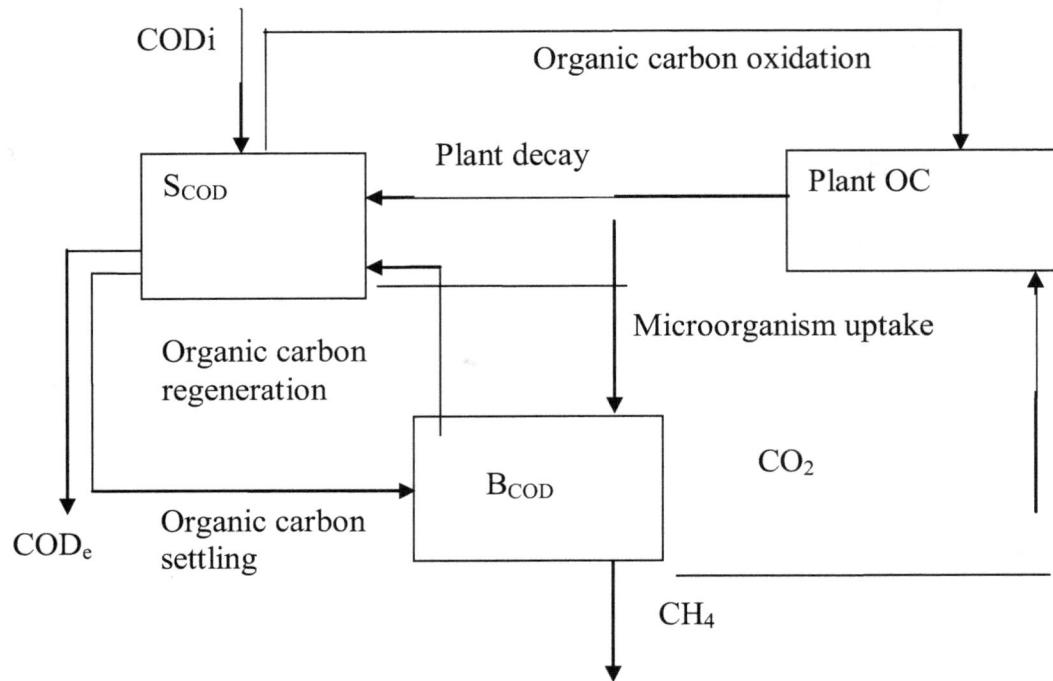

Figure 3. Conceptual model for organic matter transformation in wetland.

state variables substrate COD (S_{COD}), plant organic carbon (plant OC) and COD in benthic layer (B_{COD}), respectively.

The system was considered as a continuous plug flow hydraulic regime. The rate of change of COD concentration with time $\frac{dC}{dt}$ in wetland was based on the design assumption of the plug flow hydrodynamics and the first-order organic matter removal kinetics (USEPA, 1988) as shown in Equation (2).

$$\frac{dC}{dt} = -k_T t \qquad (2)$$

In which t = hydraulic retention time (day^{-1}); and k_T is temperature-dependent first-order reaction rate constant at wastewater temperature of T°C, which is given as

$$k_T = k_{20}\theta^{(T-20)}$$

Where k_{20}= Reaction rate constant at 20°C and θ = Temperature coefficient. Equation (3) is obtained on integrating Equation (2).

$$C_{eff} = C_{inf}e^{-k_T t} \qquad (3)$$

Where C_{eff} and C_{inf} are concentrations of COD in effluent in mg l^{-1} and influent in mg l^{-1}, respectively.

In analyzing the main influential mathematical expressions, mass balance equations of all organic materials were developed. The entire material balance that encompasses all forms of substances produced and consumed in biochemical processes, accumulations, inflows and outflows were considered. In this model, a steady state system was assumed. Equation (4) defines the overall mass balance equation for transformation and removal of organic carbon within the boundary of the water hyacinth constructed wetlands.

$$\left(\frac{dOC}{dt} = COD_{inf} - K_{ox} + K_{regCOD} + K_{dec} - K_{up} - K_{sd} - COD_{eff} \right) \qquad (4)$$

Where dOC/dt = Rate of change of organic carbon concentration (g m^{-2} day^{-1}); K_{up} = Uptake rate of organic carbon by heterotrophic bacteria (g m^{-2} day^{-1}); K_{ox} = Organic carbon oxidized to carbon dioxide (g m^{-2} day^{-1}); $K_{reg\ COD}$ = Organic carbon regenerated from the sediment (g m^{-2} day^{-1}); K_{sd} = Sedimentation rate of particulate organic carbon settling (g m^{-2} day^{-1}); K_{dec} = Plant decay rate (g m^{-2} day^{-1}); COD_{inf} = Chemical oxygen demand load in influent (g m^{-2} day^{-1}); COD_{eff} = Chemical oxygen demand load in the effluent (g m^{-2} day^{-1}).

The oxidation term (K_{ox}) in Equation (4) is the amount of organic carbon oxidized to CO_2 (day^{-1}) by heterotrophic bacteria and is influenced by the concentration of dissolved oxygen and temperature in accordance with Equation (5).

$$K_{ox} = \left[(k_{20} + k_b) \cdot \left(\theta^{(T-20)} \cdot \left(\frac{DO}{K_{DO} + DO} \right) \right) \right] S_{COD}$$

(5)

Where K_{ox} = Oxidation term (g m^{-2} day^{-1}); k_{20} = First order reaction rate constant (day^{-1}); k_b = Biofilm constant (day^{-1}); θ = temperature coefficient; DO = Dissolved oxygen concentration (g m^{-3}); K_{DO} = Half saturation rate constant for dissolved oxygen concentration (g m^{-3}); B_{COD}= COD concentration (g m^{-2}).

The substrate consumption rate by the biofilm bacteria K_{sb} is defined by Equation (6).

$$K_{sb} = k_b C_t$$

(6)

Where C_t = Substrate concentration at time t (g m^{-2}). Biofilm constant k_b was modelled using Equation (7) in accordance with Polprasert and Agarwalla (1994).

$$k_b = a_s \left(\frac{\alpha \lambda}{\alpha + \lambda} \right)$$

(7)

a_s is specific area for biofilm activity (m^2 m^{-3}) which is calculated from the sum of the surface area of the bottom, sidewalls of the wetland and the roots of the plants in accordance with Equation (8).

$$a_s = \frac{1}{h} + \frac{2}{W} + \frac{2}{l} + \frac{R_S}{h}$$

(8)

The term α (m day^{-1}) in Equation (7) is defined by Equation (9).

$$\alpha = \frac{D_s}{L_S}$$

(9)

Where D_s = Diffusivity of the substrate through the liquid (m^2 day^{-1}); L_s = the liquid sub-layer thickness (m). The liquid sub-layer thickness L_s was considered by Williamson and McCarty (1976) to consist of two layers, namely: L_o the outer liquid sub-layer and L_b the inner layer. The inner layer is considered constant with a dimension of 56 μm. Experiments on column reactor conducted by Rittmann and McCarty (1980) revealed that the liquid sub-layer thickness (L_s) ranged from 1.198×10^{-4} to 2.26×10^{-4} m for superficial flow velocities between 3.22 and 43 m day^{-1}.

Coefficient λ in Equation (7) is mathematically defined by Equation (10).

$$\lambda = \tanh(\phi) K_{fa} \frac{L_f}{\phi}$$

(10)

Where K_{fa} = First order biofilm rate constant (day^{-1}); L_f = Biofilm thickness (m) which range from 1.462×10^{-4} to 1.615×10^{-4} in water hyacinth wetlands (Polprasert and Agarwalla, 1994). The characteristic biofilm parameter (ϕ) is defined by Equation (11).

$$\phi = \sqrt{\left(\frac{k_{fa} L_f^2}{D_f} \right)}$$

(11)

The uptake rate of organic carbon by heterotrophic bacteria in Equation (4) is influenced by forcing functions such as dissolved oxygen, quality of substrate, temperature and pH within the system of growth. The uptake rate of organic carbon by heterotrophic bacteria in g m^{-2}day^{-1} is given by Equation (12).

$$K_{up} = \frac{S_{COD}}{Y_{max}} \left[\mu_{max} \left(\frac{S_{COD}}{K_{SCOD} + S_{COD}} \right) \left(\frac{DO}{K_{DO} + DO} \right) \left(\frac{K_{pH}}{K_{pH} - \left[10^{(pH_{opt} - pH)} - 1 \right]} \right) \theta^{(T-20)} \right].$$

(12)

Where μ_{max}= Maximum growth rate of heterotrophic bacteria (day^{-1}); Y_{max} = Maximum substrate utilization rate (mg biomass/mg COD); K_{pH} = Limiting value for pH; pH_{opt} = Optimum pH; K_{SCOD} = Half rate saturation constant for COD concentration (g m^{-2}); S_{COD} = COD concentration (g m^{-2}).

The decay of plants K_{dec} depends on the uptake of inorganic carbon and mortality of the plants. Plant decaying rate is modeled using first-order kinetics in accordance with Equation (13).

$$K_{dec} = P_{OC} D_{rate}$$

(13)

Where K_{dec} = Plants decay rate (g m^{-2}day^{-1}); P_{OC} = Uptake of organic carbon by plants (g m^{-2}); D_{rate} = Plant decay constant (day^{-1}).

The sedimentation term K_{sd} (g m^{-2}day^{-1}) is considered to be the sum of sedimentation in the root zone (S_{root}) and plain sedimentation in the liquid zone (S_w) and is represented by Equation (14).

$$K_{sd} = S_{root} + S_w$$

(14)

Sedimentation in the root zone (S_{root}) is defined by Equation (15). The reaction rate coefficient is based on the cylindrical collector (Logan et al., 1993) because the root of the water hyacinth plants is assumed cylindrical.

$$S_{root} = \frac{4}{\pi} \eta \frac{\alpha u_f (1 - p)}{d_c}$$

(15)

The parameter η is a single collector removal efficiency calculated from Stokes' law in accordance with O'Melia (1985) and is defined by Equation (16).

$$\eta = \frac{(\rho_p - \rho_w)gd^2{}_p}{18\mu_T}$$

(16)

Where g = Acceleration due to gravity (m s^{-2}); ρ_p = Density of particle (kg m^{-3}); ρ_w = Density of water (kg m^{-3}); d_p = Diameter of settling particle (m) which ranges from 0.5 to 40 μm (Metcalf and Eddy, 1995); α = Sticking coefficient of the particle which ranges from 0.0008 to 0.012 (Khatiwada and Polprasert, 1999); u_f = Q/A = Flow velocity of the liquid (m s^{-2}), where Q and A are flow rate and surface area; P = Porosity of the media in percentage, which varies from 95 to 96.5% (Kim and Kim, 2000); D_{co} = Diameter of collector (m) which ranges from 0.0006 to 0.003 (Reddy, 1985).

For temperature above 20°C the viscosity of water (μ) varies with temperature in accordance with Equation (17) (Weast, 1981).

$$\log \frac{\mu_T}{\mu_{20}} = \left(\frac{1.3220(20-T) - 0.0010539(T-20)^2}{T+105}\right)$$

(17)

Where μ_T = Viscosity of water at T°C (kg m^{-1} s^{-1}); μ_{20} = Viscosity of water at 20°C = 1.002×10^{-3} (kg m^{-1} s^{-1})

Plain settling in the liquid zone follows Stoke's law in accordance with Equation (18).

$$S_w = \left(\frac{\rho_p - \rho_w}{18\mu}\right)gd^2{}_p$$

(18)

The total sedimentation term (K_{sd}) in the wetland unit is then defined by Equation (19).

$$K_{sd} = \left[\frac{4}{\pi}\left(\frac{\rho_p - \rho_w}{18*\mu}\right)gd^2{}_p u_f\left(\frac{1-P}{D_{CO}}\right) + \left(\frac{\rho_p - \rho_w}{18*\mu}\right)gd^2{}_p\right]S_{COD} - (19)$$

The mass balance for COD$_{sink}$ term (g m^{-2}) is represented by Equation (20).

$$\frac{dSink_{COD}}{dt} = K_{up} + K_{sd} - K_{reg_{COD}} - K_{CO_2} - K_{CH_4}$$

(20)

The rate of methane generation (K_{CH4}) is influenced by anaerobic condition in the COD sink. The generation rate of methane is modelled using first order kinetics in accordance with Equation (21).

$$K_{CH_4} = D_m Sink_{BOD}$$

(21)

Where K_{CH4} = Generation rate of methane (g m^{-2} day^{-1}); COD$_{sink}$= Sink COD term (g m^{-2}); D_m = Generation constant for methane (day^{-1}); K_{CO2} = Release rate of carbon (day^{-1}) from the sediment in Equation (20), and is defined by Equation (22).

$$K_{CO_2} = U_{r(20)}\beta^{(T-20)}S_{BOD} * 0.531S_b$$

(22)

Where $U_{r(20)}$ = Release rate (day^{-1}); β = Arrhenius temperature constant; COD$_{sink}$ = Active bacterial biomass (g m^{-2}); S_b = Fraction of bacteria settling.

The mass balance for plant uptake rate of inorganic is defined by Equation (23).

$$\frac{dP_{OC}}{dt} = K_{OX} + K_{CO_2} - K_{dec}$$

(23)

Calibration and optimization of model parameters

STELLA 6.0.1 software was used to run the developed conceptual model that incorporated equations of different processes involved in the wetland system. Mathematical processes were used to connect the various relationships among state variables and forcing function. Equations defining processes like sedimentation, biofilm activity and growth of microorganism, regeneration of organic carbon, decay of plant biomass and oxidation of organic carbon to carbon dioxide were included into the model. The data collected from the wetland units were used as inputs to the model for model calibration. The inputs were the influent concentration of COD, dissolved oxygen concentrations, pH and temperature values that were measured on daily basis. The simulation was done using Stella II software, which integrated the model using the in-built fourth-order Runge-Kutta approximation. The best values for unknown coefficient were obtained through calibration using observed data against simulated ones. The model efficiency R^2 was calculated from Nash and Sutcliffe (1970), which is given by Equation (24).

$$R^2 = \frac{F_o - F}{F}$$

(24)

Where F_o is the sum of the difference of squares between the observed and mean of observed values while F is the sum of the difference of squares between the observed and computed values.

The conceptual diagram shown by Figure 3 was used for modeling the transformation of organic carbon in wetlands. The sensitivity analysis of model results against model inputs indicated the sensitive parameters as the maximum growth rate of microorganism bacteria (μ_{max}), temperature coefficient (θ), specific area for biofilm

Table 2. Model calibration values.

No.	Parameter	Literature range	Reference	Calibration
1.	Temperature coefficient (Θ)	1.0 to 1.1	Metcalf and Eddy (1995)	1.02
2.	Maximum growth rate of heterotrophic bacteria at 20°C, μ_{max_20}	0.18	Ferrara and Hermann (1980)	0.16
3.	Optimum pH, pH_{opt}	4.0 ~ 9.5	Barnes et al. (1981)	7.5
4.	First order biofilm rate constant K_{fa} (day^{-1})	336.6	Polprasert and Agarwalla (1994)	336.6
5.	Half rate saturation constant for DO, K_{DO} (g m^{-3})	0.1 ~ 1.0	Okabe et al. (1995)	1.0
6.	Sticking coefficient, α	0.0008 ~ 0.012	Polprasert and Khatiwada (1999)	0.008
7.	Specific area for biofilm activity, a_s (m^2 m^{-3})	5.76 ~ 20.83	Polprasert and Agarwalla (1994)	6.76
8.	Diffusivity of a substrate through a liquid, D_f (m^2 day^{-1})	5.26×10^{-5}	Polprasert and Agarwalla (1991)	5.268×10^{-5}
9.	Liquid layer thickness, L_s (m)	1.19×10^{-4} ~ 2.26×10^{-4}	Ritmann and McCarty (1980)	2.26×10^{-4}
10.	Biofilm thickness, L_f (m)	1.462×10^{-4} ~ 1.615×10^{-4}	Polprasert and Agarwalla (1994)	1.615×10^{-3}
11.	Half rate saturation constant for COD, K_{SCOD} (g m^{-3})	15 to 75	Okabe et al. (1995)	18
12.	Density of settling particle, ρ_p (kg m^{-3})	1050 to 1500	Metcalf and Eddy (1995)	1300
13.	Fraction of bacteria settling, S_b	0.05	Canale (1976)	0.05
14.	Maximum substrate utilization, Y_{max} (mg biomass/mg COD)	0.5 to 1.0	Metcalf and Eddy (1995)	1.0
15.	Limiting value for pH, K_{pH}	199 ~ 288	Mashauri and Kayombo (2002)	288
16.	First order reaction rate constant, k_{20} (day^{-1})	0.1 ~ 1.2	Reed et al. (1995)	0.1
17.	Viscosity of water, μ_{20} (kg m^{-1} s^{-1})	1.002×10^{-3}	Metcalf and Eddy (1995)	1.002×10^{-3}
18.	Diffusivity of a substrate in biofilm layer, D_f (m^2 day^{-1})	2.3×10^{-5}	Rittman and McCarty (1980)	2.38×10^{-5}
19.	Diameter of collector, D_{co} (m)	0.0006 ~ 0.003	Reddy (1985)	0.0007
20.	Regeneration rate for inorganic carbon, U_r (day^{-1})	0.09	Foree and Jewell (1970)	0.07
21.	Arrhenius temperature constant at 20°C, B	1.02 ~ 1.09	Fritz et al. (1979)	1.04
22.	Porosity, P (%)	95~ 96.5	Kim and Kim (2000)	96
23.	Generation constant for Methane, D_m (day^{-1})	-	-	0.006
24.	Settling particle diameter, d_p (μm)	0.5 ~ 40	Metcalf and Eddy (1995)	5.6 μm

activity (a_s), settling particle diameter (d_p), first order reaction rate constant (k_{20}), half rate saturation constant for COD concentration (K_{COD}), density of settling particle (ρ_p) and maximum substrate utilization (Y_{max}).

The main objective of the simulation was to predict the effect of the activities of suspended biomass and attached plant root biofilm, on transformation and removal mechanisms of organic carbon in water hyacinths constructed wetlands. The efficiency of the model was found to be 73%, which indicates that the observed data fits well with the simulated values. The values of the optimized constants and coefficients used in the model during calibration are shown in Table 2.

The system performance showed the mean observed COD effluent was 110.5 g m^{-2} while the computed effluent was 104.4 g m^{-2}. The Mean Absolute Deviations (MAD), which is the average deviations between each data value and the mean, was determined from 25 data sets collected over a two month period. MAD of the observed and the computed effluent were 16.4 and 19.4 g m^{-2}, respectively and the observed and simulated removal percentages of COD in the unit were 38.3 and 40.5%, respectively. The closeness of these values suggests that the model predicted well the organic carbon transformation mechanisms in water hyacinth constructed wetlands. Table 3 shows the COD mass balance for state variables simulated by the model. The state variables

were substrate COD, plant organic carbon and sink COD. The processes involved in transformation mechanisms were oxidation of organic carbon to carbon dioxide, plant decay, regeneration of organic carbon, uptake of organic carbon by heterotrophic bacteria, settling of organic carbon and generation of methane.

The overall mass balance was 69.2 g m^{-2} day^{-1} which is close to 67.1 g m^{-2} day^{-1}, which is the total sum of the accumulation in plant OC, sink$_{COD}$, and S$_{COD}$ in the system. This shows that the system obey the law of mass conservation. About 7.5 g m^{-2}day^{-1} of carbon dioxide was consumed by plants in the photosynthesis process (the oxidation process contributed 6.4%) while the decay process returned 6.7 gm^{-2} day^{-1} to the water body (contribution to the transformation process was 4.4%) Thus, an accumulation of 0.95 g m^{-2} day^{-1} organic carbon in the plant biomass was observed which is completely removed from the system. About 17.9 g m^{-2} day^{-1} of organic carbon was consumed by microorganism in the formation of new cellular material (contribution to the transformation was 4.2%). The uptake of organic carbon by microorganism resulted to the regeneration of 53.1 g m^{-2} day^{-1} of organic carbon back to water column (this transformation process accounted only 25.5%). About 106.4 g m^{-2} day^{-1} of organic carbon was settled down to the sediment leading to high removal of 56.5% from the system. Therefore, due to sludge accumulation anaerobic

Table 3. Transformation mechanisms of organic material in Water hyacinth wetland.

No.	Transformation process	Mass load of organic material (g m^{-2} day^{-1})
1.	Influent organic material	175.5
2.	Effluent organic material	104.4
3.	Organic material settled in the benthic layer	106.4
4.	Organic material regenerated from the benthic layer	53.1
5.	Methane released from the benthic layer	3.8
6.	Water hyacinth decay rate	6.7
7.	Organic material oxidized by heterotrophic bacteria	7.4
8.	Uptake of organic material by microorganisms	17.9
9.	Oxidative catabolism of organic material to CO_2	2.7

decomposition in the sediment released 2.7 g m^{-2} day^{-1} of CO_2 to the system and 3.8 g m^{-2} day^{-1} of CH_4 was completely removed from the system (CO_2 and CH_4 contributed 1.2 and 1.8% to the transformation process). The permanent COD removal was 26.7% and largely contributed by sedimentation, biofilm activity and to smaller extent by generation of methane.

The effects of biofilm biomass activities were studied by setting the biofilm parameter to zero in the calibrated model and re-run. The output produced the removal percentage of COD was 34.0% when the model was simulated without considering the effect of biofilm, which is lower than 40.5% when biofilm effect was taken into consideration. The mean absolute deviation (MAD) of simulated effluents with and without biofilm effect was 10.0 and 34.5 g m^{-2}, respectively. This high divergence showed the significance of the biofilm effect in transformation and removal of organic carbon through bio-oxidation mechanism as studied by Sooknar (2000). These results confirm the results of Stowell et al. (1981) who reported that water hyacinth roots provide physical support for a thick bacteria biofilm that actively degrades the organic matter.

Model applications and limitations

This mathematical model can be used to predict transformation and removal of organic material in artificial wetlands implanted with *E. crassipes*. Its applications require knowledge of influent concentration of organic material, effective surface area of *E. crassipes* roots, flow rates of domestic wastewater, density and size of settling particle, porosity of the media and biofilm activities. Other environmental and physical-chemical parameters required include pH, temperature and dissolved oxygen. Stella 6.0.1 software or any other software may be used to simulate any state variable or process provided all inputs are known.

Conclusions

From the results of this study the following conclusions

are made:

(1) Transformation of organic material in *E. crassipes* artificial wetland was governed by settling of organic solids to benthic layer (104.4 g m^{-2}day^{-1}), regeneration of soluble organic matter from particulate matter (53.1 g m^{-2}day^{-1}) and uptake of organic matter by microorganisms (17.9 g m^{-2}day^{-1}) and uptake by organic material by biofilm growth on *E. crassipes* (7.4 g m^{-2}day^{-1}). Other transformation route such as decay of *E. crassipes* (6.7 g m^{-2}day^{-1}) and release of gases such as methane (3.8 g m^{-2}day^{-1}) and carbon-dioxide (2.7 g m^{-2}day^{-1}) from benthic layer were relatively ineffective.
(2) Organic material was largely removed through net loss to sediments (64.7 g m^{-2}day^{-1}), production of methane from benthic layer (3.8 g m^{-2}day^{-1}) and net carbon removed through uptake by plants and microorganisms (3.4 g m^{-2}day^{-1}). The total mass of removed carbon amounted to 71.9 g m^{-2}day^{-1}, which is equivalent to 41% efficiency of organic carbon removal in this wetland.

Conflict of Interests

The author(s) have not declared any conflict of interests.

REFERENCES

Barnes D, Bliss PJ, Gould BW, Valentine HR (1981). Water and Wastewater Engineering Systems. Pitman Books Limited London.

Barrie A (2002). Modelling COD removal in water hyacinths constructed wetland. MSc Dissertation, Department Water Resources Engineering, University of Dar es Salaam, Tanzania.

Bigambo T, Mayo AW (2005). Nitrogen transformation in horizontal subsurface flow constructed wetlands II: Effect of biofilm. J. Phys. Chem. Earth 30:668-672.

Canale RP (1976). Modelling biochemical processes in aquatic ecosystems. Ann Arbor Science Pub. Co, Ann Arbor, Michigan.

Ferrara RA, Hermann DPF (1980). Dynamic Nutrient Cycle Model for Waste Stabilization Ponds. J. Environ. Eng. Div. Am. Soc. Civ. Eng. 106(1):37-55.

Foree EG, Jewell WJ (1970). The extent of nitrogen and phosphorus regeneration from decomposing algae. In: Advances in Water Pollution Research. Proceedings of the 5th International Conference on Water Research. Pergamon Press Ltd., London. New Zealand J.

Mar. Freshw. Res. 14(2):121-128.

Fritz JJ, Middleton AC, Meredith DD (1979). Dynamic Process Modelling of Wastewater Stabilization Ponds. J. Water Pollut. Control Fed. 51(11):2724-2742.

Gersberg RM, Elkins SR, Lyons A, Godman CR (1985). Role of aquatic plants in wastewater treatment by artificial wetlands. Water Res. 20:363-368.

Kalibbala M, Mayo AW, Asaeda T, Shilla DA (2008). Modelling faecal *streptococci* mortality in constructed wetlands implanted with *Eichhornia crassipes*. Wetland Ecol. Manage. 16:499-510.

Khatiwada NR, Polprasert C (1999). Kinetics of fecal coliform removal in constructed wetlands. Water Sci. Technol. 40(3):109-116.

Kim Y, Kim WJ (2000). Roles of water hyacinth and their roots for reducing algal concentration in the effluent for waste stabilization ponds. Water Res. 34(13):3285-3294.

Knight LR, Clarke AR, Bastian KR (2000). Treatment Wetlands as Habitat for Wildlife and Humans. 7th International Conference on Wetland Systems for Water Pollut. Control (1):37.

Logan BE, Hilbert TAA, Anold RG (1993). Removal of bacteria in laboratory filters: Models and experiments. Water Res. 27(6):955-962.

Mashauri DA, Kayombo S (2002). Application of the two Coupled Models for Water Quality Management: Facultative pond cum Constructed Wetland Models. J. Phys. Chem. Earth 27:773-781.

Mayo AW (1989). Effect of the pond depth on bacteria mortality rate, J. Environ. Eng. Div. Am. Soc. Civ. Eng. 115(5):964-977.

Mayo AW, Muraza M, Norbert J (2013). The Role of Mara River Basin Wetland in Reduction of Nitrogen Load to Lake Victoria. Int. J. Water Resour. Environ. Eng. 5(12):659-669.

Mayo AW, Bigambo T (2005). Nitrogen transformation in horizontal subsurface flow constructed wetlands I: Model development. J. Phys. Chem. Earth 30:658-667.

Mayo AW, Kalibbala M (2007). Modelling faecal coliform mortality in water hyacinths ponds. J. Phys. Chem. Earth 32:1212-1220.

Metcalf and Eddy Inc. (1995). Wastewater Engineering: Treatment, Disposal and Reuse. 3rd McGraw-Hill Ltd. New Delhi, pp. 82-1204.

Mugasha AL (1995). A study of the potential use of water hyacinth, *Eichhornia crassipes* (Mart.) Solms., in the control of chromium and lead pollution in fresh waters. MSc. Thesis, University of Dar es Salaam.

Muyodi FJ (2000). Microbiological analysis of the waters of Lake Victoria in relation to the invasion of the water hyacinth *Eichhornia crassipes* (Mart.) Solms.: A case study of the lakeshores of Mwanza Municipality. Ph.D Thesis, University of Dar es Salaam.

Nakibuule J (2013). The transformation and removal of nitrogen in horizontal sub-surface flow constructed wetlands. Final Year Project, Department of Water Resources Engineering, University of Dar es Salaam.

Nash JE, Sutcliffe JV (1970). River flow forecasting through conceptual models. J. Hydrol., 10:282-290.

Okabe S, Hirata K, Watanabe Y (1995). Dynamic changes in spatial microbial distribution in mixed population biofilm, experimental and model simulation. Water Sci. Technol. 32(8):67-74.

Okurut D (2013). Biofilm Effects on Nitrogen Transformation and Removal in Subsurface Flow Constructed Wetlands planted with *Typha Domingensis*. MSc Dissertation, Department of Water Resources Engineering, University of Dar es Salaam.

O'Melia CR (1985). Particles, pretreatment and performance in water filtration. J. Environ. Eng. Div. Am. Soc. Civ. Eng. 116(6):874-905.

Polprasert C, Agarwalla BK (1994). A Facultative Pond Incorporating Biofilm activity. Water Environ. Res. 66(5):725-732.

Polprasert C, Khatiwada NR (1997). Role of Biofilm Activity in Water Hyacinths Pond Design and Operation. Proceedings of Asian Water Quality 97, 6th IAWQ Asian-Pacific Regional Conference, Int. Association on Water Quality and Korean SOC of Water Quality. Seoul, Korea.

Reddy KR (1985). Nutrient transformations in aquatic macrophyte filters used for water purification. Purification Water Reuse 111, American Water Works Association, Denver, Colorado, pp. 660-678.

Reed SC, Crites RW, Middlebrooks EJ (1995). Natural systems for waste Management and Treatment. McGraw-Hill Book Co., New York, NY, pp. 225-262.

Rittmann BE, McCarty PL (1980). Evaluation of Steady State Biofilm Kinetics. Biotechnol. Bioeng. 22:2343-2359.

Senzia MA (2003). Modelling of nitrogen transformation and removal in subsurface flow constructed wetlands during treatment of domestic wastewater. PhD Thesis, Department Water Resources Engineering, University of Dar es Salaam, Tanzania.

Senzia AM, Mashauri DA, Mayo AW (2004). Modelling Nitrogen Transformation in Horizontal Subsurface flow Constructed Wetlands Planted with *Phragmites mauritianus*. J. Civ. Eng. Res. Pract. 1(2):1-15.

Sooknar R (2000). A review of the mechanisms of pollutant removal in water hyacinth systems. Sci. Technol. Res. J. Univ. Mauritius, Reduit, Mauritius 6:50-56.

Standard Methods for the Examination of Water and Wastewater (2012). American Public Health Association, American Water Works Associations, Water Pollution Control Federal - 22nd Edition, Washington, DC.

Stowell R, Ludwig R, Colt J, Tchobanoglous T (1981). Concepts in aquatic treatment design. J. Environ. Eng. Div. Am. Soc. Civ. Eng. 112:885-894.

U.S Environmental Protection Agency (1988). Design Manual. Constructed Wetlands and Aquatic plant Systems for Municipal Waste Treatment. EPA/625/1-88/002, Cincinnati, OH 45268, pp. 2-53.

Vymazal J (2010). Constructed wetlands for wastewater treatment. Water 2:503-549.

Water 21 (2000). Constructed wetlands: A global technology. Magazines of International Water Association, June, pp. 57-58.

Weast RC (1981). Handbook of Chemistry and Physics 61st Ed. CRC Press, Boca Raton, Florida.

Williamson KJ, McCarty PL (1976). A Model of Substrate Utilization by Bacteria Films. J. Water Pollut. Control Fed. 48(1):9-24.

Yi Q, Hur C, Kim Y (2009). Modelling nitrogen removal in water hyacinth ponds receiving effluent from waste stabilization ponds. Ecol. Eng. 35(1):75-84.

24

An integrated process investigation of self-sustaining incineration in a novel waste incinerator: Drying, pyrolysis, gasification, and combustion of sludge-plastic blends

Jianzhong Zhu[1,2] , Lieqiang Chen[3], Jun Fang[4] and Buchang Shi[5]

[1]Key Laboratory of Integrated Regulation and Resource Development on Shallow Lakes, Ministry of Education HOHAI University, HOHAI University, Nanjing, PR China 210098.
[2]School of Environmental Science and Environmental Engineering, Nanjing, PR China 210098.
[3]College of Chemistry and Chemical Engineering, South China University of Technology, Guangzhou, PR China 510006.
[4]Delon Hampton and Associates, OMAP Program at District of Columbia Water and Sewer Authority, Washington DC, USA 20032.
[5]Department of Chemistry Western Kentucky University, Bowling Green, USA 42101.

Utilization of waste-water sludge is one of the most difficult processes of environment protection because of the high moisture and contents of harmful substances. Incineration is the effective method of utilization of such waste mater. To deal with these issues more effectively, a novel sludge incineration technology was developed in this study. The results showed sludge drying, pyrolysis, gasification and incineration were achieved as a spectrum of combustion by a pilot screw model incinerator in one step. In practice, the primary chamber of this technology actually acted as both gasifier for organic matter and vitrifying reactor for ashes, and the combustion process was mainly completed in the secondary chamber. The pre-dried sludge blended with plastic successfully realized self-sustaining incineration in the incinerator, and the incineration temperature reached about 1000°C with complete combustion. The study primarily demonstrated co-incineration might be realized by mixing sludge and waste plastic in the novel incinerator, some adverse effects due to single sludge or waste plastic combustion may be avoided in co-incineration process in this way.

Key words: Incinerator, sewage sludge, waste plastic, self-sustaining incineration.

INTRODUCTION

In China, more than 600 wastewater treatment plants are currently operating, and consequently, about 5000 million kg of dry sewage sludge are produced annually. Furthermore, according to national planning, the municipal wastewater treatment rate is expected to increase to 40 to 50% by 2010, which will result in a considerable increase in sewage sludge production (Wang, 1997; Shanghai City Government, 2004; Tang and Zhao, 2005; Jiang et al., 2010). The growing shortage of suitable land for disposal sites strengthens the preference for sludge incineration in China. With the increasing cost of sewage sludge disposal which has

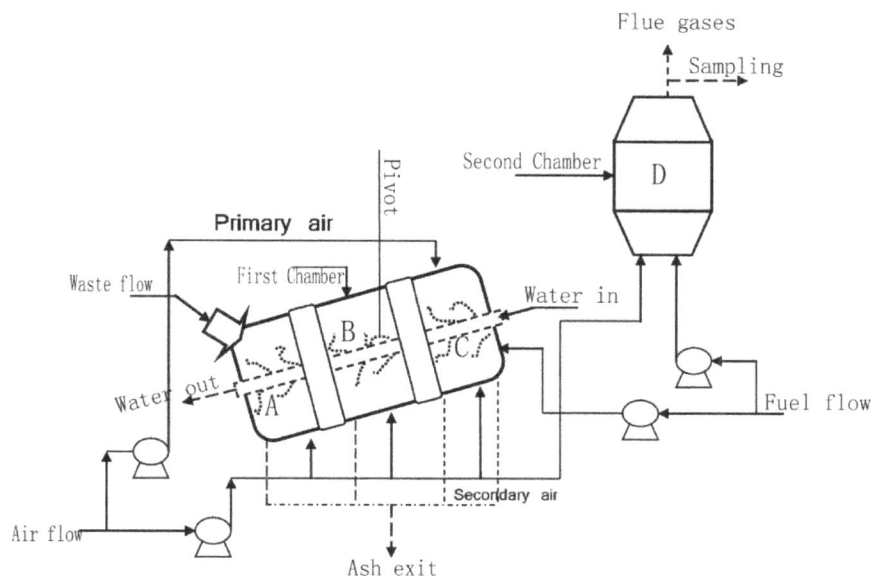

Figure 1. Flow chart of the new model screw incinerator.

posed a serious problem (Holmes et al., 1993; Kiely, 1997; Jimenez et al., 2004; NEPA, 1999; Zhang et al., 2002; Lowe, 1993). It may be expected that the role of incineration as a route for sludge disposal will also increase.

At the same time, waste plastic has become an environmental problem. For sustainable development, new ways of environmentally friendly waste plastic recycling have been of interest, and among them, the use of waste plastics as a supplemental fuel with coal has attracted interest (Jan and Weiss, 1996; Arapaima, 1996; Kim et al., 2002), plastics has greater fuel energy content than the sludge wastes. Therefore, co-incineration technology is a promising technique that offers a number of economical and environmental benefits, developments in national and international biomass waste recycling and reuse programmes have led to the process engineering design of an integrated unit.

The objective of this work is to investigate the operations in a novel integral type waste incinerator during the co-combustion of blends consisting of sludge and waste plastic, as an effort to improve and optimize the successful combustion with several processing variables, to obtain information about the running conditions, emissions and ashes of the biomass waste incinerator and evaluate the technological feasibility of the novel incinerator.

EXPERIMENTAL APPARATUS AND MATERIALS

Experimental setup

A process flowchart of the experimental apparatus is presented in

Figure 1. The pilot screw model incinerator consists of two chambers. A primary chamber is 50 cm inside diameter and 5.0 m total length, a secondary chamber formed of two deformed cubes is 80 cm total height. The two chambers were connected with a transformative column integrated a whole and a long throat on the primary chamber for feeding. The primary chamber, a rotating cylindrical reactor with inner rotating helix fins, where drying (A zone), pyrolysis and gasification (B zone), combustion of sludge-plastic wastes and vitrification of residues occur (C zone), the primary chamber of the furnace is inclined, adjustable and rotatable at a low speed. Pre-dried solid waste can be turned and conveyed by screw, and consequently passed through varying temperature zones where drying, volatilising, combusting and cooling can take place. A hollow pivot with helix fins located in center of the primary chamber, which may regulate temperatures with cooling water, the helix fins are used to screw the waste forward and transfer the heat; the secondary combustion chamber (D zone), a deformed cube reactor with a fuel nozzle, where the yielded flue gases in the primary chamber were burnt out to completely destroy toxic organic compounds contained in the fuel gas.

For the newly installed pilot incineration unit, the ignition process was needed. At first, pre-drying blends was feed into the grate, and then gasoline was sprayed onto the waste blends. The flame-thrower was fixed on the top of the primary chamber ignited. Simultaneously, the auxiliary fuel was injected into the secondary chamber and ignited by another flame-thrower fixed on the top of the secondary chamber. After about 30~60 min, sludge-plastic waste was added. It took about 1 h to keep the combustion temperature higher than 850°C in the secondary chamber and about 2~3 h to make various parameters stable in the primary chamber.

Materials and analysis

The sewage sludge used in this study was obtained from an urban wastewater plant located in Suzhou and Guangzhou (China), which was treated with chemical softening and mechanical dehydration. The sewage sludge was dried and grounded into fine particles

Table 1. Main Characteristics of sludge and waste plastic used in the experiment.

Item	Proximate analysis (wt %)		Ultimate analysis (dry basis, wt %)	
Sludge	Volatile matter	59.23	C	41.87
			H	6.00
	Fixed carbon	8.63	N	5.79
			O	24.44
	Ash	39.14	S	1.83
			Cl	0.085
	Heating value (MJ/kg)	15.38	Others	19.985
Plastic	Volatile matter	80.12	C	59.14
			H	2.12
	Fixed carbon	0.023	N	0.40
			O	7.70
	Ash	19.91	S	0.61
			Cl	23.40
	Heating value (MJ/kg)	41.23	Others	6.63

before using in the experiments. The Poly (Vinyl Chloride) (PVC) used in this study was commercially available by Kaida waste recycle company, China (Its diameter <400 mm). Ultimate analyses of samples were determined using a CHN-600 LECO elemental analyzer. Table 1 shows the proximate analysis and ultimate analysis of test wastes, respectively. Thermo-gravimetric analyses (TGA) were carried out with the prepared samples using a LECO TGA-500. K-thermocouples were used to measure the temperatures and flue gas analyzer (Germany, Test 0330-2) was used to monitor gas concentrations, nine points in the primary chamber and three points in secondary chamber. For the nine points in the primary chamber, one was located on the top of the furnace body in each zone, other two points on the middle of both sides of the furnace body in each zone, respectively. For the secondary chamber, the three monitor points are located on the bottom, middle and top of the furnace wall, respectively. The raw waste and residues were characterized by scanning electron microscopy (SEM) to evaluate the alterations of surface morphology by the incineration.

RESULTS AND DISCUSSION

Pyrolysis characterization of waste

Figure 2 shows TGA results of sludge, waste plastic and their blends. For the sludge, three different behaviors were conducted in range 170 to 220°C, 300 to 400°C and 420 to 500°C. A primary weight loss was apparently taken place within range 170 to 220°C, two weight losses in range 300~500°C showed short transitions from volatilization to burn-off in sludge combustion, therefore, the two processes may be considered to take place parallel to each other (Gomez-Rico et al., 2005). Two main loss stages were observed in the PVC result. The primary weight loss within the range of 280 to 300°C was mainly due to HCl elimination in PVC and a polyenes structure formation (Makherjee and Gupta, 1983; et al.,

Szakacs 2004), the second weight loss stage within the range of 450 to 500°C was due to the thermal degradation of carbon chain of PVC that produced flammable volatiles (Troitskii and Troitskaya, 1999). For the blend sample, it showed that for the combustion, there was no marked separation between the drying ending and volatilizing beginning at temperatures, blend sample shortly burned after the volatilizing.

Operation parameters of self-sustaining incineration

Due to variations of the characteristics of the feed waste, complete combustion conditions including continuous changeable incineration temperature and adjustable air excess ratio were observed during the process of the co-incineration. At the beginning of test, the temperature of the primary chamber was heated to 100°C with gas fuel, thereafter, the semi-dried sludge (40 to 50 wt% dry matter) and waste plastic powder blends were fed into the incinerator by means of the screw type feeder. Supply air and gas fuel were distributed through ducts and kept at a constant feed rate until the pyrolysis gases could be automatically combusted in the secondary chamber, after, the gas fuel tank was closed.

Air is one of the most important parameter in combustion processes. The air amount is approximately proportional to the calorific value of fuel. A change of fuel composition may result in a change of the air amount that must be supplied to the chambers. The parameters on co-incineration of sludge and waste PVC blends were discussed in this section, which might show combustion situation of the incinerator. Firstly, keep feed and supply air a constant value, frequently monitored temperature changes in each zone, their results are shown in Figure 3,

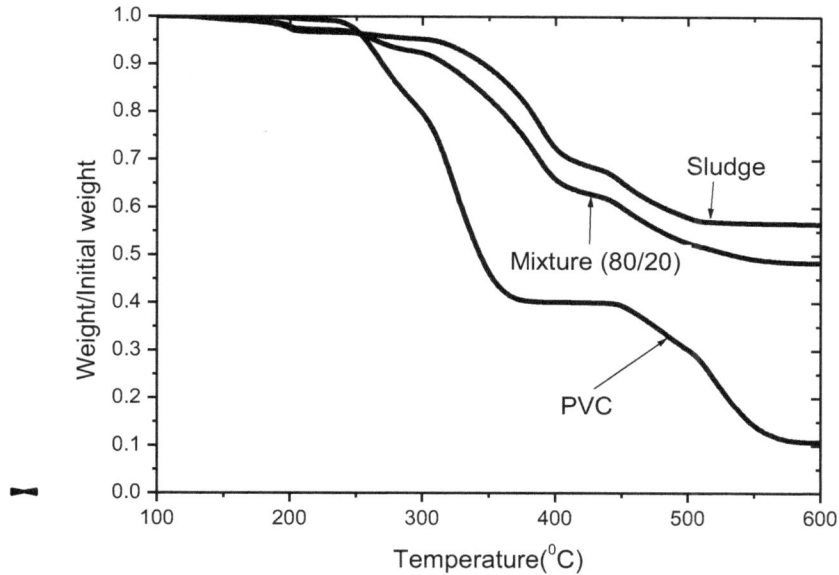

Figure 2. TGA profiles of the waste fuels.

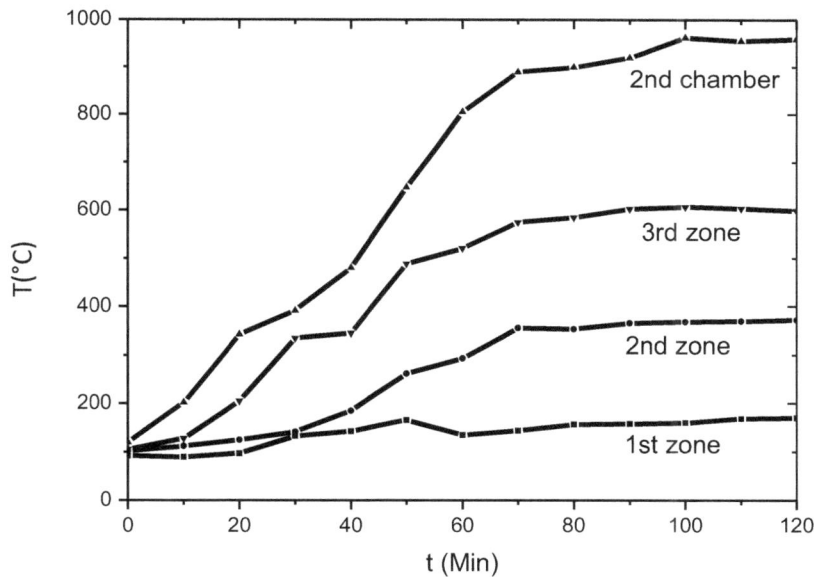

Figure 3. Distribution of temperatures in incinerator.

which shows the ranges of temperature distribution: 100~180°C in drying zone, 200~300°C in gasifying zone, and 400~600°C in combusting zone.

When the lighter wastes were screwed to the third segment with the helix fins, the furnace temperature rose up rapidly, and reached 400°C. Here, a heat balance was held among the newly ingoing wastes, the fume and the incineration. Table 2 shows that relation between the supply air and incineration temperatures in the

incinerator. Experimental results showed the amount of supply air in the primary chamber was too small to complete incineration, the heat could not realize auto-incineration in the secondary chamber. Certainly, the secondary incineration could not be obtained for enough air in the third zone and less combustible gases in the secondary chamber. Incinerating temperatures and the total air excess ratio showed a negative and positive correlation in four zones.

Table 2. Effect of feed air on combustion.

Primary chamber						Secondary chamber		
First sect		Second sect		Third sect		Chamber		Out-flue
Air (m³/h)	T(°C)	Air (m³/h)	T(°C)	Air (m³/h)	T (°C)	Air (m³/h)	T(°C)	T(°C)
2.2	97.3	5.2	110.4	8.1	403.2	14	446.4	441.4
1.8	110.5	5.5	130.3	9.2	430.9	15	615.4	595.2
1.4	140.5	1.6	265.2	6	448.1	10	929.1	723.0
1.2	85.6	2.4	210.6	6	519.3	10	697.5	691.2
1.2	109.7	3.2	188.0	6	641.4	10	855.5	819.5
1	127.2	2.5	228.5	8	331.2	10	844.9	804.3
1	153.0	3	300.4	8	591.0	10	1020.1	875.1

The changes of the combustion temperatures are caused by the varies of oxygen contents in the air excess ratios, only when the total amounts of the oxygen in the supplied air can be reached to balance, and the produced heating energy cannot be flowed away due to a higher air excess ratio. Then the incinerating temperatures might be constantly maintained in each part in the whole system (Inguanzo et al., 2002). Another influencing factor of auto-incineration is that a right distribution of the supplied air in the primary chamber can burst violently out the combustible gases in the 3rd zone of the primary chamber. The operation conditions were optimized during the experiments: First stage air ratio: 0.5~0.7; Feed rate: 6~8.5 kg/h; Screw speed: 2~3r/min.

Incinerator temperatures with combustion

All operating conditions were kept constant except the temperatures in the experiments. At the beginning, the incinerator was heated, and upon attaining the temperature of 100°C, then, the blend samples were fed into the incinerator. The average temperatures of the nine points in the primary chamber and three points in the secondary chamber were calculated. Figure 3 shows experimental curves corresponding to drying, pyrolysis and combustion in dynamic runs. The drying curve showed the demanded temperature characteristic within a 150-200°C range. Based on the sludge dry studies (McGhee, 1991), the sludge moisture was characterized as free moisture which could be removed during a constant drying rate period. The blend wastes appeared to have some differences compared with single waste for pyrolysis property, sludge and waste plastic had a significantly high content of volatiles, its volatile carbon was up to 80% (Heo et al., 2000).

The pyrolysis in the secondary zone and combustion in the third zone showed approximately consistent results with the designed requirements. The blast starts of volatilization and the combustion beginning lead to a rapid increase of the chamber temperature within 70 min, consistent with the TGA results at 400 and 600°C,

respectively. For the secondary chamber, the rapidly pyrolyzed gases cannot be completely combusted for having not enough time and air in the primary chamber, so the larger quantity of combustible gases flowed into the secondary chamber, subsequently auto-thermal combustion was realized successfully within a high temperature (about 1000°C), which definitely might destroy pollutants effectively in such a waste incinerator (Nottrodt et al., 1990).

Waste blend ratio with combustion

The energy is required for the evaporation of the moisture in the sludge. If the net energy is not sufficient for auto-thermal combustion, then supplementary fuel must be supplied the use of recycled plastic as a secondary substitute fuel for its high energy efficiency (Oral et al., 2005; Werle and Wilk, 2010). Figure 4 shows the plastic requirements as the supplementary fuel to maintain auto-thermal incineration. At 6 wt. % of PVC in the mixed waste, the average temperatures in the two combustion chambers were about 563.5 and 839.5°C, respectively. This began to maintain auto-thermal incineration, which means that the incinerator operating conditions were at acceptable levels. The blended waste, which is mixed with PVC and sludge, has the higher heating value as the higher co-incineration percentage. For example, the co-incineration percentage was 7 wt.%, the low heating value of the mixed waste was 17.19 MJ/kg, the average temperatures in the two combustion chambers were about 601.2 and 899.8°C, respectively. But, when the co-incineration percentages were 10 and 12 wt.%, the low heating value of the mixed waste was 17.83 and 18.48 MJ/kg, respectively. The average temperatures in the two combustion chambers were about 569.2 and 809.8, 435.7 and 646.8°C.

Therefore, as the co-incineration percentage increased, the combustion chamber temperature decreased and dropped to less than 435.7 and 646.8°C, respectively. The result shows that sludge, and 6 to 10% waste plastic blends might maintain auto-thermal incineration, however

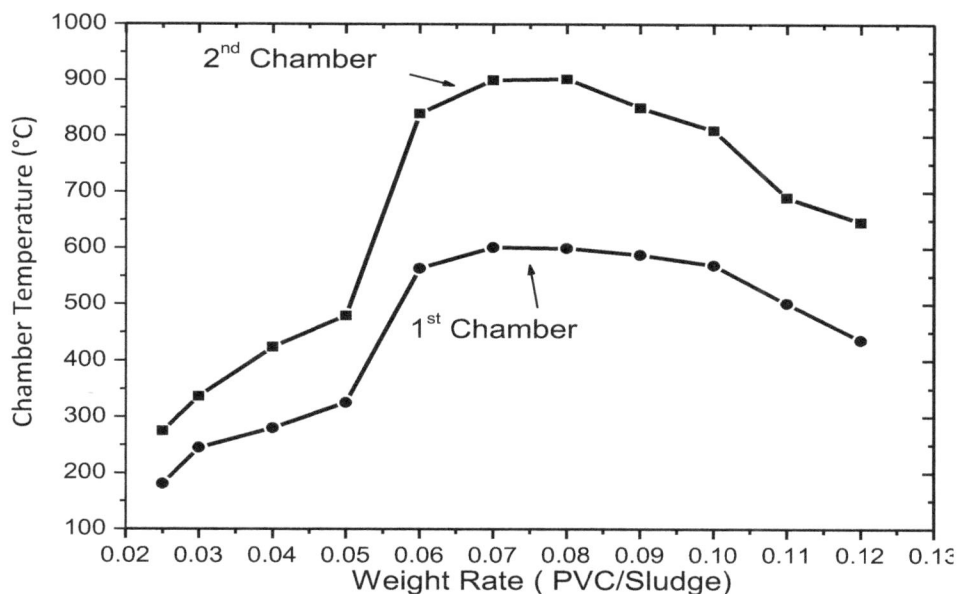

Figure 4. Blend ratio of waste fuels with combustion.

Table 3. Effect of feed rate on combustion (Gas: Vol.-%).

Feed (kg/h)	5	6	7	8	9
T (°C)	495.4	523.1	567.3	532.7	483.4
Burnout (%)	87.4	86.9	86.4	86.0	76.3
O_2	10.2	8.5	8.3	7.9	5.4
CO_2	6.2	7.3	8.8	8.9	9.2
CO	0.94	1.0	1.4	1.9	2.1

the overfull waste plastic in blend wastes would agglomerate with showing a burnt and sintered surface. Whereas the inside remains were unburned, because plastic materials were generally melt from surface to core due to low thermal conductivity, the problem of the "sticky phase" must be solved in the combustion systems (Oral et al., 2005). The amount of air is approximately proportional to the calorific value of the mixing waste fuel. For air flow below the stoichiometric requirement an oxygen deficient combustion regime is established within the waste matrix. This regime is characterized by low reaction temperatures, favors endothermic pyrolysis. The stoichiometry controls the reaction temperature, a PVC enhancement leads to a decrease in temperature for a constant total air.

Feed and screw rates with combustion

Incineration experiments were conducted in keeping constant operating parameters except variable feed rate in the incinerator, their temperatures, burnout rates and

O_2 CO_2; CO vol.% were measured in each round at the third zone, the CO concentration is one of important parameters in combustion processes indicating the efficiency (Shi et al., 2009). The results are shown in Table 3. Experimental results suggested the temperature played a key index for the auto-thermal combustion, automatic combustion was difficult in happening when the temperature of the third zone was lower than 500°C. When the mixed waste was gradually supplied to the incinerator with a increase rate, the chamber temperatures and the CO contents were increased with the enhancement of the mixed waste, the burnout rate was maintained at similar levels, then the feed rate reached at 9 kg/h, the chamber temperatures in the 3rd zone was under 500°C, the co-combustion of the variations of the feed rate have also affected the emissions of gases and burnout rates. So, an optimal value of feed rate was set at 6 to 8.5 kg/h. For the rotation speed, keeping constant operating parameters except variable screw rates in the incinerator, the temperatures, burnout rates and O_2, CO_2; CO vol.% were measured in each round at the third zone, the experimental

Table 4. Effect of screw speed on combustion (Gas: Vol.-%).

Rotation speed (rpm)	Temperature(°C)	Burnout (%)	O_2	CO_2	CO
2	524.8	79.4	7.2	6.3	1.6
3	589.0	84.3	6.5	7.2	0.4
4	435.9	69.9	7.8	5.7	2.8

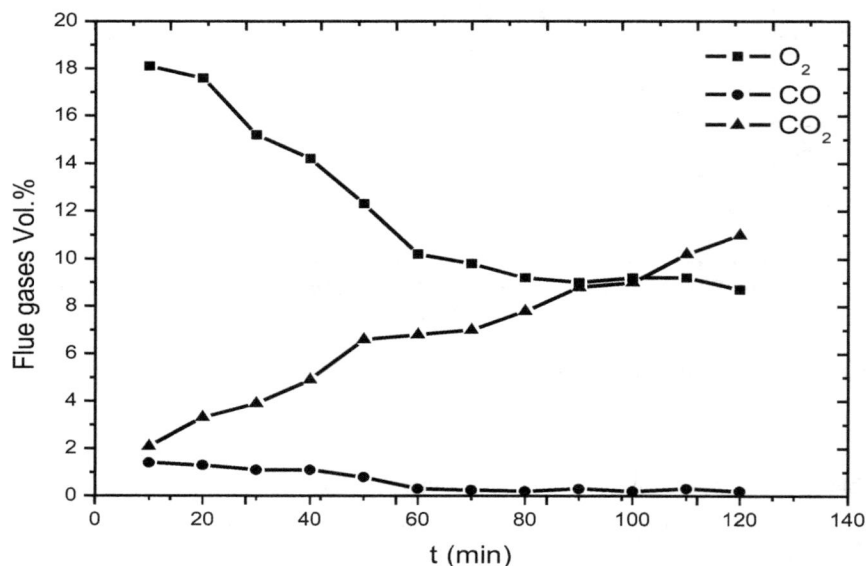

Figure 5. Variations of O_2, CO_2 and CO compositions in the secondary chamber.

Table 5. Combustion status in the incinerator (Mean; Gas: vol.-%).

Item	Zones			2nd chamber	Flue gas
	1st	2nd	3rd		
T(°C)	172.3	343.4	589.5	974.6	646.5
O_2	5.8	6.7	7.2	10.8	9.7
CO_2	2.3	4.6	6.2	8.9	8.2
CO	1.1	2.3	5.0	0.14	0.13

results, as shown in Table 4. With no change in the quantities of the mixed waste, the burnout rates and combustion temperatures were increased as the increase of the rotation speed from 2 to 3 rpm. If the rotation speed continued to increase to 4 rpm, the CO content was sharply increased to 2.8 vol. %, and the combustion temperature was only 435.9°C. This indicated that a rate in 4 r/min could decrease the temperatures and burnout rates in the third zone, and a significant change in O_2, CO_2 and CO values, so an optimal screw rate was set at 2~3 r/min.

Combustion efficiency

The O_2, CO_2 and CO concentrations were inspected in different parts of the incinerator for indicating combustion. In general, CO concentration can be used as a surrogate indicator for combustion efficiency. The results are shown in Figure 5 and Table 5. Curves in Figure 5 show the combusting situation according to the changes of O_2, CO_2 and CO concentrations in the secondary chamber. The CO_2 generation at the O_2 expense directly shows the combustion reactions, the profiles lead to an insight of the combusting process in the incinerator, oxygen concentrations were decreased with an increase of CO_2 composition, the CO behavior was associated with the imperfect combustion of hydrocarbons under a constantly supplied air. The measured temperatures in different sections showed that the mixtures were heated up slowly as they passed through from the primary zone to the third

Table 6. Evaluation of the designed combustion zones (rate, wt %).

	Zones		Combustibility (Fly ash)
Drying 1st	Volatilization 2nd	Burnout 3rd	
64.3%	69.8%	91.2%	7.45%

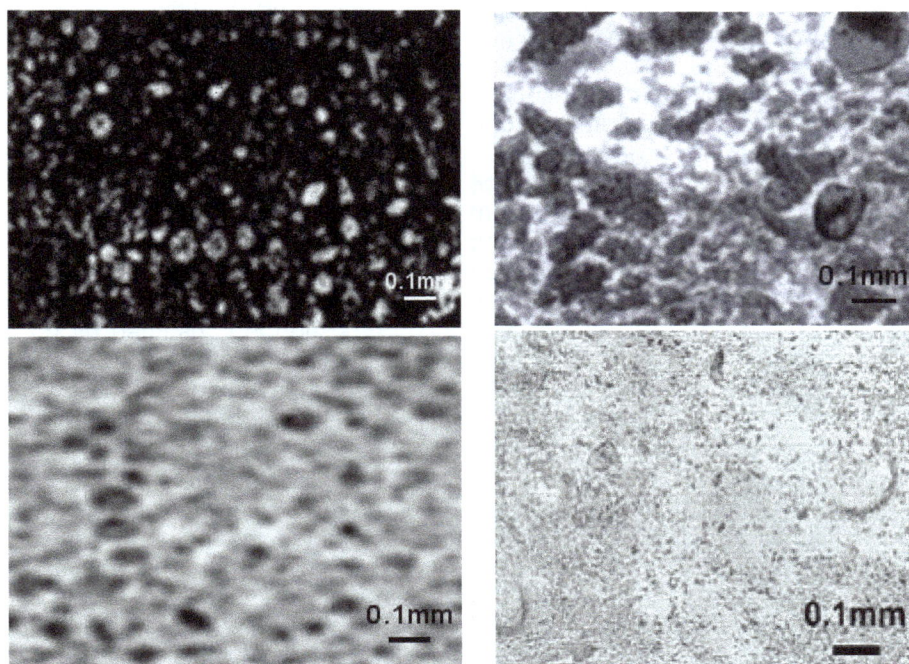

Figure 6. SEM micrographs of raw fuels and residues (In order: Raw wastes, Residues in volatile zone, Residues in burnout zone and Fly ash in secondary chamber).

zone. Table 5 shows constitution (%) of flue gases, under a constant total air excess ratio in the whole system, if the combustible gases from the primary chamber were incompletely incinerated and the formation of CO increased.

The emission of CO gradually decreased with enhancing the combustion efficiency. The results show that the residual O_2 was about 9.7 vol. %, CO, 0.13 and CO_2, 8.2vol. %. The contents of O_2 and CO_2 were slightly related to the waste feed period, but that of CO was more stabile, the combustion temperature in the secondary chamber was 974.6°C and a low CO emission was achieved. To understand drying, volatilizating and burnout in the incinerator, the overall combustion was analyzed with a selected case; the evaluated efficiency of co-combustion with corresponding measurement is presented in Table 6. In the heat balance system, the increase in char formation is the main reason for the reduction in temperature with the co-incineration. The residual oxygen increase under a constant total excess air can also be explained as being due to this large increase in the unburned char of waste. With the co-incineration of 7 wt.% PVC and 93 wt.% sludge waste,

the unburned char fractions were 8.8 wt.% in the 3rd zone and 7.4 wt.% in fly ash from the secondary chamber, respectively, which was a low combustibility in fly ash.

The results showed a high efficient incineration and also indicated a successful process of drying and volatile release. Experiments provided the detected results with SEM on raw wastes and residues in the test, as shown in Figure 6. In contrast, the changes might be clearly observed among the raw wastes, residues in volatile zone, residues in burnout zone and fly ash in the secondary chamber, the obvious changes were shown from physical shapes to colors in the whole combustion process.

Pollutant control with incinerator characteristics

Although, combustion procedure may lead to the production of compounds like polychlorinated dibenzo-p-dioxins and dibenzofurans (PCDD/Fs) (Wang et al., 1999). Two different process technologies have been adopted for waste incineration to reduce pollutants emission, namely, rotary kilns or thermal oxidizers. For instant, Dioxin emissions could be reduced below 0.1 ng I-TEQ/m^3 with rotary kilns (Rappe et al., 1992; McKay, 2002).

To minimize these possibilities for dioxin formation, the optimum design of an incinerator must pay particular attention to the four cornerstones of high destruction efficiency–temperature, time, turbulence and excess oxygen, such that PCDDs and PCDFs were certainly destroyed after adequate residence times and temperatures above 800°C (WHO, 1986; Vogg and Stieglitz, 1986). These might exert a different influence on the PCDD/Fs formation. Furnace temperature is one of the key variables that determine the combustion condition in incinerator, and it exerted a large influence on PCDD/Fs formation in a waste incineration system (Hatanaka et al., 2000; Lenoir et al., 1991). High furnace temperature, which intended to promote combustion reactions, was generally recommended to reduce their release. It was generally agreed that combustion temperatures of 850°C and a gas residence time of 2s or 1000°C and a gas residence time of 1 s were necessary for total destruction of PCDD/Fs (McKay, 2002).

Figure 3 shows the results of the temperature change during the combustion, the temperature of third zone in the primary chamber reached 600°C, and that of the secondary chamber reached 1000°C, it was certainly effective to keep the temperature of the secondary combustion chamber high enough to reduce the PCDD/Fs release. Three slag exits and the upper part of the secondary chamber were designed to minimize contact between flue gas and ash, which definitely inhibited the PCDD/Fs formation in this incinerator. Poor combustion was shown to result in relatively high emissions of dioxin formation or other pollutants, some studies reported a correlation of CO emission to PCDD/PCDF concentration during combustion processes (Yde et al., 1994; Sato et al., 1994), the O_2, CO_2 and CO values were parameters indicating high quality of combustion in the incineration chamber. Further studies (Raghunathan and Gullett, 1996; Ogawa et al., 1996; Lindbauer et al., 1992; Gullett and Raghunathan, 1997) have shown the presence of sulphur dioxide reduces the level of PCDD/F formation during incineration processes. A number of mechanisms have been proposed but the basic concept was that sulphur (Verhust et al., 1996) could scavenge the chlorine molecule in the presence of moisture producing SO_3 and HCl and also sulphur might block the activity of metallic catalysts in the ash thus reducing their PCDD/Fs activity formation (Chen et al., 1997).

The municipal sewage sludge taken as potential inhibitors of PCDD/PCDF formation has been investigated by some researchers (Ruokojarvi et al., 2001; Mininni et al., 2004; Caneghem et al., 2010). Addition of waste plastic to sludge waste changed the fuel composition (S/Cl ratio). The increase of S/Cl ratio in wastes could inhibit formation of chlorinated pollutants by

generating SO_2 which could convert the chlorine formed via the Deacon reaction back to HCl (Fullana et al., 2004; Hagenmaler et al., 1987), this slowed chlorination process as HCl was a much poorer chlorination agent than chlorine. Werther and Ogada (1999) suggested that the relative lower PCDD/Fs concentration (in comparison with municipal waste incineration) in sewage sludge was perhaps due to the high concentration of sulfur in the sewage sludge. There are competitive reactions with transition metals between sulfur oxides and chlorides, thus, the formation of transition metal sulfides may reduce the produce of chlorinated pollutants. We are combining removal of pollutants in integrated processes and conducting a series of experiments to evaluate the effects of additives ($CaSO_3$, $Ca(OH)_2$) on pollutant formations (HCl SO_2 and Dioxin). Several cleaning steps are being integrated into the incinerator to test the cleaning efficiency. The chlorine generated from PVC combustion is consumed by reaction with the SO_2 from sludge in the fume gases, add additional adsorbents to capture heavy metals as well as dioxins/furans and other harmful components (Fullana et al., 2004; Hagenmaler et al., 1987; Skodras et al., 2007; Pandelova et al., 2007).

The cleaning process based on the reactions among Cl_2, SO_2, $CaCO_3$ and $Ca(OH)_2$ according to self-sustaining incineration in the waste incinerator has been investigated and it is possible to study the relations among HCl, SO_2 and dioxin in the incinerator in the future. It is possible to study the relations among HCl, SO_2 and dioxin in the incinerator in the future.

Conclusions

A pilot screw incinerator with two chambers was designed and constructed for sludge incineration. The co-combustion possibility of using waste plastics as a source of secondary fuel was available in the incinerator, the experimental results showed self-sustaining combustion was successfully realized by drying, pyrolysing and combusting process of pre-dry sludge with waste plastic blends, the success of this process was critically dependent upon the optimization of operating conditions. This strongly suggests that the combustibility of sludge with waste plastic could be improved by controlling these variables such as temperature, air-excess and so on. The designed incinerator can be effective to keep the temperature of the secondary combustion zone high enough to reduce pollutant release. On the other hand, the combined designs (2 s residence time, incline, screw feed and multi-exits) might result in reducing the pollutant concentration with completing combustion.

ACKNOWLEDGEMENTS

The authors wish to thank the Scientific Research Starting Foundation for Returned Overseas Chinese Scholars,

Ministry of Education, China (Grant No. 1061-51200312), the National Science Fund for Distinguished Young Scholars (Grant No. 50925932) and the Fundamental Research Funds for the Central Universities, China (Grant No. B11020157) for the financial supports.

REFERENCES

Arapaima T (1996). Development of a new scrap melting process based on massive coal and plastics injection, in Proceedings of the third International Iron Making Congress pp. 314-321.

Caneghem JV, Block C, Vermeulen I, Brecht AV, Royen PV, Jaspers M, Wauters G, Vandecasteele C (2010). Mass balance for POPs in a real scale fluidized bed combustor co-incinerating automotive shredder residue. J. Hazard. Mater. 181:827-835.

Chen JC, Wey MY, Yan MH (1997). Theoretical and experimental study of metal Capture During Incineration Process. Theoretical and Experimental Study of Metal Capture During Incineration Process. J. Environ. Eng. 123(11):1100-1106.

Fullana A, Conesa JA, Font R, Sidhu S (2004). Formation and Destruction of Chlorinated Pollutants during Sewage Sludge Incineration. Environ. Sci. Technol. 38: 2953-29538.

Gomez-Rico MF, Font R, Fullana A (2005). Thermo-gravimetric study of different sewage sludge and their relationship with the nitrogen content. J. Anal. Appl. Pyrol. 74:421-428.

Gullett BK, Raghunathan K (1997). Observations on the effect of process parameters on dioxin/furan yield in municipal waste and coal systems. Chemosphere 34:1027-1032.

Hagenmaler H, Kraft M, Brunner H, Haag R (1987). Catalytic Effects of Fly Ash from Waste Incineration Facilities on the Formation and Decomposition of Polychlorinated Dibenzo-p -dioxins and Polychlorinated Dibenzofurans. Environ. Sci. Technol. 21:1080-1084.

Hatanaka T, Imagawa T, Takeuchi M (2000). Formation of PCDD/Fs in artificial solid waste incineration in a laboratory-scale fluidized-bed reactor: Influence of contents and forms of chlorine sources in high-temperature combustion. Environ. Sic. Technol. 34:3920-3924.

Heo NH, Baek CY, Yim CH (2000). Analysis of furnace conditions with waste plastics injection into blast furnace. J. Korean Inst. Resour. Recyc. 9:23-30.

Holmes G, Singh BR, Theodore L (1993). Handbook of environmental management and technology. New York: John Wiley & Sons Inc.

Inguanzo M, Domínguez A, Menéndez JA, Blanco CG, Pis JJ (2002). On the pyrolysis of sewage sludge: The influence of pyrolysis conditions on solid, liquid and gas fractions. J. Anal. Appl. Pyrolysis 63(1):209-222.

Janz J, Weiss W (1996). Injecting waste plastics into the blast furnace of steelwork. In Proceedings of the Third International Iron Making Congress.114-119.

Jiang J, Du X, Yang S (2010). Analysis of the combustion of sewage sludge-derived fuel by a thermogravimetric method in China. Waste Manage. 30:1407-1413.

Jimenez B, Barrios JA, Mendez JM, Diaz J (2004). Sustainable sludge management in developing countries. Water Sci. Technol. 49(10):251-258.

Kiely G (1997). Environmental engineering, Maidenhead: McGraw-Hill Publishing Co.

Kim DG, Shin SS, Son SM, Choy KS, Ban BC (2002). Waste plastics as supplemental fuel in the blast furnace process: Improving combustion efficiencies. J. Hazard. Mater. B94:213-222.

Lenoir D, Kaune A, Hutzinger O, Mutzenich G, Horch K (1991). Influence of operating parameters and fuel type on PCDD/PCDF emissions from fluidized bed incinerator. Chemosphere 23:1491-1500.

Lindbauer RL, Wurst F, Prey T (1992). Combustion dioxin suppression in municipal solid waste incineration with sulfur additives. Chemosphere 25:1409-1414.

Lowe P (1993). The development of a sludge disposal strategy for Hong Kong territories. J. Inst. Water Environ. Manage. 7:350-353.

Makherjee AK, Gupta A (1983). Graft Copolymerization of Vinyl Monomers onto Polypropylene. J. Macromol. Sci. Part A. 19(7):1069-1099.

McGhee TJ (1991). Water supply and sewerage, New York: McGraw-Hill.

McKay G (2002). Dioxin characterization, formation and minimization during municipal solid waste (MSW) incineration: Review. Chem. Eng. J. 86:343-368.

Mininni G, Sbrilli A, Guerriero E, Rotatori M (2004). Dioxins and furans in pilot scale incineration tests of sewage sludge spiked with organic chlorine. Chemosphere 54:1337-1350.

NEPA (1999). Annual Report on the Environmental Situation in China 1999, Beijing (in Chinese): National Environmental Protection Agency of China (NEPA).

Nottrodt A, Duwel U, Ballschmiter K (1990). The influence of increased excess air on the formation of PCDD/PCDF in a municipal waste incineration plant. Chemosphere 20:1847.

Ogawa H, Orita N, Horaguchi M, Suzuki T, Okada M, Yasuda S (1996). Dioxin reduction by sulfur component addition. Chemosphere 32:151-157.

Oral J, Sikula J, Puchy R (2005). Energy utilization from industrial sludge processing. Energy 30:1343-1352.

Pandelova M, Lenoir D, Schramm KW (2007). Inhibition of PCDD/F and PCB formation in co-combustion. J. Hazard. Mater. 149:615-618.

Raghunathan K, Gullett BK (1996). Role of sulfur in reducing PCDD and PCDF formation. Environ. Sci. Technol. 30:1827-1834.

Rappe C, Lindstrom G, Hansson M, Andersson K (1992). Levels of PCDDs and PCDFs in cow's milk and worker's blood collected in connection with a hazardous waste incinerator in Sweden. Organohalogen Comp. 9:199-202.

Ruokojarvi P, Aatamila M, Tuppurainen K, Ruuskanen J (2001). Effect of urea on fly ash PCDD/F concentrations in different particle sizes. Chemosphere 43:757-762.

Sato Y, Shizuma M, Sasaki M, Futamura O (1994). Technique for reduction of dioxin emission in waste incinerators. Organohalogen Com. 19:389-393.

Shanghai City Government (2004). Future Plan for Sewage Sludge Treatment and Management in 2020, Shanghai: Report by Sewage Bureau.

Shi DZ, Tang XJ, Wu WX, Fang J, Shen CF, McBride MB, Chen YX (2009). Effect of MSW Source-Classified Collection on Polycyclic Aromatic Hydrocarbons in Residues from Full-Scale Incineration in China. Water Air Soil Pollut. 198:347-358.

Skodras G, Palladas A, Kaldis SP, Sakellaropoulos GP (2007). Cleaner co-combustion of lignite–biomass–waste blends by utilising inhibiting compounds of toxic emissions. Chemosphere 67:S191-S197.

Szakacs T, Ivan B, Kupai J (2004). Thermal stability of cationically allylated poly(vinyl chloride) and poly(vinyl chloride-co-2-chloropropene) copolymer. Poly. Degrad. Stabil. 85(3):1029-1033.

Tang X, Zhao L (2005). The development of sludge disposal strategy. (in Chinese, with English abstract) Environ. Sci. Manage. 30:68-70, 90.

Troitskii BB, Troitskaya LS (1999). Degenerated branching of chain in poly(vinyl chloride) thermal degradation. Eur. Poly. J. 35(12):2215-2224.

Verhust D, Buekens A, Spancer PJ, Erikson G (1996).Thermodynamic behavior of metal chlorides and sulfates under the conditions of incineration sources. Environ. Sci. Technol. 30:50-56.

Vogg H, Stieglitz L (1986). Thermal behavior of PCDD/PCDF in fly ash from municipal incinerators. Chemosphere 15:1373-1378.

Wang KS, Chiang KY, Lin SM, Tsai CC, Sun CJ (1999). Effects of chlorides on emissions of toxic compounds in waste incineration: study on partitioning characteristics of heavy metal. Chemosphere 38:1833-1849.

Wang MJ (1997). Land application of sewage sludge in China. Sci. Total Environ. 197(1-3):149-160.

Werle S, Wilk RK (2010). A review of methods for the thermal utilization of sewage sludge: The Polish perspective. Expt. Therm. Fluid Sci. 34:387-395.

Werther J, Ogada T (1999). Sewage sludge combustion. Prog. Energy Combust. Sci. 25:55-116.

WHO (1986). Dioxins and furans from municpal incinerators, Geneva:

World Health Organisation.

Yde Y, Morimoto S, Morioka S, Uji S, Furubayashi K (1994). Organohalogen compounds emission control on flue gas from MSW incineration plants. Organohalogen Comp. 19:395-400.

Zhang CS, Wang LJ, Shen WR (2002). Characteristics of sludge from sewage discharge channels of tianjin, China. Water, Air Soil Pollut. 134:239-254.

Water quality index of fresh water streams feeding Wular Lake, in Kashmir Himalaya, India

Sayar Yaseen, Ashok K. Pandit and Javaid Ahmad Shah

Aquatic Ecology Laboratory, Department of Environmental Science University of Kashmir, Srinagar 190006, J&K (India).

The quality of drinking water is of vital concern for human health and life. The present investigation was aimed at assessing the water quality index (WQI) of five fresh water streams feeding Wular lake. Analysis of the data revealed that the WQI values ranged from a minimum of 45.4 to a maximum of 48.9. Among the study sites, Makdhoomyari stream showed higher values of WQI while as lowest was shown by Madhumati. Pearson matrix revealed that conductivity showed significant positive correlation with total dissolved solids ($P < 0.01$, $r = 0.807$) and total alkalinity ($P < 0.01$, $r = 0.635$). Total hardness was found to bear strong positive correlation with calcium ($P < 0.01$, $r = 0.819$), while temperature maintained inverse relationship with dissolved oxygen ($P < 0.01$, $r = 0.78$). The results support that the water parameters are desirable and the water quality of these streams falls under Category I based on water quality index values. Bray Curtis similarity dendrogram depicted that Erin and Gurura streams had maximum (99%) similarity; while as lowest similarity was observed between Makdhoomyari and Ashtungu streams (85.98%). The present finding revealed that these streams need immediate attention to prevent them from further deterioration.

Key words: Water quality index, Wular lake, Pearson correlation, Kashmir, Himalaya.

INTRODUCTION

The fresh water streams in Kashmir Himalaya are the potable sources of water for the region but unfortunately due to their exploitation for various purposes like drinking, domestic, agriculture, hydropower, etc. These vital resources are getting not only degraded but also polluted as the human population grows. Comparatively, little work has been done on the stream ecosystems of the Kashmir valley and it is only very recently some work have been conducted on the physicochemical and biological aspects of streams (Rashid et al., 2006; Bhat et al., 2011; Hussain and Pandit, 2011). Further, very recently, WQI was applied to evaluate the water quality of Vishav stream in Kashmir (Hamid et al., 2013). However, no substantial work on WQI has been carried out on the incoming streams of Wular lake, the largest fresh water lake in Indian subcontinent.

MATERIALS AND METHODS

Description of the study area and sites

Wular Lake is the largest fresh water body in the Indian Sub-continent is located 34 km northwest of Srinagar city. Geographically, the lake is situated at an altitude of 1,580 m (a.m.s.l), between 34°16′-34°20′N latitudes and 74°33′-74°44′E longitudes and covers an area of 189 km^2 (Shah et al., 2014). For

Table 1. Location of five sampling sites around Wular lake.

Site	Sampling sites	Latitude E	Longitude N	Elevation
I	Makhdoomyari	34°17′39.8″	74°37′-29.6″	1586
II	Gurura	34°22′37.6″	74°40′-21.7″	1598
III	Erin	34°24′26.5″	74°39′10.2 ″	1599
IV	Madhumati	34°25′30.7″	74°37′45.1″	1594
V	Ashtang	34°24′30.3″	74°32′23.6″	1585

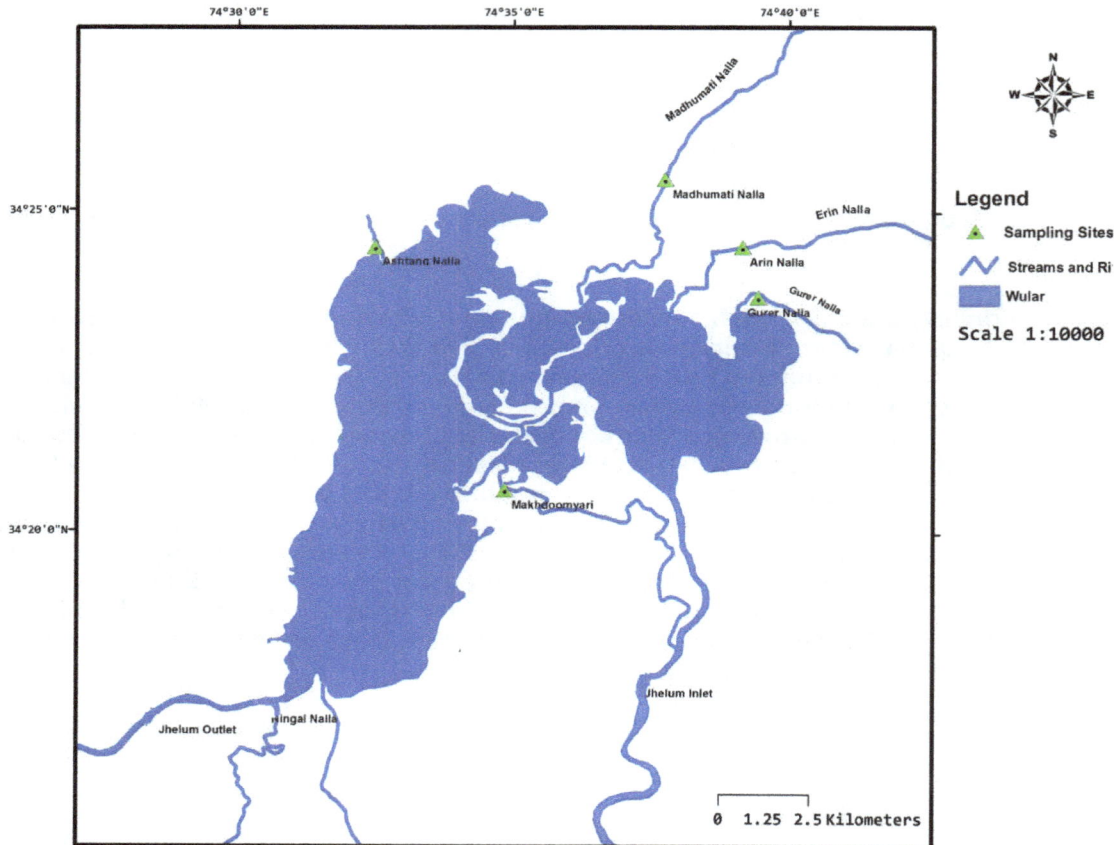

Figure 1. Location map of the study area with sampling stations.

the present investigation, five streams namely Makhdoomyari, Gurura, Erin, Madhumati and Ashtungu were selected which directly drain into the Wular lake for the assessment of water quality and calculation of Water Quality Index (Table 1 and Figure 1).

Sample collection and analysis

Surface water samples were collected in clean polyethylene bottles for the analysis of various physico-chemical parameters on monthly basis from December, 2011 to November, 2012. Parameters like water temperature and pH were measured on spot by means of a mercury thermometer and digital pH meter. The remaining parameters were analysed in the laboratory as per the standard methods of APHA (2005).

Statistical analysis

Statistical analysis were performed using SPSS (statistical version 16 for windows 7, SPSS and Chicago, IL,USA). The relation between various study sites were calculated by another software programme PAST (statistical version 1.93 for windows 7).

Calculation of Water Quality Index (WQI)

Eight water parameters were considered for calculation of water quality index (Tiwari and Manzor, 1988; Mohanta and Patra, 2000; Kesharwani et al., 2004; Padmanabha and Belagalli., 2005):

Water Quality I index (WQI) = $\Sigma q_i w_i$

Table 2. Physico chemical characteristics and water quality index at Site I.

Parameter	WHO Standard	Wi (unit weight)	Mean (±) S.D	WIQI
Temperature (°C)	......		14.1±8.2	-
pH	8.5	0.319	7.8±0.3	17.01227
EC (µS/cm)	750	0.0036	232.3±80.8	0.111979
TDS (mg/l)	500	0.0054	155.6±54.1	0.112509
DO (mg/l)	5	0.5423	9.1±1.6	31.06837
Total alkalinity (mg/l)	200	0.0136	91±18.3	0.164498
Total hardness (mg/l)	300	0.009	133±41.3	0.16028
Calcium (mg/l)	75	0.0362	53.3±17.6	0.25693
Chloride (mg/l)	250	0.0108	10.4±4.2	0.01504
Nitrate (mg/l)	45	0.0603	0.76±0.33	0.00612
Water Quality Index ∑WiQi = 48.91				

Table 3. Physico chemical characteristics and water quality index at Site II.

Parameter	WHO Standard	Wi (unit weight)	Mean (±) S.D	WiQi
Temperature (°C)	......		13.4±7.9	-------
Ph	8.5	0.319	7.8±0.3	17.01227
EC (µS/cm)	750	0.0036	214.3±64.8	0.103302
TDS (mg/l)	500	0.0054	143.5±43.4	0.10376
DO (mg/l)	5	0.5423	9.3±1.6	29.94038
Total alkalinity (mg/l)	200	0.0136	91.6±21.1	0.165582
Total hardness (mg/l)	300	0.009	124.2±37.3	0.149675
Calcium (mg/l)	75	0.0362	45.2±12.8	0.217884
Chloride (mg/l)	250	0.0108	10.1±3.6	0.014606
Nitrate (mg/l)	45	0.0603	0.80±0.4	0.006427
Water Quality Index ∑WiQi = 47.7				

Where qi (water quality rating) = 100 x (Va- Vi) / (Vs-Vi), when Va = actual value present in the water sample, Vi = ideal value (0 for all parameters except pH and dissolved oxygen which are 7.0 and 14.6 mg/L respectively). Vs = standard value. If quality rating qi = 0 means complete absence of pollutants, While 0< qi < 100 implies that, the pollutants are within the prescribed standard; When qi > 100 implies that, the pollutants are above the standards:

Wi (unit weight) = K / Sn

$$K(constant) = 1 + \cfrac{1}{\frac{1}{Vs1} + \frac{1}{Vs2} + \frac{1}{Vs3} + \frac{1}{Vs4} + \cdots \frac{1}{Vsn}}$$

Where Sn = 'n' number of standard values.

RESULTS AND DISCUSSION

The annual average values of various physico-chemical properties and water quality index of five streams are presented in Tables 2 to 6 and in Figure 2. During the present study, the annual mean water temperature ranged from 12.8 to 14.8°C, with highest temperature being recorded at Site I and lowest at Site III. The higher temperature during the summer season can be attributed probably to high atmospheric temperature, low relative humidity (Sinha et al., 2004; Ayoade et al., 2006; Atobatele and Ugwumba, 2008). pH is an important parameter in water quality assessment as it influences many biological and chemical processes within a water body (Gray, 1999; Shah and Pandit, 2013). In this present study, pH value ranged between 7.7 at Site V to 7.9 at Site III. In the majority of studies conducted on freshwater ecosystems, the pH values are generally reported between 6 and 9 (Kamran et al., 2003).

Specific conductivity in aquatic ecosystems depends on ionic concentration or dissolved in organic substances. It can also be used to give a rough estimate of the total amount of dissolved solids in water. In the present study, the highest conductivity of 232 µScm[-1] was registered at Site I as against the lowest of 123.7 µScm[-1] being recorded at Site V. The high conductivity values recorded at Site I can be due to excessive use of agricultural fertilizers (Clenaghan et al., 1998).

A total dissolved solid is very useful parameter

Table 4. Physico chemical characteristics and water quality index at Site III.

Parameter	WHO Standard	Wi (unit weight)	Mean (±) S.D	WiQi
Temperature (°C)	……		12.8±7.2	
pH	8.5	0.319	7.9±0.3	19.14
EC (µs/cm)	750	0.003615	215.6±70.2	0.103929
TDS (mg/l)	500	0.005423	144.4±47.0	0.104411
DO (mg/l)	5	0.5423	9.8±1.3	27.115
Total alkalinity (mg/l)	200	0.013558	90.4±22.6	0.163413
Total hardness (mg/l)	300	0.009038	118.2±35.5	0.142444
Calcium (mg/l)	75	0.036153	44.1±15.9	0.212582
Chloride (mg/l)	250	0.010846	9.1±2.9	0.01316
Nitrate (mg/l)	45	0.060256	0.80±0.4	0.006427
Water Quality Index ∑Wi Qi = 47.00				

Table 5. Physico chemical characteristics and water quality index at Site IV.

Parameter	WHO Standard	Wi (unit weight)	Mean (±) S.D	WiQi
Temperature (°C)	……		13.6±7.7	-------
pH	8.5	0.319	7.7±0.2	14.88773
EC (µs /cm)	750	0.003615	123.7±39.6	0.059629
TDS (mg/l)	500	0.005423	142.1±56.4	0.102748
DO (mg/l)	5	0.5423	9.1±1	31.06837
Total alkalinity (mg/l)	200	0.013558	94.3±18.8	0.170463
Total hardness (mg/l)	300	0.009038	123.7±39.6	0.149072
Calcium (mg/l)	75	0.036153	47.7±13.1	0.229935
Chloride (mg/l)	250	0.010846	11.7± 5	0.01692
Nitrate (mg/l)	45	0.060256	0.6±0.2	0.00482
Water Quality Index ∑WiQi = 46.68				

Table 6. Physico chemical characteristics and water quality index at Site V.

Parameter	WHO Standard	Wi (unit weight)	Mean (±) S.D	WiQi
Temperature (°C)	-	-	13.2 ±7.7	-
pH	8.5	0.319	7.9±0.4	19.14
EC (µs /cm)	750	0.003615	211.3 ± 68.5	0.101856
TDS (mg/l)	500	0.005423	141.6 ± 45.9	0.102386
DO (mg/l)	5	0.5423	10 ± 1.6	25.54233
Total alkalinity (mg/l)	200	0.013558	95.5 ± 22.4	0.172632
Total hardness (mg/l)	300	0.009038	117.3 ± 22.3	0.14136
Calcium (mg/l)	75	0.036153	42.9 ± 13.6	0.206797
Chloride (mg/l)	250	0.010846	8.9 ± 3.8	0.012871
Nitrate (mg/l)	45	0.060256	0.7 ± 0.3	0.005624
Water Quality Index ∑WiQi = 45.43				

describing chemical constituents of the water and can be in general related to the edaphic factor that contributes to the productivity within the water body (Goher, 2002). In the present study, total dissolved solids fluctuated between a high of 155.6 mg/L at Site I and a low of 82.9 mg/L at Site V. The concentration of total dissolved solids tends to be higher at Site I due to increased siltation caused by surface run-off (Shinde, 2011). Dissolved

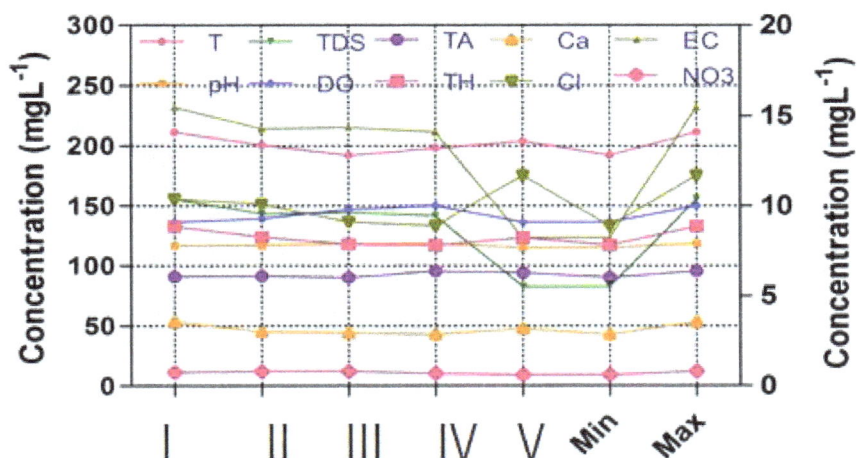

Figure 2. Spatial variation of annual average physico chemical characteristics of water.

oxygen is of paramount importance in all aquatic ecosystems as it regulates most of metabolic processes of organism and also the community architecture as a whole (Hussain and Pandit, 2011). The highest amount of dissolved oxygen (10 mg/L) was noted at Site IV while the lowest dissolved oxygen of 9.1 mg/L was evinced at Site V.

Alkalinity of water is the capacity to neutralize strong acids and is primarily a function of carbonate, bicarbonate and hydroxide content, being formed due to the dissolution of carbondioxide in water (Dallas and Day, 2004). During the present investigation, the maximum alkalinity of 95 mg/L was registered again at Site IV as against the minimum of 90.4 mg/L, being registered at Site III. The lower alkalinity values in the present investigation may be attributed to the high flow discharge (Harlow, 2003).

Hardness usually includes only Ca^{2+} and Mg^{2+} ions expressed in the terms of equivalent $CaCO_3$ (Das and Singh, 1996). During the entire study period, the highest value of total hardness (133 mg/L) were maintained at Site I as against the lowest of 117.3 mg/L at Site IV. The slightly higher hardness values may be attributed to the increased mobilization of elements (calcium and magnesium) from subsurface ground (Badrakh et al., 2008). Calcium is present in all waters as Ca^{2+} and is readily dissolved from rocks rich in calcium minerals, particularly as carbonates and sulphates, especially limestone and calcite (Chapman, 1996). In the present study the values of calcium hardness ranged between a high of 53.3 mg/L at Site I and a low of 42.9 mg/L at Site IV. The higher concentration of Ca^{2+} at Site I is the direct attribute of the lithology of catchment area as suggested by the findings of (Jaiswala et al., 2009). Chloride is chemically and biologically unreactive and it occurs naturally in all types of water. It enters surface waters, with the weathering of some sedimentary rocks (mostly

rock salt deposits) and from industrial and sewage effluents, and agricultural and road run-off (Link and Inman, 2003). In the present study, the levels of chloride fluctuated between 11.7 mg/L at Site V and 8.4 mg/L at Site IV. The highest chloride concentrations at Site V may be explained on the account of the increasing anthropogenic activities (Mooers and Alexander, 1994).

Nitrate-nitrogen is known to be a vital nutrient for growth, reproduction, and the survival of organisms. In the present investigation, the concentration of nitrate could not show marked fluctuations and ranged from (0.6 mg/l) to (0.8 mg/l). Higher levels of nitrate were obtained at Sites II and III, while as lower levels were registered at Site V. High nitrate levels (>1 mg L^{-1}) are not good for aquatic life (Kilham, 1990). Further, the fluctuations noticed in the concentration of nitrate may be attributed to increased agricultural runoff and sewage contamination (Ali et al., 1999).

Water quality index (WQI)

The concept of water quality index (WQI) is based on the comparison of water quality parameters with respect to regulatory standard (Khan et al., 2003). WQI is defined as a rating that reflects the composite influence of different water quality parameters. WQI is calculated from the point of the suitability of surface waters for variety of uses including human consumption (Cude, 2001; Atulegwu and Njoku, 2004).

Analysis of the data revealed that the WQI values for the streams fluctuated from a high of 48.9 at Site I and to a low of 45.4 at Site IV with an average value of 47.1 ± 1.3 (Figure 3). Higher values of WQI obtained at Site I may be attributed to the fact that it drains out through the large catchment surrounded by huge tracts of agriculture fields and also has severe anthropogenic stresses as

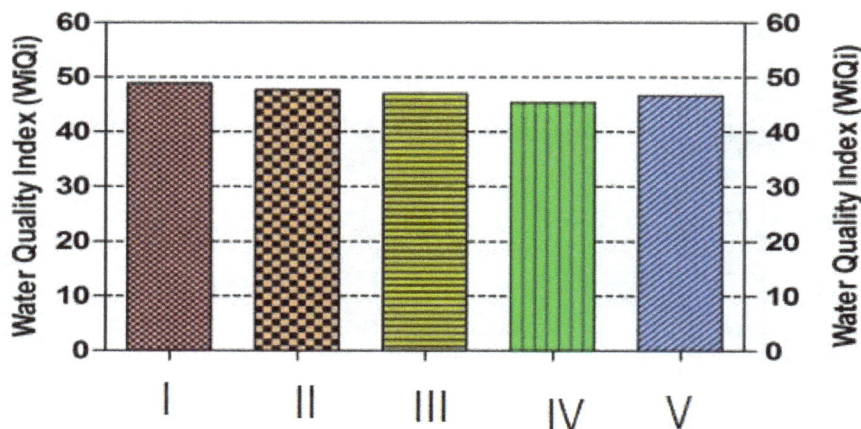

Figure 3. Variation of WQI of five study sites.

compared to other sites. On the other hand, lower WQI values obtained at Site IV may be on the account of its glacier fed nature, being surrounded by catchments of dense forests and meadows (Kanth and Hassan, 2012). The findings of the present study revealed that the water quality parameters are desirable and the water quality of these stream falls under Category I (that is, slightly polluted based on water quality index values) as per (Sinha et al., 2004). Yet, they need immediate attention as they are main sources of water to Wular lake.

Relationship among hydrological parameters

In the present study, water temperature showed negative correlation with almost all the parameters. The parameters depicting highly significant negative correlation with water temperature were dissolved oxygen ($P < 0.01$, $r = -0.784$), (Gurumahum et al., 2002; Idowu et al., 2013), nitrate ($P < 0.01$, $r = -0.820$) (Shah and Pandit, 2012), total dissolved solids ($P < 0.01$, $r = -0648$) and total hardness ($P < 0.01$, $r = -0.676$). Conductivity revealed highly significant positive correlation with total alkalinity ($P < 0.01$, $r = 0.635$). However, the highly significant positive correlation between conductivity and total dissolved solids was evident from the results which was proved statistically ($P < 0.01$, $r = 0.807$) (Heydari and Abbasi, 2013). The most significant positive correlation of total dissolved solids was recorded with total alkalinity ($P < 0.01$, $r = 0.765$) and chloride ($P < 0.01$, $r = 0.655$). Total hardness was found to bear strong positive correlation with calcium ($P < 0.01$, $r = 0.819$) (Kumar and Sinha, 2010) pH was the only exception which could not depict strong positive correlation with any of the parameters. The relationship among hydrological attributes has been also diagrammatically shown in Figure 4.

Bray Curtis similarity analysis shows that Sites II and III have maximum (99%) similarity, while as lowest similarity was observed between Sites I and V (Figure 5). This may

be attributed to the fact that the former two sites are very close to each other and both the streams have almost same origin but drain through different watersheds of almost similar catchment, Sites I and V are totally dissimilar due to the fact that Site I is an inlet of river Jhelum, perennial in nature and drains out through large catchment while as site V is a stream having very low flow and thus has large quantity of pollutants due to the absence of strong dilution effect also carries out with itself the whole domestic sewage of villages present in its banks immediately into the Wular lake, hence has great anthropogenic stress.

Conclusions

From the observations, it may be concluded that among five streams, the water quality of all the streams is slightly polluted based on water quality index. The study revealed that these streams are experiencing initial stage of anthropogenic stress pollution and needs immediate attention of aquatic ecologists. Further, the results of the study could be helpful in the management of the lake for its water quality, fisheries and recreation. The data obtained could also form baseline and reference point while assessing further changes that might be caused by nature or man in the lake.

Conflict of Interest

The authors have not declared any conflict of interest.

ACKNOWLEDGEMENTS

This work is part of the Ph.D. research of the first author Sayar Yaseen for which he is indebted to Director, Centre of Research for Development (CORD) and Head,

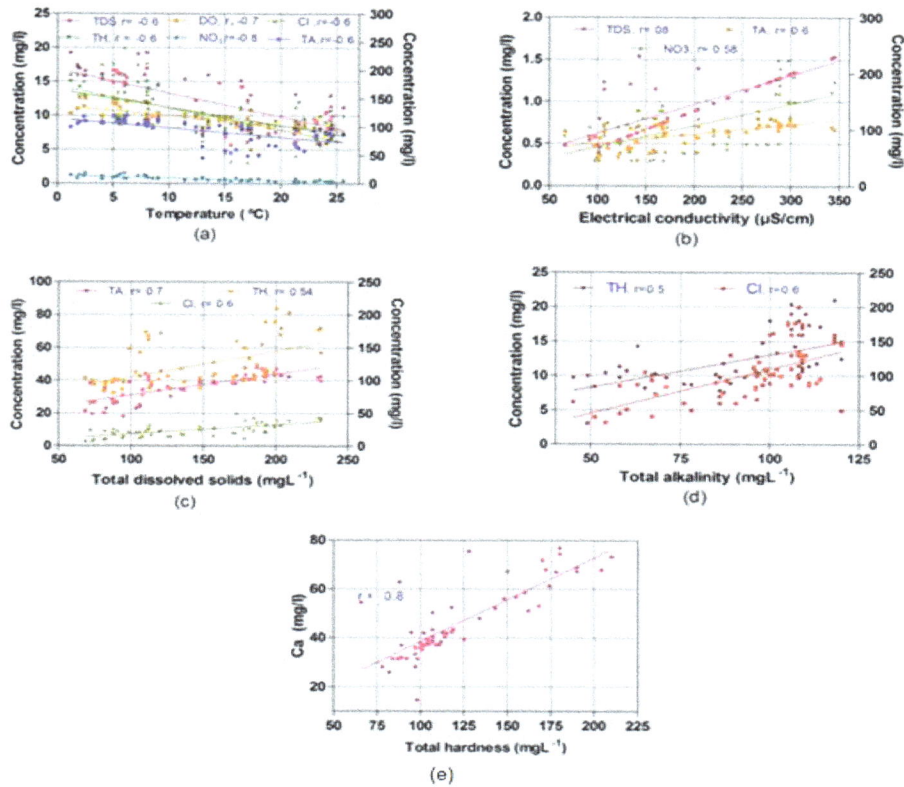

Figure 4. Nature of relationship between physico-chemical parameters of water.

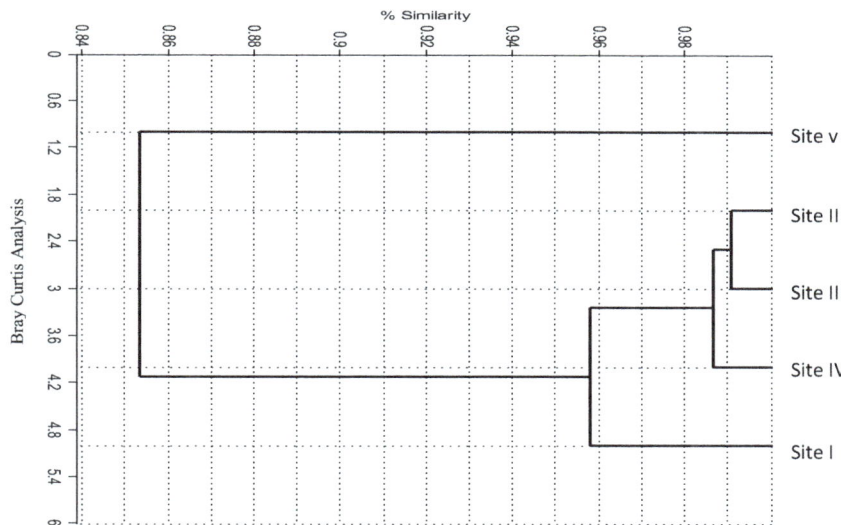

Figure 5. Bray Curtis similarity analysis of study sites.

Department of Environmental Sciences, University of Kashmir for providing necessary laboratory facilities.

REFERENCES

Ali MB, Tripathi RD, Rai UN, Pal A, Singh SP (1999). Physicochemical

characteristics and pollution level of lake Nainital (U.P., India): role of macrophytes and phytoplankton in bio monitoring and phyto remediation of toxic metal ions. Chemosphere 39(12):2171-2182.
APHA (2005). Standard Methods for the Examination of Water and Waste Water, 21st Ed. Amer. Pub. Health Assoc. Inc., Washington D.C.
Atobatele OE, Ugwumba OA (2008). Seasonal Variation in the physicochemistry of a small tropical reservoir (Aiba Reservoir, Iwo,

Osun, Nigeria). Afr. J. Biotechnol. 7(12)1962-1971.

Atulegwu PU, Njoku JD (2004). The impact of biocides on the water quality. Int. Res. J. Eng. Sci. Technol. 1:47-52.

Ayoade AA, Fagade SO, Adebisi AA (2006). Dynamics of limnological features of two manmade lakes in relation to fish production. Afr. J. Biotechnol. 5(10):1013-1021.

Badrakh A, Tserendorj T, Vanya D, Dalaijamts C, Shinee E, Chultemdorji T, Hagan R, Govind S (2008). A study of the quality and hygienic conditions of spring water in Mongolia. J. Water Health, 6(1):141-148.

Bhat SU, Sofi AH, Yaseen T, Pandit AK, Yousuf AR (2011). Macroinvertebrate community from Sonamarg streams of Kashmir Himalaya. Pak. J. Biol. Sci. 14(3):182-194.

Chapman D (1996). Water Quality Assessments: A Guide to the Use of Biota, Sediments and Water in Environmental Monitoring, 2nd ed. UNESCO, London.

Clenaghan C, Giller PSO, Halloran J, Hernan R (1998). Stream macro invertebrate communities in a conifer afforested catchment: Relationships to physico-chemical and biotic factors. Fresh Water Biol. 40:175-193.

Cude C (2001). Oregon water quality index: A tool for evaluating water quality management effectiveness. J. Am. Water Resour. Assoc. 37:125-137.

Dallas FH, Day JA (2004). The Effect of Water Quality Variables on Aquatic Ecosystems: Review. Report to the Water Research Commission, WRC Report No. TT224/04.

Das BK, Singh M (1996). Water chemistry and control of weathering of Pichola Lake, Udaipur District, Rajasthan, India. Enviromental Geology, 27: 184-190

Goher MEM (2002). Chemical studies on the precipitation and dissolution of some chemical element in Lake Qarun, Ph.D. Thesis Fac. of Sci, Al-Azhar University, Egypt.

Gray NF (1999). Water Technology: an Introduction for Environmental Scientists and Engineers, 1st Ed., Arnold Publishers, London.

Gurumahum SD, Daimari P, Goswami BS, Sakar A, Choudhury M (2000). Physico chemical qualities of water and plankton of selected rivers in Meghalaya. J. Inland Fish Soc. India, 34:36-42.

Hamid A, Dar NA, Bhat SU, Pandit AK (2013). Water Quality Index: A Case Study of Vishav Stream, Kulgam, Kashmir. Int. J. Environ. Bioenergy 5(2):108-122.

Harlow R (2003). Stream biomonitoring using aquatic macroinvertebrates in small catchments: a case study in Emigrant Creek, north-east NSW. Honours thesis, School of Environmental Science and Management, Southern Cross University, Lismore, NSW, Australia.

Heydari MM, Abasi A (2013). Correlation study and regression analysis of drinking water quality in Kashan City, Iran. Middle-East J. Sci. Res. 13(9):1238-1244.

Hussain QA, Pandit AK (2011). Hydrology, geomorphology and rosgen classification of Doodhganga stream in Kashmir Himalaya, India. Int. J. Water Resour. Environ. Eng. 3(3):57-65.

Idowu EO, Ugwumba AA, Edward JB, Oso JA (2013). Study of the Seasonal Variation in the Physico- Chemical Parameters of a Tropical Reservoir Greener. J. Phys. Sci. 3(4):142-148

Jaiswala MK, Bhat MI, Bali B, Ahmad S, Chen GY (2009). Luminescence characteristics of quartz and feldspar from tectonically uplifted terraces in Kashmir Basin, Jammu and Kashmir, India. Radiation Measurements (44):523-528.

Kamran TM, Abdus S, Muhammed L, Tasveer Z (2003). Study of the seasonal variations in the physicochemical and biological aspects of Indus River Pakistan. Pak. J. Biol. Sci. 6(21):1795-1801.

Kanth TA, Hassan Z (2012). "Morphometric analysis and prioritization of watersheds for soil and water resource management in Wular catchment using geo- spatial tools". Int. J. Geol. Earth Environ. Sci. 2(1):30-41

Kesharwani S, Mandoli AK, Dube KK (2004). Determination of water quality index (WQI) of Amkhera pond of Jabalpur city. M. Ntl. J. Life Sci. 1:61-66.

Khan F, Husain T, Lumb A (2003). Water quality evaluation and trend analysis in selected watersheds of the Atlantic Region of Canada. Environ. Monit. Assess. 88:221-242.

Kilham P (1990). Mechanisms controlling the chemical composition of lakes and rivers: Data from Africa. Limnol. Oceanogr. 35:80-83.

Kumar N, Sinha DK (2010). Drinking water quality management through correlation studies among various physicochemical parameters: A case study. Int. J. Environ. Sci. (1): 2:253-259.

Link M, Inman D (2003). Ground water monitoring at livestock waste control facilities in Nebraska. Nebraska Department of Environmental Quality, pp. 1-14.

Mohanta BK, Patra AK (2000). Studies in the water quality index of river Sanamachhakananda at Keonjargarh, Orissa, India. Pollut. Res. 19:377-385.

Mooers HD, Alexander J (1994). Contribution of spray irrigation of wastewater to groundwater contamination in the Karst of Southeastern Minnesota, USA. J. Hydrogeol. 2:34-44.

Padmanabha B, Belagali SL (2005). Comparative study on the water quality index of four lakes in the Mysore city. IJEP 25:873-876.

Rashid HU, Pandit AK (2006). Food preferences of the brown trout (Salmo trutta L.) in relation to the benthic macroinvertebrates of River Sindh, Kashmir valley. Indian J. Environ. Ecoplan. 12(1):9-16.

Shah JA, Pandit AK (2012). Physico-chemical characteristics of water in Wular lake-A Ramsar Site in Kashmir Himalaya. Int. J. Geol. Earth Environ. Sci. 2(2):257-265.

Shah JA, Pandit AK, Shah GM (2014). Spatial and temporal variations of nitrogen and phosphorus in wular lake leading to eutrophication. Ecologia 4(2):44-55.

Shah JA, Pandit AK (2013). Relationship between physicochemical limnology and Crustacearn Community in Wular Lake of Kashmir Himalaya. Pak. J. Biol. Sci. 16(19):976-983.

Shinde SE, Pathan TS, Raut KS, Sonawane DL (2011). Studies on the Physico-chemical Parameters and Correlation Coefficient of Harsool-savangi Dam, District Aurangabad, India Middle-East J. Scientific Res. 8 (3):544-554.

Sinha DK, Saxena S, Saxena R (2004). Water quality index for Ram Ganga river at Mordabad. Pollut. Res. 23:527-531.

Tiwari TN, Manzoor A (1988). Water quality index for Indian rivers. In Ecology and pollution in Indian rivers (Ed.: R.K. Trivedy). Ashish Publishing House, New Delhi. pp. 271-286.

Permissions

The contributors of this book come from diverse backgrounds, making this book a truly international effort. This book will bring forth new frontiers with its revolutionizing research information and detailed analysis of the nascent developments around the world.

We would like to thank all the contributing authors for lending their expertise to make the book truly unique. They have played a crucial role in the development of this book. Without their invaluable contributions this book wouldn't have been possible. They have made vital efforts to compile up to date information on the varied aspects of this subject to make this book a valuable addition to the collection of many professionals and students.

This book was conceptualized with the vision of imparting up-to-date information and advanced data in this field. To ensure the same, a matchless editorial board was set up. Every individual on the board went through rigorous rounds of assessment to prove their worth. After which they invested a large part of their time researching and compiling the most relevant data for our readers.

The editorial board has been involved in producing this book since its inception. They have spent rigorous hours researching and exploring the diverse topics which have resulted in the successful publishing of this book. They have passed on their knowledge of decades through this book. To expedite this challenging task, the publisher supported the team at every step. A small team of assistant editors was also appointed to further simplify the editing procedure and attain best results for the readers.

Apart from the editorial board, the designing team has also invested a significant amount of their time in understanding the subject and creating the most relevant covers. They scrutinized every image to scout for the most suitable representation of the subject and create an appropriate cover for the book.

The publishing team has been an ardent support to the editorial, designing and production team. Their endless efforts to recruit the best for this project, has resulted in the accomplishment of this book. They are a veteran in the field of academics and their pool of knowledge is as vast as their experience in printing. Their expertise and guidance has proved useful at every step. Their uncompromising quality standards have made this book an exceptional effort. Their encouragement from time to time has been an inspiration for everyone.

The publisher and the editorial board hope that this book will prove to be a valuable piece of knowledge for researchers, students, practitioners and scholars across the globe.

List of Contributors

Falah A. Almottiri
Civil Engineering Department, College of Technological Studies (Kuwait) P.O. Box: 34 Ardia, 13136 Kuwait

Falah M. Wegian
Civil Engineering Department, College of Technological Studies (Kuwait) P.O. Box: 34 Ardia, 13136 Kuwait

Rouzbeh Abbassi
Faculty of Engineering and Applied Science, Memorial University of Newfoundland, St.John's, NL, Canada, A1B 3X5, Canada

Faisal Khan
Faculty of Engineering and Applied Science, Memorial University of Newfoundland, St.John's, NL, Canada, A1B 3X5, Canada

Kelly Hawboldt
Faculty of Engineering and Applied Science, Memorial University of Newfoundland, St.John's, NL, Canada, A1B 3X5, Canada

S. M. Al-Rawi
Environment Research Center (ERC), Mosul University, Mosul, Iraq

A. Binesh
Physics Department, Payam Nour University, Fariman, Iran

H. Arabshahi
Physics Department, Ferdowsi University of Mashhad, Mashhad, Iran

Vipul Shinde
Agricultural and Food Engineering Department, Indian Institute of Technology, Kharagpur, India-721302, India

K. N. Tiwari
Agricultural and Food Engineering Department, Indian Institute of Technology, Kharagpur, India-721302, India

Manjushree Singh
Agricultural and Food Engineering Department, Indian Institute of Technology, Kharagpur, India-721302, India

M. Kasebele
Department of Environmental Engineering, University of Dodoma, Dodoma, Tanzania

W. J. S. Mwegoha
School of Environmental Science and Technology, Ardhi University, Dar es Salaam, Tanzania

M. I Oladapo
Department of Applied Geophysics, Federal University of Technology, Akure, Nigeria

O. O Adeoye-Oladapo
Department of Physics, Adeyemi College of Education, Ondo, Nigeria

F. S Adebobuyi
Groundwater and Geophysical Services (GGS) Limited, Lagos, Nigeria

Majid Dabbaghian Amiry
Urban Planning, Geographic Department, Piam Noor University, Sari, Iran

A. A. Mohammadi
Department of Watershed Management, Science and Research Branch, Islamic Azad University, Tehran, Iran

Adimasu Woldesenbet Worako
Department of Water Resources and Irrigation Management, Dilla University College of Agriculture and Natural Resources, Ethiopia

Rouhollah Soltani Goharrizi
Collage of Natural Resources, Department Of Environment, Islamic Azad University, Bandar Abbas, Iran

Fazlollah Soltani
Department of Civil Engineering, Faculty of Engineering, Graduate University of Advanced Technology, Kerman, Iran

Bahador Abolpour
Department of Chemical Engineering, Faculty of Engineering, Shahid Bahonar University of Kerman, Jomhoori Blvd., Post Code 76175, Kerman, Iran

Kuforiji Titilope Shakirat
Department of Zoology, Faculty of Science, Lagos State University (LASU), P.O. BOX LASU 001, Ojo, Lagos, Nigeria

Ayandiran Tolulope Akinpelu
Department of Pure and Applied Biology, Faculty of Science, Ladoke Akintola University of Technology, Ogbomoso, Oyo State, Nigeria

Sudipta Dey
Department of Biotechnology, Heritage Institute of Technology, Anandapur, Chowbaga Road, PO: Kolkata, PIN:700107, West Bengal, India

Somnath Mukherjee
Professor, Environmental Engineering Division, Civil Engineering Department, Jadavpur University, Raja S. C. Mallic Road, Kolkata, PIN: 700032, West Bengal, India

Sonya Dimitrova
University of Architecture, Civil Engineering and Geodesy, 1 Christo Smirnenski blvd., 1046 Sofia, Bulgaria

Nadejda Taneva
University of Architecture, Civil Engineering and Geodesy, 1 Christo Smirnenski blvd., 1046 Sofia, Bulgaria

Kapka Bojilova
University of Architecture, Civil Engineering and Geodesy, 1 Christo Smirnenski blvd., 1046 Sofia, Bulgaria

Vesela Zaharieva
University of Architecture, Civil Engineering and Geodesy, 1 Christo Smirnenski blvd., 1046 Sofia, Bulgaria

Svetlana Lazarova
University of Architecture, Civil Engineering and Geodesy, 1 Christo Smirnenski blvd., 1046 Sofia, Bulgaria

Mariana Koleva
University of Architecture, Civil Engineering and Geodesy, 1 Christo Smirnenski blvd., 1046 Sofia, Bulgaria

Rumen Arsov
University of Architecture, Civil Engineering and Geodesy, 1 Christo Smirnenski blvd., 1046 Sofia, Bulgaria

Tony Venelinov
University of Architecture, Civil Engineering and Geodesy, 1 Christo Smirnenski blvd., 1046 Sofia, Bulgaria

O. W. Obot
Department of Mechanical Engineering, Faculty of Engineering, University of Uyo, Nigeria

C. N. Anyakwo
Department of Metallurgical and Materials Engineering, Federal University of Technology, Owerri, Nigeria

A.Geethakarthi
Department of Civil Engineering, VIT University, Vellore 632014, India

B. R. Phanikumar
Department of Civil Engineering, VIT University, Vellore 632014, India

I. D. Hussaini
Adamawa State College of Agriculture, P. M. B. 2088 Ganye, Adamawa State, Nigeria

B. Aliyu
Modibbo Adama University of Technology, Yola Adamawa State, Nigeria

A. A Bassi
Adamawa State College of Agriculture, P. M. B. 2088 Ganye, Adamawa State, Nigeria

S. I Abubakar
Federal Polytechnic Bali, Taraba State, Nigeria

M. Aminu
Adamawa State College of Agriculture, P. M. B. 2088 Ganye, Adamawa State, Nigeria

Tomètin A. S. Lyde
Laboratoire de Chimie Inorganique et de l'Environnement (LACIE), Université d'Abomey-Calavi, 01 BP 526, Cotonou (Bénin) République du Bénin

Mama Daouda
Laboratoire de Chimie Inorganique et de l'Environnement (LACIE), Université d'Abomey-Calavi, 01 BP 526, Cotonou (Bénin) République du Bénin

Sagbo Etienne
Laboratoire de Chimie Inorganique et de l'Environnement (LACIE), Université d'Abomey-Calavi, 01 BP 526, Cotonou (Bénin) République du Bénin
Laboratoire d'Hydrologie Appliquée (LHA) Université d'Abomey-Calavi, République du Bénin

Fatombi K. Jacques
Laboratoire d'Expertise et de Recherche en Chimie de l'Eau et de l'Environnement (LERCEE) Faculté des Sciences et Techniques/ Université d'Abomey-Calavi, 01 BP 526, Cotonou, Bénin

Aminou W.Taofiki
Laboratoire d'Expertise et de Recherche en Chimie de l'Eau et de l'Environnement (LERCEE) Faculté des Sciences et Techniques/ Université d'Abomey-Calavi, 01 BP 526, Cotonou, Bénin

Bawa L. Moctar
Laboratoire de Chimie de l'Eau, Faculté Des Sciences, Université de Lomé, BP 1515, Lomé, Togo

Emmanuel Okoh Agyemang
Energy Systems Engineering Department, Koforidua Polytechnic, P. O. Box KF981, Koforidua, Ghana

Esi Awuah
Civil Engineering Department, Kwame Nkrumah University of Science and Technology (KNUST) Private Mail Bag, Kumasi, Ghana

Lawrence Darkwah
Chemical Engineering Department, Kwame Nkrumah University of Science and Technology (KNUST) Private Mail Bag, Kumasi, Ghana

Richard Arthur
Energy Systems Engineering Department, Koforidua Polytechnic, P. O. Box KF981, Koforidua, Ghana

Gabriel Osei
Mechanical Engineering Department, Koforidua Polytechnic, P. O. Box KF981, Koforidua, Ghana

N. O Adebisi
Department of Earth Sciences, Faculty of Science, Olabisi Onabanjo University, Ago Iwoye, Ogun State, Nigeria

O. S Oluwafemi
Department of Chemistry, Cape-Peninsula University of Technology, P.O.Box 652, Capetown, 8000, Western Cape, South Africa

S. P Songca
Department of Chemistry, Walter Sisulu University, Private Bag X1, Mthatha, 5117, South Africa

I. Haruna
Ikorodu Local Government, Ikorodu, Lagos State, Nigeria

H. M. Rasel
Department of Civil Engineering, Rajshahi University of Engineering and Technology, Rajshahi, Bangladesh

M. R. Hasan
Institute of Environmental Science, University of Rajshahi, Bangladesh

S. C. Das
Bangladesh Water Development Board (BWDB), Dhaka, Bangladesh

R. P. Verma
CSIR-Central Institute of Mining and Fuel Research, Barwa Road, Dhanbad-826015, India
Department of Environmental Science and Engineering, Indian School of Mines, Dhanbad-826004, India

R. Mandal
CSIR-Central Institute of Mining and Fuel Research, Barwa Road, Dhanbad-826015, India

S. K. Chaulya
CSIR-Central Institute of Mining and Fuel Research, Barwa Road, Dhanbad-826015, India

P. K. Singh
Department of Environmental Science and Engineering, Indian School of Mines, Dhanbad-826004, India

A.K. Singh
CSIR-Central Institute of Mining and Fuel Research, Barwa Road, Dhanbad-826015, India

G. M. Prasad
CSIR-Central Institute of Mining and Fuel Research, Barwa Road, Dhanbad-826015, India

Bupe Mwanza
Harare Institute of Technology, P. O. Box BE277 Belvedere Harare

Anthony Phiri
Harare Institute of Technology, P. O. Box BE277 Belvedere Harare

Aloyce W. Mayo
Department of Water Resources Engineering, University of Dar es Salaam, Tanzania

Jianzhong Zhu
Key Laboratory of Integrated Regulation and Resource Development on Shallow Lakes, Ministry of Education HOHAI University, HOHAI University, Nanjing, PR China 210098.School of Environmental Science and Environmental Engineering, Nanjing, PR China 210098

Lieqiang Chen, Jun Fang
College of Chemistry and Chemical Engineering, South China University of Technology, Guangzhou, PR China 510006

Jun Fang
Delon Hampton and Associates, OMAP Program at District of Columbia Water and Sewer Authority, Washington DC, USA 20032

Buchang Shi
Department of Chemistry Western Kentucky University, Bowling Green, USA 42101

Sayar Yaseen
Aquatic Ecology Laboratory, Department of Environmental Science University of Kashmir, Srinagar 190006, J&K (India)

Ashok K. Pandit
Aquatic Ecology Laboratory, Department of Environmental Science University of Kashmir, Srinagar 190006, J&K (India)

Javaid Ahmad Shah
Aquatic Ecology Laboratory, Department of Environmental Science University of Kashmir, Srinagar 190006, J&K (India)

www.ingramcontent.com/pod-product-compliance
Lightning Source LLC
Chambersburg PA
CBHW080632200326
41458CB00013B/4598